Advances in Computational Fluid Dynamics

Advances in Computational Fluid Dynamics

Baldev Singh Chikara

RANDOM

RANDOM PUBLICATIONS

NEW DELHI (INDIA)

Advances in Computational Fluid Dynamics

ISBN 978-93-5111-574-8

Published in 2015 in India by

RANDOM PUBLICATIONS

4376-A/4B, Gali Murari Lal, Ansari Road
New Delhi-110 002
Phone : +9111-43580356, 011-23289044, 011-43142548
e-mail: sales@randompublications.com,
info@randompublications.com, randomexports@gmail.com

Reprinted 2026

Type Setting by : Friends Media, Delhi-110089

Preface

The use of computational fluid Dynamics to simulate and predict fluid flows, heat transfer and associated phenomena continues to grow throughout many engineering disciplines.

Computational fluid dynamics, usually abbreviated as CFD, is a branch of fluid mechanics that uses numerical methods and algorithms to solve and analyze problems that involve fluid flows. Computers are used to perform the calculations required to simulate the interaction of liquids and gases with surfaces defined by boundary conditions. With high-speed supercomputers, better solutions can be achieved. Ongoing research yields software that improves the accuracy and speed of complex simulation scenarios such as transonic or turbulent flows. Initial experimental validation of such software is performed using a wind tunnel with the final validation coming in full-scale testing, e.g. flight tests.

Fluid dynamics is the sub-discipline of fluid mechanics dealing with fluids (liquids and gases) in motion. It has several subdisciplines itself, including aerodynamics (the study of gases in motion) and hydrodynamics (the study of liquids in motion). Fluid dynamics has a wide range of applications, including calculating forces and moments on aircraft, determining the mass flow rate of petroleum through pipelines, predicting weather patterns, understanding nebulae in interstellar space and reportedly modelling fission weapon detonation. Some of its principles are even used in traffic engineering, where traffic is treated as a continuous fluid.

Fluid dynamics offers a systematic structure that underlies these practical disciplines and that embraces empirical and semi-empirical laws, derived from flow measurement, used to solve practical problems. The solution of a fluid dynamics problem typically involves calculation of various properties of the fluid, such as velocity, pressure, density, and temperature, as functions of space and time.

The book is aimed at graduate students, researchers, engineers and physicists involved in fluid computations. An up-to-date account is given of the present state of the art of numerical methods employed in computational fluid dynamics.

I would like to thank my team for standing beside me throughout my career and writing this book. My special thanks go to "Random Publications" who have published the book.

–Baldev Singh Chikara

Contents

1

Computational Fluid Dynamics

INTRODUCTION

Computational fluid dynamics, usually abbreviated as CFD, is a branch of fluid mechanics that uses numerical methods and algorithms to solve and analyze problems that involve fluid flows. Computers are used to perform the calculations required to simulate the interaction of liquids and gases with surfaces defined by boundary conditions. With high-speed supercomputers, better solutions can be achieved. Ongoing research yields software that improves the accuracy and speed of complex simulation scenarios such as transonic or turbulent flows. Initial experimental validation of such software is performed using a wind tunnel with the final validation coming in full-scale testing, *e.g.* flight tests.

BACKGROUND AND HISTORY

The fundamental basis of almost all CFD problems are the Navier–Stokes equations, which define any single-phase (gas or liquid, but not both) fluid flow. These equations can be simplified by removing terms describing viscous actions to yield the Euler equations. Further simplification, by removing terms describing vorticity yields the full potential equations. Finally, for small perturbations in subsonic andsupersonic flows (not transonic or hypersonic) these equations can be linearized to yield the linearized potential equations.

Historically, methods were first developed to solve the linearized potential equations. Two-dimensional (2D) methods, using conformal transformations of the flow about a cylinder to the flow about an airfoil were developed in the 1930s.

One of the earliest type of calculations resembling modern CFD are those by Lewis Fry Richardson, in the sense that these calculations used finite differences and divided the physical space in cells. Although they failed dramatically, these calculations, together with Richardson's book "Weather prediction by numerical process", set the basis for modern CFD and numerical meteorology. In fact, early CFD calculations during the 1940s using ENIAC used methods close to those in Richardson's 1922 book. The computer power available paced development of three-dimensional methods. Probably the first work using computers to model fluid flow, as governed by the Navier-Stokes equations, was performed at Los Alamos National Lab, in the T3 group. This group was led by Francis H. Harlow, who is widely considered as one of the pioneers of CFD. From 1957 to late 1960s, this group developed a variety of numerical methods to simulate transient two-dimensional fluid flows, such as Particle-in-cell method (Harlow, 1957), Fluid-in-cellmethod (Gentry, Martin and Daly, 1966), Vorticity stream function method (Jake Fromm, 1963), and Marker-and-cell method (Harlow and Welch, 1965). Fromm's vorticity-stream-function method for 2D, transient, incompressible flow was the first treatment of strongly contorting incompressible flows in the world.

The first paper with three-dimensional model was published by John Hess and A.M.O. Smith of Douglas Aircraft in 1967. This method discretized the surface of the geometry with panels, giving rise to this class of programmes being called Panel Methods. Their method itself was simplified, in that it did not include lifting flows and hence was mainly applied to ship hulls and aircraft fuselages.

The first lifting Panel Code (A230) was described in a paper written by Paul Rubbert and Gary Saaris of Boeing Aircraft in 1968. In time, more advanced three-dimensional Panel Codes were developed at Boeing (PANAIR, A502), Lockheed (Quadpan), Douglas (HESS), McDonnell Aircraft (MACAERO), NASA(PMARC) and Analytical Methods (WBAERO, USAERO and VSAERO). Some (PANAIR, HESS and MACAERO) were higher order codes, using higher order distributions of surface singularities, while others (Quadpan, PMARC, USAERO and VSAERO) used single singularities on each surface panel. The advantage of the lower order codes was that they ran much faster on the computers of the time. Today, VSAERO has grown to be a multi-order code and is the most widely used programme of this class. It has been used in the development of many submarines, surface ships, automobiles, helicopters, aircraft, and more recently wind turbines. Its sister code, USAERO is an unsteady panel method that has also been used for modeling such things as high speed trains and racing yachts. The NASA PMARC code from an early version of VSAERO and a derivative of PMARC, named CMARC, is also commercially available.

In the two-dimensional realm, a number of Panel Codes have been developed for airfoil analysis and design. The codes typically have a boundary layer analysis included, so that viscous effects can be modeled. Professor Richard Eppler of the University of Stuttgart developed the PROFILE code, partly with NASA funding, which became available in the early 1980s. This was soon followed by MIT Professor Mark Drela's XFOIL code. Both PROFILE and XFOIL incorporate two-dimensional panel codes, with coupled boundary layer codes for airfoil analysis work. PROFILE uses a conformal transformation method for inverse airfoil design, while XFOIL has both a conformal transformation and an inverse panel method for airfoil design.

An intermediate step between Panel Codes and Full Potential codes were codes that used the Transonic Small Disturbance equations. In particular, the three-dimensional WIBCO code, developed by Charlie Boppe of Grumman Aircraft in the early 1980s has seen heavy use.

Developers turned to Full Potential codes, as panel methods could not calculate the non-linear flow present at transonic speeds. The first description of a means of using the Full Potential equations was published by Earll Murman and Julian Cole of Boeing in 1970. Frances Bauer, Paul Garabedian and David Korn of the Courant Institute at New York University (NYU) wrote a series of two-dimensional Full Potential airfoil codes that were widely used, the most important being named Programme H. A further growth of Programme H was developed by Bob Melnik and his group at Grumman Aerospace as Grumfoil. Antony Jameson, originally at Grumman Aircraft and the Courant Institute of NYU, worked with David Caughey to develop the important three-dimensional Full Potential code FLO22 in 1975. Many Full Potential codes emerged after this, culminating in Boeing's Tranair (A633) code, which still sees heavy use.

The next step was the Euler equations, which promised to provide more accurate solutions of transonic flows. The methodology used by Jameson in his three-dimensional FLO57 code (1981) was used by others to produce such programmes as Lockheed's TEAM programme and IAI/Analytical Methods' MGAERO programme. MGAERO is unique in being a structured cartesian mesh code, while most other such codes use structured body-fitted grids (with the exception of NASA's highly successful CART3D code,Lockheed's SPLITFLOW code and Georgia Tech's NASCART-GT). Antony Jameson also developed the three-dimensional AIRPLANE code which made use of unstructured tetrahedral grids.

In the two-dimensional realm, Mark Drela and Michael Giles, then graduate students at MIT, developed the ISES Euler programme (actually a suite of programmes) for airfoil design and analysis. This code first became available in 1986 and has been further developed to design, analyze and optimize single or multi-element airfoils, as the MSES programme.MSES sees wide use throughout the world. A derivative of MSES, for the design and analysis of airfoils in a cascade, is MISES, developed by Harold "Guppy" Youngren while he was a graduate student at MIT.

The Navier–Stokes equations were the ultimate target of developers. Two-dimensional codes, such as NASA Ames' ARC2D code first emerged. A number of three-dimensional codes were developed (ARC3D, OVERFLOW, CFL3D are three successful NASA contributions), leading to numerous commercial packages.

METHODOLOGY

In all of these approaches the same basic procedure is followed.

- During preprocessing
- The geometry (physical bounds) of the problem is defined.
- The volume occupied by the fluid is divided into discrete cells (the mesh). The mesh may be uniform or non-uniform.
- The physical modeling is defined – for example, the equations of motion + enthalpy + radiation + species conservation
- Boundary conditions are defined. This involves specifying the fluid behaviour and properties at the boundaries of the problem. For transient problems, the initial conditions are also defined.
- The simulation is started and the equations are solved iteratively as a steady-state or transient.
- Finally a postprocessor is used for the analysis and visualization of the resulting solution.

DISCRETIZATION METHODS

The stability of the selected discretisation is generally established numerically rather than analytically as with simple linear problems. Special care must also be taken to ensure that the discretisation handles discontinuous solutions gracefully. The Euler equations and Navier–Stokes equations both admit shocks, and contact surfaces.

Some of the discretisation methods being used are:

FINITE VOLUME METHOD

The finite volume method (FVM) is a common approach used in CFD codes, as it has an advantage in memory usage and solution speed, especially for large problems, highReynolds number turbulent flows, and source term dominated flows (like combustion).

In the finite volume method, the governing partial differential equations (typically the Navier-Stokes equations, the mass and energy conservation equations, and the turbulence equations) are recast in a conservative form, and then solved over discrete control volumes. This discretisation guarantees the conservation of fluxes through a particular control volume. The finite volume equation yields governing equations in the form,

$$\frac{\partial}{\partial t}\iiint Q\,dV + \iint F\,dA = 0,$$

where Q is the vector of conserved variables, F is the vector of fluxes (see Euler equations or Navier–Stokes equations), V is the volume of the control volume element, and A is the surface area of the control volume element.

FINITE ELEMENT METHOD

The finite element method (FEM) is used in structural analysis of solids, but is also applicable to fluids. However, the FEM formulation requires special care to ensure a conservative solution. The FEM formulation has been adapted for use with fluid dynamics governing equations. Although FEM must be carefully formulated to be conservative, it is much more stable than the finite volume approach. However, FEM can require more memory and has slower solution times than the FVM.

In this method, a weighted residual equation is formed:

$$R_i = \iiint W_i Q \, dV^e$$

where R_i is the equation residual at an element vertex i_i, Q is the conservation equation expressed on an element basis, W_i is the weight factor, and V^e is the volume of the element.

FINITE DIFFERENCE METHOD

The finite difference method (FDM) has historical importance and is simple to programme. It is currently only used in few specialized codes, which handle complex geometry with high accuracy and efficiency by using embedded boundaries or overlapping grids (with the solution interpolated across each grid).

$$\frac{\partial Q}{\partial t} + \frac{\partial F}{\partial x} + \frac{\partial G}{\partial y} + \frac{\partial H}{\partial z} = 0$$

where Q is the vector of conserved variables, and F, G, and H are the fluxes in the x, y, and z directions respectively.

SPECTRAL ELEMENT METHOD

Spectral element method is a finite element type method. It requires the mathematical problem (the partial differential equation) to be cast in a weak formulation. This is typically done by multiplying the differential equation by an arbitrary test function and integrating over the whole domain. Purely mathematically, the test functions are completely arbitrary - they belong to an infinitely dimensional function space. Clearly an infinitely dimensional function space cannot be represented on a discrete spectral element mesh. And this is where the spectral element discretization begins. The most crucial thing is the choice of interpolating and testing functions. In a standard, low order FEM in 2D, for quadrilateral elements the most typical choice is the bilinear test or interpolating function of the form $v(x, y) = ax + by + cxy + d$. In a spectral element method however, the interpolating and test functions are chosen to be polynomials of a very high order (typically *e.g.* of the 10th order in CFD applications). This guarantees the rapid convergence of the method.

Furthermore, very efficient integration procedures must be used, since the number of integrations to be performed in a numerical codes is big. Thus, high order Gauss integration quadratures are employed, since they achieve the highest accuracy with the smallest number of computations to be carried out. At the time there are some academic CFD codes based on the spectral element method and some more are currently under development, since the new time-stepping schemes arise in the scientific world. You can refer to the C-CFD website to see movies of incompressible flows in channels simulated with a spectral element solver or to the Numerical Mechanics (see bottom of the page) web site to see a movie of the lid-driven cavity flow obtained with a compeletely novel unconditionally stable time-stepping scheme combined with a spectral element solver.

BOUNDARY ELEMENT METHOD

In the boundary element method, the boundary occupied by the fluid is divided into a surface mesh.

HIGH-RESOLUTION DISCRETIZATION SCHEMES

High-resolution schemes are used where shocks or discontinuities are present. Capturing sharp changes in the solution requires the use of second or higher-order numerical schemes that do not introduce spurious oscillations. This usually necessitates the application of flux limiters to ensure that the solution is total variation diminishing.

TURBULENCE MODELS

In computational modeling of turbulent flows, one common objective is to obtain a model that can predict quantities of interest, such as fluid velocity, for use in engineering designs of the system being modeled. For turbulent flows, the range of length scales and complexity of phenomena involved in turbulence make most modeling approaches prohibitively expensive; the resolution required to resolve all scales involved in turbulence is beyond what is computationally possible. The primary approach in such cases is to create numerical models to approximate unresolved phenomena. This section lists some commonly-used computational models for turbulent flows.

Turbulence models can be classified based on computational expense, which corresponds to the range of scales that are modeled versus resolved (the more turbulent scales that are resolved, the finer the resolution of the simulation, and therefore the higher the computational cost). If a majority or all of the turbulent scales are not modeled, the computational cost is very low, but the tradeoff comes in the form of decreased accuracy.

In addition to the wide range of length and time scales and the associated computational cost, the governing equations of fluid dynamics contain a non-linear convection term and a non-linear and non-local pressure gradient term. These nonlinear equations must be solved numerically with the appropriate boundary and initial conditions.

REYNOLDS-AVERAGED NAVIER–STOKES

Reynolds-averaged Navier-Stokes (RANS) equations are the oldest approach to turbulence modeling. An ensemble version of the governing equations is solved, which introduces new *apparent stresses* known as Reynolds stresses. This adds a second order tensor of unknowns for which various models can provide different levels of closure. It is a common misconception that the RANS equations do not apply to flows with a time-varying mean flow because these equations are 'time-averaged'. In fact, statistically unsteady (or non-stationary) flows can equally be treated. This is sometimes referred to as URANS. There is nothing inherent in Reynolds averaging to preclude this, but the turbulence models used to close the equations are valid only as long as the time over which these changes in the mean occur is large compared to the time scales of the turbulent motion containing most of the energy.

RANS models can be divided into two broad approaches:

BOUSSINESQ HYPOTHESIS

This method involves using an algebraic equation for the Reynolds stresses which include determining the turbulent viscosity, and depending on the level of sophistication of the model, solving transport equations for determining the turbulent kinetic energy and dissipation. Models include k-å (Launder and Spalding), Mixing Length Model (Prandtl), and Zero Equation Model (Cebeci and Smith). The models available in this approach are often referred to by the number of transport equations associated with the method. For example, the Mixing Length model is a "Zero Equation" model because no transport equations are solved; the $k-\in$ is a "Two Equation" model because two transport equations (one for k and one for $\in$) are solved.

REYNOLDS STRESS MODEL (RSM)

This approach attempts to actually solve transport equations for the Reynolds stresses. This means introduction of several transport equations for all the Reynolds stresses and hence this approach is much more costly in CPU effort.

LARGE EDDY SIMULATION

Large eddy simulation (LES) is a technique in which the smallest scales of the flow are removed through a filtering operation, and their effect modeled using subgrid scale models. This allows the largest and most important scales of the turbulence to be resolved, while greatly reducing the computational cost incurred by the smallest scales. This method requires greater computational resources than RANS methods, but is far cheaper than DNS.

DETACHED EDDY SIMULATION

Detached eddy simulations (DES) is a modification of a RANS model in which the model switches to a subgrid scale formulation in regions fine enough for LES calculations. Regions near solid boundaries and where the turbulent length scale is less than the maximum grid dimension are assigned the RANS mode of solution. As the turbulent length scale exceeds the grid dimension, the regions are solved using the LES mode. Therefore the grid resolution for DES is not as demanding as pure LES, thereby considerably cutting down the cost of the computation. Though DES was initially formulated for the Spalart-Allmaras model (Spalart et al., 1997), it can be implemented with other RANS models (Strelets, 2001), by appropriately modifying the length scale which is explicitly or implicitly involved in the RANS model. So while Spalart-Allmaras model based DES acts as LES with a wall model, DES based on other models (like two equation models) behave as a hybrid RANS-LES model. Grid generation is more complicated than for a simple RANS or LES case due to the RANS-LES switch. DES is a non-zonal approach and provides a single smooth velocity field across the RANS and the LES regions of the solutions.

DIRECT NUMERICAL SIMULATION

Direct numerical simulation (DNS) resolves the entire range of turbulent length scales. This marginalizes the effect of models, but is extremely expensive. The computational cost is proportional to Re^3. DNS is intractable for flows with complex geometries or flow configurations.

COHERENT VORTEX SIMULATION

The coherent vortex simulation approach decomposes the turbulent flow field into a coherent part, consisting of organized vortical motion, and the incoherent part, which is the random background flow. This decomposition is done using wavelet filtering. The approach has much in common with LES, since it uses decomposition and resolves only the filtered portion, but different in that it does not use a linear, low-pass filter. Instead, the filtering operation is based on wavelets, and the filter can be adapted as the flow field evolves. Farge and Schneider tested the CVS method with two flow configurations and showed that the coherent portion of the flow exhibited the $-\frac{40}{39}$ energy spectrum exhibited by the total flow, and corresponded to coherent structures (vortex tubes), while the incoherent parts of the flow composed homogeneous background noise, which exhibited no organized structures. Goldstein and Oleg applied the FDV model to large eddy simulation, but did not assume that the wavelet filter completely eliminated all coherent motions from the subfilter scales. By employing both LES and CVS filtering, they showed that the SFS dissipation was dominated by the SFS flow field's coherent portion.

PDF METHODS

Probability density function (PDF) methods for turbulence, first introduced by Lundgren, are based on tracking the one-point PDF of the velocity, $fv(v; x, t)dv$, which gives the probability of the velocity at point x being between v and $v + dv$. This approach is analogous to the kinetic theory of gases, in which the macroscopic properties of a gas are described by a large number of particles. PDF methods are unique in that they can be applied in the framework of a number of different turbulence models; the main differences occur in the form of the PDF transport equation. For example, in the context of large eddy simulation, the PDF becomes the filtered PDF. PDF methods can also be used to describe chemical reactions, and are particularly useful for simulating chemically reacting flows because the chemical source term is closed and does not require a model. The PDF is commonly tracked by using Lagrangian particle methods; when combined with large eddy simulation, this leads to a Langevin equation for subfilter particle evolution.

VORTEX METHOD

The vortex method is a grid-free technique for the simulation of turbulent flows. It uses vortices as the computational elements, mimicking the physical structures in turbulence. Vortex methods were developed as a grid-free methodology that would not be limited by the fundamental smoothing effects associated with grid-based methods. To be practical, however, vortex methods require means for rapidly computing velocities from the vortex elements – in other words they require the solution to a particular form of the N-body problem (in which the motion of N objects is tied to their mutual influences). A breakthrough came in the late 1980s with the development of the fast multipole method (FMM), an algorithm by V. Rokhlin (Yale) and L. Greengard (Courant Institute). This breakthrough paved the way to practical computation of the velocities from the vortex elements and is the basis of successful algorithms. They are especially well-suited to simulating filamentary motion, such as wisps of smoke, in real-time simulations such as video games, because of the fine detail achieved using minimal computation.

Software based on the vortex method offer a new means for solving tough fluid dynamics problems with minimal user intervention. All that is required is specification of problem geometry and setting of boundary and initial conditions. Among the significant advantages of this modern technology;

- It is practically grid-free, thus eliminating numerous iterations associated with RANS and LES.

- All problems are treated identically. No modeling or calibration inputs are required.
- Time-series simulations, which are crucial for correct analysis of acoustics, are possible.
- The small scale and large scale are accurately simulated at the same time.

VORTICITY CONFINEMENT METHOD

The vorticity confinement (VC) method is an Eulerian technique used in the simulation of turbulent wakes. It uses a solitary-wave like approach to produce a stable solution with no numerical spreading. VC can capture the small scale features to within as few as 2 grid cells. Within these features, a non-linear difference equation is solved as opposed to the finite difference equation. VC is similar to shock capturing methods, where conservation laws are satisfied, so that the essential integral quantities are accurately computed.

LINEAR EDDY MODEL

The Linear eddy model is a technique used to simulate the convective mixing that takes place in turbulent flow. Specifically, it provides a mathematical way to describe the interactions of a scalar variable within the vector flow field. It is primarily used in one-dimensional representations of turbulent flow, since it can be applied across a wide range of length scales and Reynolds numbers. This model is generally used as a building block for more complicated flow representations, as it provides high resolution predictions that hold across a large range of flow conditions.

TWO-PHASE FLOW

The modeling of two-phase flow is still under development. Different methods have been proposed lately. The Volume of fluid method has received a lot of attention lately, for problems that do not have dispersed particles, but the Level set method and front tracking are also valuable approaches . Most of these methods are either good in maintaining a sharp interface or at conserving mass . This is crucial since the evaluation of the density, viscosity and surface tension is based on the values averaged over the interface. Lagrangian multiphase models, which are used for dispersed media, are based on solving the Lagrangian equation of motion for the dispersed phase.

SOLUTION ALGORITHMS

Discretization in the space produces a system of ordinary differential equations for unsteady problems and algebraic equations for steady problems. Implicit or semi-implicit methods are generally used to integrate the ordinary differential equations, producing a system of (usually) non-linear algebraic equations. Applying a Newton or Picard iteration produces a system of linear equations which is non-symmetric in the presence of advection and indefinite in the presence of incompressibility. Such systems, particularly in 3D, are frequently too large for direct solvers, so iterative methods are used, either stationary methods such as successive overrelaxation or Krylov subspace methods. Krylov methods such as GMRES, typically used with preconditioning, operate by minimizing the residual over successive subspaces generated by the preconditioned operator.

Multigrid has the advantage of asymptotically optimal performance on many problems. Traditional solvers and preconditioners are effective at reducing high-frequency components of the residual, but low-frequency components typically require many iterations to reduce. By operating on multiple scales, multigrid reduces all components of the residual by similar factors, leading to a mesh-independent number of iterations.

For indefinite systems, preconditioners such as incomplete LU factorization, additive Schwarz, and multigrid perform poorly or fail entirely, so the problem structure must be used for effective preconditioning. Methods commonly used in CFD are the SIMPLE and Uzawa algorithms which exhibit mesh-dependent convergence rates, but recent advances based on block LU factorization combined with multigrid for the resulting definite systems have led to preconditioners that deliver mesh-independent convergence rates.

UNSTEADY AERODYNAMICS

CFD made a major break through in late 70s with the introduction of LTRAN2, a 2-D code to model oscillating airfoils based on transonic small perturbation theory by Ballhaus and associates. It uses a Murman-Cole switch algorithm for modeling the moving shock-waves. Later it was extended to 3-D with use of a rotated difference scheme by AFWAL/Boeing that resulted in LTRAN3.

NUMERICAL METHODS IN FLUID MECHANICS

Fluid motion is governed by the Navier–Stokes equations, a set of coupled and non-linear partial differential equations derived from the basic laws of conservation of mass, momentum and energy. The unknowns are usually the velocity, the pressure and density and temperature. The analytical solution of this equation is impossible hence scientists resort to laboratory experiments in such situations. The answers delivered are, however, usually qualitatively different since dynamical and geometric similitude are difficult to enforce simultaneously between the lab experiment and the prototype. Furthermore, the design and construction of these experiments can be difficult (and costly), particularly for stratified rotating flows. Computational fluid dynamics (CFD) is an additional tool in the arsenal of scientists. In its early days CFD was often controversial, as it involved additional approximation to the governing equations and raised additional (legitimate) issues. Nowadays CFD is an established discipline alongside theoretical and experimental methods. This position is in large part due to the exponential growth of computer power which has allowed us to tackle ever larger and more complex problems.

DISCRETIZATION

The central process in CFD is the process of discretization, *i.e.*. the process of taking differential equations with an infinite number of degrees of freedom, and reducing it to a system of finite degrees of freedom. Hence, instead of determining the solution everywhere and for all times, we will be satisfied with its calculation at a finite number of locations and at specified time intervals. The partial differential equations are then reduced to a system of algebraic equations that can be solved on a computer. Errors creep in during the discretization process. The nature and characteristics of the errors must be controlled in order to ensure that:

- We are solving the correct equations (consistency property)
- That the error can be decreased as we increase the number of degrees of freedom (stability and convergence).

Once these two criteria are established, the power of computing machines can be leveraged to solve the problem in a numerically reliable fashion. Various discretization schemes have been developed to cope with a variety of issues. The most notable for our purposes are: finite difference methods, finite volume methods, finite element methods, and spectral methods.

FINITE DIFFERENCE METHOD

Finite difference replace the infinitesimal limiting process of derivative calculation:

$$\lim_{\Delta x \to 0} f'(n) = \frac{f(x+\Delta x) - f(x)}{\Delta x}$$

with a finite limiting process,*i.e.*.

$$f'(x) = \frac{f(x+\Delta x) - f(x)}{\Delta x} + O(\Delta x)$$

The term O(_x) gives an indication of the magnitude of the error as a function of the mesh spacing. In this instance, the error is halfed if the grid spacing, _x is halved, and we say that this is a first order method. Most FDM used in practice are at least second order accurate except in very special circumstances. Finite Difference method is still the most popular numerical method for solution of PDEs because of their simplicity, efficiency and low computational cost. Their major drawback is in their geometric inflexibility which complicates their applications to general complex domains. These can be alleviated by the use of either mapping techniques and/or masking to fit the computational mesh to the computational domain.

FINITE ELEMENT METHOD

The finite element method was designed to deal with problem with complicated computational regions. The PDE is first recast into a variational form which essentially forces the mean error to be small everywhere. The discretization step proceeds by dividing the computational domain into elements of triangular or rectangular shape. The solution within each element is interpolated with a polynomial of usually low order. Again, the unknowns are the solution at the collocation points. The CFD community adopted the FEM in the 1980s when reliable methods for dealing with advection dominated problems were devised.

SPECTRAL METHOD

Both finite element and finite difference methods are low order methods, usually of 2nd " 4th order, and have local approximation property. By local we mean that a particular collocation point is affected by a limited number of points around it. In contrast, spectral method have global approximation property. The interpolation functions, either polynomials or trigonomic functions are global in nature. Their main benefits is in the rate of convergence which depends on the smoothness of the solution (*i.e.*. how many continuous derivatives does it admit). For infinitely smooth solution, the error decreases exponentially, *i.e.*. faster than algebraic. Spectral methods are mostly used in the computations of homogeneous turbulence, and require relatively simple geometries. Atmospheric model have also adopted spectral methods because of their convergence properties and the regular spherical shape of their computational domain.

FINITE VOLUME METHOD

Finite volume methods are primarily used in aerodynamics applications where strong shocks and discontinuities in the solution occur. Finite volume method solves an integral form of the governing equations so that local continuity property do not have to hold.

COMPUTATIONAL COST

The CPU time to solve the system of equations differs substantially from method to method. Finite differences are usually the cheapest on a per grid point basis followed by the finite element method and spectral method. However, a per grid point basis comparison is a little like comparing apple and oranges. Spectral methods deliver more accuracy on a per grid point basis than either FEM or FDM. The comparison is more meaningful if the question is recast as "what is the computational cost to achieve a given error tolerance?".

The problem becomes one of defining the error measure which is a complicated task in general situations.

APPLICATION OF FINITE DIFFERENCES TO ORDINARY DIFFERENTIAL EQUATIONS

Let us consider the problem of solving the following partial differential equation:

$$u_t + cu_x = vu_{xx},\ 0 \le x \le L$$

The two independent variables are *t* for time and *x* for space. Because of the periodicity, it is sensible to expand the unknown function in a Fourier series:

$$u = \sum_{k=-\infty}^{\infty} \hat{u}_n(t) r^{ik_n x}$$

The Fourier functions form what is called an orthonormal basis. The Fourier series can be differentiated term by term to get an expression for the derivatives of u. The governing equation for the Fourier amplitude is

$$\frac{d\hat{u}}{dt} = -(ick + vk^2)\hat{u}$$

The solution to this simple ODE is:

$$u = \hat{u}_0 e^{kt}$$

FORWARD EULER APPROXIMATION

$$\frac{u^{n+1} - u^n}{\Delta t} \approx ku^n$$

Equation is an explicit approximation to the original differential equation since no information about the unknown function at the future time $(n + 1)_t$ has been used on the right hand side of the equation. In order to derive the error committed in the approximation we rely again on Taylor series.

BACKWARD DIFFERENCE

This is an example of an implicit method since the unknown $u(n + 1)$ has been used in evaluating the slope of the solution on the right hand side; this is not a problem to solve for $u(n + 1)$ in this scalar and linear case. For more complicated situations like a non-linear right hand side or a system of equations, a non-linear system of equations may have to be inverted.

DISCRETIZATION OF NAVIER–STOKES EQUATIONS

Discretization of the Navier–Stokes equations is a reformulation of the equations in such a way that they can be applied to computational fluid dynamics. Several methods of discretization can be applied.

FINITE VOLUME METHOD

INCOMPRESSIBLE FLOW

We begin with the incompressible form of the momentum equation. The equation has been divided through by the density ($P = p/\tilde{n}$) and density has been absorbed into the body force term.

$$\frac{\partial u_i}{\partial t} + \frac{\partial u_i u_j}{\partial x_j} = -\frac{\partial P}{\partial x_i} + v\frac{\partial^2 u_i}{\partial x_j \partial x_j} + f_i$$

The equation is integrated over the control volume of a computational cell.

$$\iiint_v \left[\frac{\partial u_i}{\partial t} + \frac{\partial u_i u_j}{\partial x_j}\right] dV = -\iiint_v \left[\frac{\partial P}{\partial x_i} + v\frac{\partial^2 u_i}{\partial x_j \partial x_j} + f_i\right] dV$$

The time-dependent term and the body force term are assumed constant over the volume of the cell. The divergence theorem is applied to the advection, pressure gradient, and diffusion terms.

$$\frac{\partial u_i}{\partial t}V + \iint_A u_i u_j n_j A = -\iint_A P n_i dA + \iint_A v\frac{\partial u_i}{\partial x_j} n_j dA + f_i V$$

where n is the normal of the surface of the control volume and V is the volume. If the control volume is a polyhedron and values are assumed constant over each face, the area integrals can be written as summations over each face.

$$\frac{\partial u_i}{\partial t}V + \sum_{nbr}(u_i u_j n_j A)_{nbr} = -\sum_{nbr}(P n_i A)_{nbr} + \sum_{nbr}\left(v\frac{\partial u_i}{\partial x_j} n_j A\right)_{nbr} + f_i V$$

where the subscript *nbr* denotes the value at any given face.

TWO-DIMENSIONAL UNIFORMLY-SPACED CARTESIAN GRID

For a two-dimensional Cartesian grid, the equation can be expanded to

$$\frac{\partial u_i}{\partial t}\Delta x \Delta y - (u_i u \Delta y)_w + (u_i u \Delta y)_e - (u_i v \Delta x)_s + (u_i v \Delta x) =$$

$$-(P n_i \Delta y)_w + (P n_i \Delta y)_e - (P n_i \Delta x)_s + (P n_i \Delta x)_n$$

$$-\left(v\frac{\partial u_i}{\partial x}\Delta y\right)_w + \left(v\frac{\partial u_i}{\partial x}\Delta y\right)_e - \left(\frac{\partial u_i}{\partial y}\Delta x)\right)_s + \left(v\frac{\partial u_i}{\partial y}\Delta x\right)_n + f_i$$

On a staggered grid, the x-momentum equation is

$$\frac{\partial u}{\partial t}\Delta x \Delta y - (uu\Delta y)_w + (uu\Delta y)_e - (uu\Delta y)_w + (uu\Delta x)_n =$$

$$+(P\Delta y)_w - (P\Delta y)_e - \left(v\frac{\partial u_i}{\partial x}\Delta y\right)_w + \left(v\frac{\partial u_i}{\partial x}\Delta y\right)_e - \left(v\frac{\partial u_i}{\partial y}\Delta x)\right)_s + \left(v\frac{\partial u_i}{\partial y}\Delta x\right)_n + f_x$$

and the y-momentum equation is

$$\frac{\partial u}{\partial t}\Delta x\Delta y-(uu\Delta y)_w+(uu\Delta y)_e-(uu\Delta y)_w+(uu\Delta x)_n=$$

$$+(P\Delta)_s-(P\Delta x)_n-\left(v\frac{\partial v}{\partial x}\Delta y\right)_w+\left(v\frac{\partial v}{\partial x}\Delta y\right)_e-\left(v\frac{\partial v}{\partial y}\Delta x)\right)_s+\left(v\frac{\partial v}{\partial y}\Delta x\right)_n+f_y$$

The goal at this point is to determine expressions for the face-values for u, v, and P and to approximate the derivatives using finite difference approximations. For this example we will use backward difference for the time derivative and central difference for the spacial derivatives. For both momentum equations, the time derivative becomes

$$\frac{\partial u_i}{\partial x}=\frac{u_i^n-u_i^{n-1}}{\Delta t}$$

where n is the current time index and $\ddot{A}t$ is the time-step. As an example for the spacial derivatives, derivative in the west-face diffusion term in the x-momentum equation becomes

$$\left(\frac{\partial u}{\partial x}\right)_w=\frac{u_{I,J}-u_{I-1,J}}{\Delta x}$$

where I and J are the indices of the x-momentum cell of interest.

2

Systematic Structure of Fluid Dynamics

Fluid dynamics is the sub-discipline of fluid mechanics dealing with fluids (liquids and gases) in motion. It has several subdisciplines itself, including aerodynamics (the study of gases in motion) and hydrodynamics (the study of liquids in motion). Fluid dynamics has a wide range of applications, including calculating forces and moments on aircraft, determining the mass flow rate of petroleum through pipelines, predicting weather patterns, understanding nebulae in interstellar space and reportedly modelling fission weapon detonation. Some of its principles are even used in traffic engineering, where traffic is treated as a continuous fluid.

Fluid dynamics offers a systematic structure that underlies these practical disciplines and that embraces empirical and semi-empirical laws, derived from flow measurement, used to solve practical problems. The solution of a fluid dynamics problem typically involves calculation of various properties of the fluid, such as velocity, pressure, density, and temperature, as functions of space and time.

EQUATIONS OF FLUID DYNAMICS AND AERODYNAMICS

The foundational axioms of fluid dynamics are the conservation laws, specifically, conservation of mass, conservation of momentum (also known as Newton's second law and third law), and conservation of energy. These are based on classical mechanics and are modified in quantum mechanics and general relativity. They are expressed using the Reynolds transport theorem.

In addition to the above, fluids are assumed to obey the continuum assumption. Fluids are composed of molecules that collide with one another and solid objects. However, the continuum assumption considers fluids to be continuous, rather than discrete. Consequently, properties such as density, pressure, temperature, and velocity are taken to be well-defined at infinitely small points, and are assumed to vary continuously from one point to another. The fact that the fluid is made up of discrete molecules is ignored.

For fluids which are sufficiently dense to be a continuum, do not contain ionized species, and have velocities small in relation to the speed of light, the momentum equations for Newtonian fluids are the Navier-Stokes equations, which are non-linear differential equations that describe the flow of a fluid whose stress depends linearly on velocity and on pressure. The unsimplified equations do not have a general closed-form solution, so they are only of use in computational fluid dynamics or when they can be simplified. The equations can be simplified in a number of ways, all of which make them easier to solve. Some of them allow appropriate fluid dynamics problems to be solved in closed form. In addition to the mass, momentum and energy conservation equations, a thermodynamical equation of state giving the pressure as a function of other thermodynamic variables for the fluid is required to completely specify the problem.

An example of this would be the perfect gas equation of state:

$$p = \frac{\rho R_u T}{M}$$

where p is pressure, ρ is density, R_u is the gas constant, M is the molecular mass and T is temperature.

COMPRESSIBLE VS INCOMPRESSIBLE FLOW

All fluids are compressible to some extent, that is changes in pressure or temperature will result in changes in density. However, in many situations the changes in pressure and temperature are sufficiently small that the changes in density are negligible. In this case the flow can be modelled as an incompressible flow. Otherwise the more general compressible flow equations must be used.

Mathematically, incompressibility is expressed by saying that the density ρ of a fluid parcel does not change as it moves in the flow, *i.e.*

$$\frac{D\rho}{Dt} = 0,$$

where D/Dt is the convective derivative. This additional constraint simplifies the governing equations, especially in the case when the fluid has a uniform density. For flow of gases, to determine whether to use compressible or incompressible fluid dynamics, the Mach number of the problem is evaluated. As a rough guide, compressible effects can be ignored at Mach numbers below approximately 0.3. For liquids, whether the incompressible assumption is valid depends on the fluid properties (specifically the critical pressure and temperature of the fluid) and the flow conditions (how close to the critical pressure the actual flow pressure becomes). Acoustic problems always require allowing compressibility, since sound waves are compression waves involving changes in pressure and density of the medium through which they propagate.

VISCOUS VS INVISCID FLOW

Viscous problems are those in which fluid friction has significant effects on the solution. The Reynolds number can be used to evaluate whether viscous or inviscid equations are appropriate to the problem. Stokes flow is flow at very low Reynolds numbers, such that inertial forces can be neglected compared to viscous forces.

On the contrary, high Reynolds numbers indicate that the inertial forces are more significant than the viscous (friction) forces. Therefore, we may assume the flow to be an inviscid flow, an approximation in which we neglect viscosity at all, compared to inertial terms. This idea can work fairly well when the Reynolds number is high, even if certain problems, such as those involving boundaries, may require that viscosity be included. Viscosity often cannot be neglected near boundaries because the no-slip condition can generate a region of large strain rate (a Boundary layer) which enhances the effect of even a small amount of viscosity, generating vorticity. Therefore, to calculate net forces on bodies (such as wings) we should use viscous equations. As illustrated by d'Alembert's paradox, a body in an inviscid fluid will experience no force. The standard equations of inviscid flow are the Euler equations. Another often used model, especially in computational fluid dynamics, is to use the Euler equations far from the body and the boundary layer equations, which incorporate viscosity, close to the body.

The Euler equations can be integrated along a streamline to get Bernoulli's equation. When the flow is everywhere irrotational and inviscid, Bernoulli's equation can be used throughout the field.

STEADY VS UNSTEADY FLOW

When all time derivatives of a flow field vanish, the flow is considered steady. Otherwise, it is called unsteady. Whether a particular flow is steady or unsteady, can depend on the chosen frame of reference. For instance, laminar flow over a sphere is steady in the frame of reference that is stationary with respect to the sphere. In a frame of reference that is stationary with respect to the background flow, the flow is unsteady.

Turbulent flows are unsteady by definition. A turbulent flow can, however, be statistically stationary. According to Pope, Turbulent Flows, Cambridge university press, page 75:

The random field U(x,t) is statistically stationary if all statistics are invariant under a shift in time. This roughly means that all statistical properties are constant in time. Often, the mean field is the object of interest, and this is constant too in a statistically stationary flow. Steady flows are often more tractable than otherwise similar unsteady flows. The governing equations of a steady problem have one dimension less (time) than the governing equations of the same problem without taking advantage of the steadiness of the flow field. Although strictly unsteady flows, time-periodic problems can often be solved by the same techniques as steady flows. For this reason, they can be considered to be somewhere between steady and unsteady.

LAMINAR VS TURBULENT FLOW

Turbulence is flow dominated by recirculation, eddies, and apparent randomness. Flow in which turbulence is not exhibited is called laminar. It should be noted, however, that the presence of eddies or recirculation does not necessarily indicate turbulent flow—these phenomena may be present in laminar flow as well. Mathematically, turbulent flow is often represented via Reynolds decomposition, in which the flow is broken down into the sum of a steady component and a perturbation component.

It is believed that turbulent flows obey the Navier-Stokes equations. Direct numerical simulation (DNS), based on the incompressible Navier-Stokes equations, makes it possible to simulate turbulent flows with moderate Reynolds numbers (restrictions depend on the power of computer and efficiency of solution algorithm). The results of DNS agree with the experimental data. Most flows of interest have Reynolds numbers too high for DNS to be a viable option, given the state of computational power for the next few decades. Any flight vehicle large enough to carry a human ($L > 3$ m), moving faster than 72 km/h (20 m/s) is well beyond the limit of DNS simulation ($Re = 4$ million). Transport aircraft wings (such as on an Airbus A300 or Boeing 747) have Reynolds numbers of 40 million (based on the wing chord). In order to solve these real life flow problems, turbulence models will be a necessity for the foreseeable future. Reynolds-averaged Navier-Stokes equations combined with turbulence modeling provides a model of the effects of the turbulent flow, mainly the additional momentum transfer provided by the Reynolds stresses, although the turbulence also enhances the heat and mass transfer. Large eddy simulation also holds promise as a simulation methodology, especially in the guise of detached eddy simulation (DES), which is a combination of turbulence modeling and large eddy simulation.

NEWTONIAN VS NON-NEWTONIAN FLUIDS

Sir Isaac Newton showed how stress and the rate of change of strain are very close to linearly related for many familiar fluids, such as water and air. These Newtonian fluids are modeled by a coefficient called viscosity, which depends on the specific fluid. However, some of the other materials, such as emulsions and slurries and some visco-elastic materials (eg. blood, some polymers), have more complicated *non-Newtonian* stress-strain behaviours. These materials include *sticky liquids* such as latex, honey, and lubricants which are studied in the sub-discipline of rheology.

OTHER APPROXIMATIONS

There are a large number of other possible approximations to fluid dynamic problems. Some of the more commonly used are listed below.

- The Boussinesq approximation neglects variations in density except to calculate buoyancy forces. It is often used in free convection problems where density changes are small.

- Lubrication theory exploits the large aspect ratio of the domain to show that certain terms in the equations are small and so can be neglected.
- Slender-body theory is a methodology used in Stokes flow problems to estimate the force on, or flow field around, a long slender object in a viscous fluid.
- The shallow-water equations can be used to describe a layer of relatively inviscid fluid with a free surface, in which surface gradients are small.
- Darcy's law is use for flow in porous media, and works with variables averaged over several pore-widths.
- In rotating systems, the quasi-geostrophic approximation assumes an almost perfect balance between pressure gradients and the Coriolis force. It is useful in the study of atmospheric dynamics.

EULER'S FORMULA

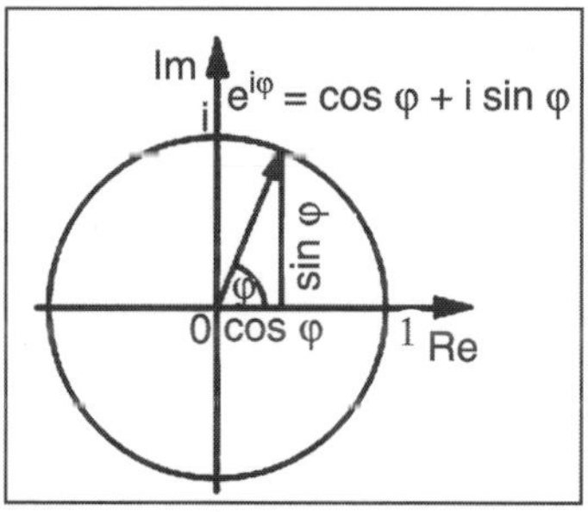

Euler's formula, named after Leonhard Euler, is a mathematical formula in complex analysis that shows a deep relationship between the trigonometric functions and the complex exponential function. (Euler's identity is a special case of the Euler formula.)

Euler's formula states that, for any real number x,

$$e^{ix} = \cos(x) + i \sin(x)$$

where,

e is the base of the natural logarithm

i is the imaginary unit

cos and sin are trigonometric functions.

Richard Feynman called Euler's formula "our jewel" and "the most remarkable formula in mathematics".

History

Euler's formula was proven for the first time by Roger Cotes in 1714 in the form,

$$\text{Ln}\,(\cos(x) + 0\sin(x)) = ix$$

(where "ln" means natural logarithm, *i.e.* log with base *e*).

It was Euler who published the equation in its current form in 1748, basing his proof on the infinite series of both sides being equal. Neither of these men saw the geometrical interpretation of the formula: the view of complex numbers as points in the complex plane arose only some 50 years later. Euler considered it natural to introduce students to complex numbers much earlier than we do today. In his elementary algebra text book, *Elements of Algebra*, he introduces these numbers almost at once and then uses them in a natural way throughout.

RHEOLOGY OF FLUIDS

For most fluids, viscosity is not a constant but varies with the shear rate. Such fluids are called rate dependent. To check for rate dependent effects, a fluid is tested for shear stress or viscosity at a number of shear rates on the same viscometer. From this data the rheological constants from the Bingham and Power Law Models can be obtained or when this data is plotted the result is called a viscosity profile. A few examples are shown below.

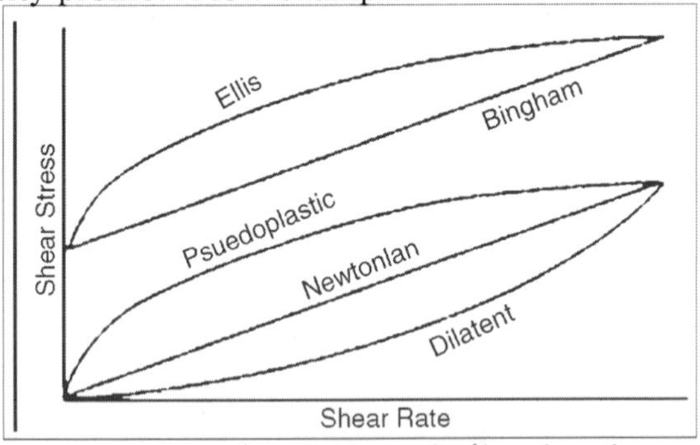

In a few systems, such as in water and oils, the shear stress and shear rate are in direct proportion. Such fluids have a constant viscosity. They are termed Newtonian after Newton, who first postulated their existence.

Most polymers are pseudoplastic. In this case, increased shear rate causes a progressive decrease in viscosity and inversely a decrease in shear rate will result in an increase in viscosity due to alignment of the long polymer chains along the flow lines. A few rare concentrated suspensions have been found to be dilatent.

These fluids increase in viscosity with increasing shear. Some fluids have a certain critical shear stress that must be exceeded before flow can begin. This critical shear stress is called the "yield value". If the fluid has essentially Newtonian flow after the yield value is exceeded, it is called a Bingham Plastic fluid. This model has been used extensively for circulating fluids in the oil industry.

Another model that has gained acceptance for pseudoplastic fluids is the Ostwald-deWaele-Nutting or Power Law model. If the fluid has a yield value and also follows the pseudoplastic model, it is referred to as an Ellis fluid and can be described by a simplified Hershel-Bulkley model. The Power Law and Hershel-Bulkley model have often been used to describe polymer solutions.

Besides the rate dependent effects discussed above, fluids can also exhibit time dependent effects. In these cases, the viscosity of the fluid tends to drop with continued shear at a constant rate. The following curve illustrates the two possible time dependent effects.

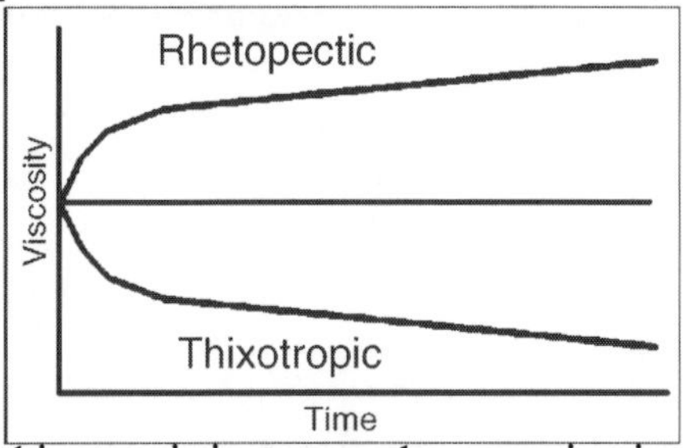

Fluids which are thixotropic in nature decrease in viscosity with time. Some polymers are thixotropic to a certain degree. In certain rare situations, highly concentrated suspensions can display rheopectic flow.

In these unusual cases, the viscosity increases with continued shear at a given rate. These models of rate and time dependent effects are designed to provide assistance in understanding fluid flow. Different models may suite the same fluid under various conditions. There are always a variety of possible explanations for any given situation.

Numerous viscometers are available for measuring the properties of a fluid. If it is desired to measure the viscosity of a sample and obtain data in basic rheological parameters (shear stress and shear rate) a viscometer that operates using one of three basic flow types; cone and plate, Colette, or capillary, must be used.

Empirical viscometers, not based on these flow types are widely used but the data from these can not be reduced to fundamental rheological measurements. Some of the more common types of viscometers are listed in Application of these principles to real fluid systems is not difficult. one must have an understanding of the critical property (or properties) of a system.

For example, a suspension that is designed to remain homogeneous must have sufficient viscosity at low shear rates in order to prevent solids settling. On the other hand, some applications require that the viscosity at high shear rates be as low as possible to reduce friction pressure losses. It is recognized that measurements at room temperature and pressure do not truly represent the fluid under down hole conditions, but extensive field studies have resulted in a high degree of success in predicting a fluids performance from this data. g system. It provides a useful addition to the "normal" steady shear measurements.

VISCO ELASTICITY

The distinction between solid and liquid materials is not as clear as the preceding would indicate. Most materials have some solid like character as well as some viscous properties. It is easy to visualize the appearance differences between viscous liquids and gelled or cross-linked solids.

These fluids are characterized by taking the viscosity of liquids and the gel strength of solids. Many different viscometers and gel testers are sold, depending on the industry.

A problem arises when one tries to characterize a fluid that behaves like both a viscous liquid and a gel. Such fluids are called viscoelastic fluids. They can be pumped easily (although they have the appearance of a viscous liquid), but they are able to suspend small solid particles indefinitely (so have the characteristic of an elastic solid). In fact, molten polymers are viscoelastic to different degrees.

A way of quantifying the viscoelastic properties of materials is by measurement of their elastic modulus and viscous modulus. It is convenient to think of these as vector quantities located 90 degrees apart. Vector quantities have a magnitude (which is indicated by the length of the vector) as well as a direction. When the rheology is measured with a viscometer in steady shear, only the resultant vector of the viscous and elastic modulus is measured, this is usually called the shear stress.

Shear stress divided by the shear rate is called complex viscosity. Both of these fluids would have the same steady shear or complex viscosity since the resultant vector of the two moduli is the same magnitude.

However, viscoelastic measurement reveals that they are quite different in their nature. Much more the "viscosity" of the elastic fluid is due to its elastic modulus while the viscous fluid derives most of its "viscosity" from its viscous modulus. The elastic fluid is much better at suspending particles or stabilizing a foam. The viscous fluid will slow down the rate of settling but the primary effect is an increase in friction pressure. If the solution has a significant elastic component, it will act like a solid and may actually be able to prevent settling.

These differences can only be revealed through viscoelastic measurements and performance comparisons. There are laboratory instruments called rheometers that can simultaneously measure the viscous and gel nature of a fluid.

The rheometer used in this report is made by Rheometrics. It has a cup and bob configuration much like the Fann. Because the rheometer operates in an oscillating mode by applying a sinusoidally varying strain or deformation to the sample, it does not generate shear rate data in units of sec-1. The technically correct designation for its shear rate measurements called "frequency" are in terms of radians per second. Fortunately frequency in radians/second closely correlates with shear rate in sec-1 from the rotational viscometer.

The rheometer measures the resistance to the oscillation, and the phase delay of the resistance. A perfect elastic gel would have no phase delay, while a perfect viscous liquid would have a 90 degree phase delay.

All materials, have a phase angle somewhere between 0 and 90 degrees. The computer attached to the rheometer can qualify both the viscous nature, called the viscous modulus, and the gel nature, called the elastic modulus or elasticity.

The instrument also determines the vector sum of these two moduli which is the complex viscosity. While it is undoubtedly a sophisticated technique, viscoelastic measurements can add a great deal of understanding of today's polymers.

MEASURE OF RESISTANCE TO FLOW

VISCOSITY

The *viscosity* is a measure of resistance to flow. It we tip over a glass of water on the table, the water will spill out before we can stop it.

If we tip over a jar of honey, we probably can set it upright again before much honey flow out; this is because the honey has much greater resistance to flow, more viscosity, than water. A more precise definition of viscosity is possible in terms of following experiment:

Consider two long, solid plates separated by a thin film of fluid . If we slide the upper plate steadily in the x direction with velocity V_0, a force is needed to overcome the friction in the fluid between the plates. This force is different for different velocities, plate sizes, fluids, and distances between the plates. We can eliminate the effect of different plate sizes, however, by measuring the force per unit area of the plate, which we define as the shear stress τ.

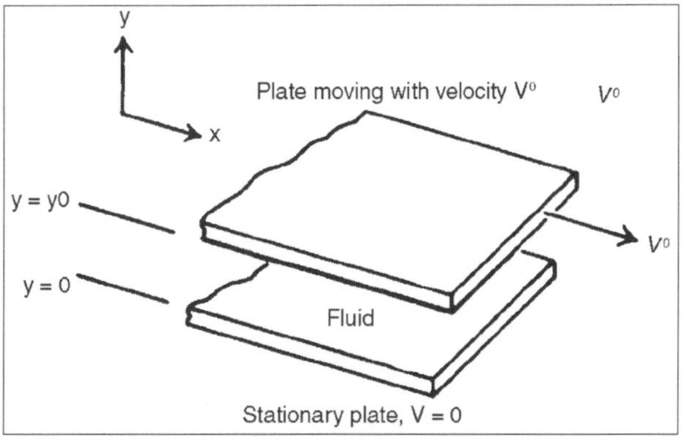

Fig. The Sliding Plate Experiment.

It has been demonstrated experimentally that at low values of V_0 the velocity profile in the fluid between the plates is linear, *i.e.*.,

$$V = \frac{V_o y}{y_o}$$

so that,

$$\frac{dV}{dy} = \frac{V_o}{y_o}$$

It has also been demonstrated experimentally that for most fluids the results of this experiment can be shown most conveniently on a plot τ versus dV / dy. As shown here, dV / dy is simply is velocity divided by a distance. In more complex geometries, it is the limiting value of such a ratio at a point. It is commonly called the *shear rate*, *rate of strain*, and *rate of shear deformation*, which all means exactly the same thing. Four different kinds of curve are shown as experiment results in the figure.

All four are observed in nature. The behaviour most common in nature is that represented by the straight line through the origin. This line is called *Newtonian* because it is described by Newton's law of viscosity,

$$\iota = \upsilon \frac{dV}{dy}$$

This equation says that the shear stress τ is linearly proportional to the velocity gradient dV / dy. It is also the definition of viscosity, because we can rearrange it as,

$$\upsilon \frac{\iota}{dV/dy}$$

Here μ is called the *viscosity* or the coefficient of *viscosity*.

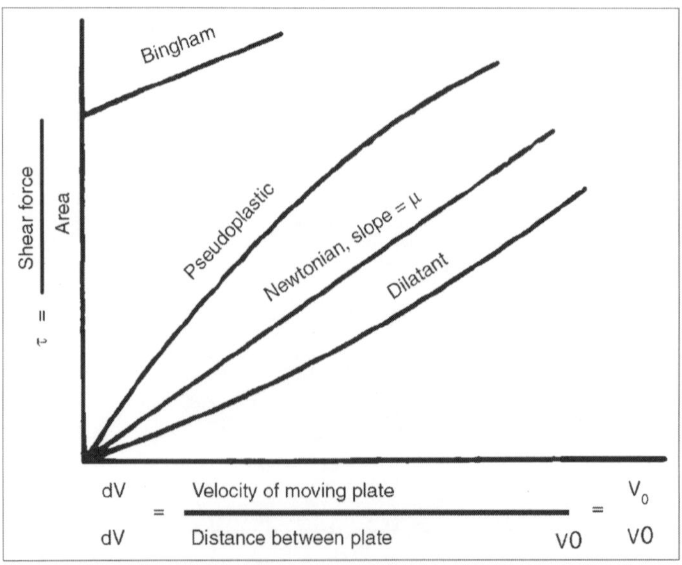

Fig. Possible outcomes of the Sliding-plate Experiment at Constant Temperature and Pressure.

For fluids such as air, the value of μ is very low; therefore, their observed behaviour is represented in Fig above by a straight line through the origin, very close to the dV/dy axis. For fluid such as corn syrup, the value of μ is very large, and the straight line through the origin is close to the τ axis.

Fluids that exhibit this behaviour in the sliding plate experiment (*i.e.*., fluids that obey Newton's law of viscosity) are called *Newtonian fluids*.

All others are called *non-newtonian fluids*. Which fluids are Newtonian? All gases are Newtonian. All liquids for which we can write a simple chemical formula are Newtonian, such as water, benzene, ethyl alcohol, carbon tetrachloride, and hexane. Most solutions of simple molecules are Newtonian, such as aqueous solutions of inorganic salts and of sugar. Which fluids are non-newtonian? Generally, non-newtonian fluids are complex mixtures: slurries, pastes, gels, polymer solutions, etc. Most non-newtonian fluids are composed of molecules or particles that are much larger than water molecules, such as the sand grains in a mud or the collagen molecules in gelatin, which are thousands or millions of times larger than water molecules.

In discussing non-newtonian fluids, we must agree on what we mean by viscosity. If we retain the definition given by Eqn. then the viscosity can no longer be considered a constant independent of *dV/dy* for a given temperature, but must bc considered a function of *dV/dy*. This is shown in figure below.

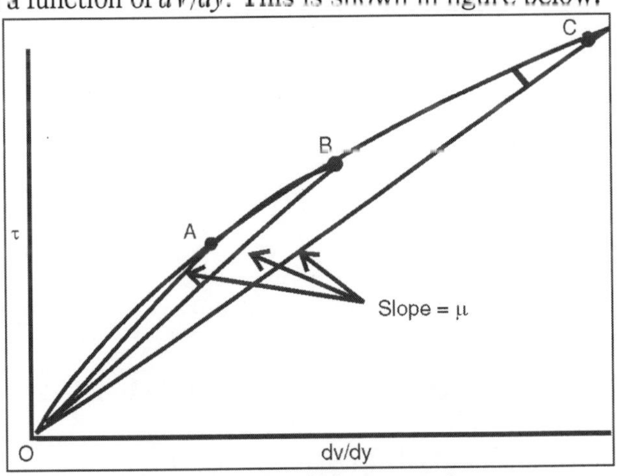

Fig. Viscosity of Nonnewtonian Fluids at Constant Temperature

Here each of lines OA, OB, and OC have slope μ, so the viscosity is sdecreasing with increasing *dV/dy*.

Using this definition, we can observe that there are three common types of nonnewtonian fluid:

- *Bingham fluids*: Sometimes called *bingham plastics*, resist a small shear stress indefinitely but flow easily under larger shear stresses. One may say that at low stresses the viscosity is infinite and that at higher stresses the viscosity decreases with increasing velocity gradient. Examples are bread dough, tooth paste, jellies, and some slurries.

- *Pseudoplastic fluids:* Show a viscosity that decreases with increasing velocity gradient. Examples are most slurries, muds, polymer solutions, solutions of natural gums, and blood.
- *Dilatant fluids:* Show a viscosity that increases with increasing velocity gradient. They are uncommon, but starch suspensions behave in this way.

The basic unit of viscosity is the poise, where 1P = 1g / (cm.s) = 0.1 Pa.s = 6.72×10^{-2} lbm / (ft.s). It is widely used for materials such as high-polymer solutions and molten polymers.

However, it is too large a unit for most common fluids. By sheer coincidence the viscosity of pure water at about 68°F is 0.01 P; for that reason the common unit of viscosity in the United States is the centipoises, where

$$1\ \text{cP} = 0.01\ \text{P} = 0.01\ \text{g} / (\text{cm.s}) = 6.72 \times 10^{-4} \text{lb}_m / (\text{ft.s}) = 0.001 \text{Pa.s.}$$

Hence, the viscosity of a fluid, expressed in centipoises, is the same as the ratio of its viscosity to that of water at room temperature.

Kinematic Viscosity

In many engineering problems, viscosity appears only in the relation of viscosity divided by density. Therefore, to save writing, we define,

$$\text{Kinematic Viscosity} = \upsilon = \frac{\mu}{\rho}$$

The most common unit of kinematic viscosity is the centistokes (cSt):

$$1\,\text{cSt} = \frac{1\,\text{cP}}{1\,\text{g}/\text{cm}^3} = 1.08 \times 10^{-5}\,\frac{\text{ft}^2}{\text{s}} = 10^{-6}\,\frac{\text{m}^2}{\text{s}}$$

(The kinematic viscosity of water at room temperature is 1cSt.) To avoid confusion over which viscosity is being used, some writers refer to the viscosity μ as the absolute viscosity.

NEWTONIAN KINETIC ENERGY

KINETIC ENERGY OF RIGID BODIES

In classical mechanics, the kinetic energy of a "point object" (a body so small that its size can be ignored), or a non-rotating rigid body, is given by the equation $E_k = \frac{1}{2}mv^2$ where m is the mass and v is the speed of the body.

For example - one would calculate the kinetic energy of an 80 kg mass traveling at 18 meters per second (40 mph) as,

$$\frac{1}{2} \cdot 80 \cdot 18^2 = 12{,}960\,\text{joules.}.$$

Note that the kinetic energy increases with the square of the speed. This means, for example, that if you are traveling twice as fast, you will have four times as much kinetic energy. As a result of this, a car traveling twice as fast requires four times as much distance to stop.

Thus, the kinetic energy can be calculated using the formula:

$$E_k = \frac{1}{2}mv^2$$

For the *translational kinetic energy* of a body with constant mass *m*, whose center of mass is moving in a straight line with speed *v*, as seen above is equal to

$$E_t = \frac{1}{2}mv^2$$

where:

m is mass of the body

v is speed of the center of mass of the body.

Thus kinetic energy is a relative measure and no object can be said to have a unique kinetic energy. A rocket engine could be seen to transfer its energy to the rocket ship or to the exhaust stream depending upon the chosen frame of reference. But the total energy of the system, *i.e.*. kinetic energy, fuel chemical energy, heat energy etc, will be conserved regardless of the choice of measurement frame.

The kinetic energy of an object is related to its momentum by the equation:

$$E_k = \frac{p^2}{2m}$$

Derivation and Definition

The work done accelerating a particle during the infinitesimal time interval *dt* is given by the dot product of *force* and *displacement*:

$$F \cdot dx = F \cdot vdt = \frac{dp}{dt} \cdot vdt = v \cdot dp = v \cdot d(mv)$$

Applying the product rule we see that:

$$d(v \cdot v) = (dv) \cdot v + v \cdot (dv) = 2(v \cdot dv)$$

Therefore (assuming constant mass), the following can be seen:

$$v \cdot d(mv) = \frac{m}{2}d(v \cdot v) = \frac{m}{2}dv^2 = d\left(\frac{mv^2}{2}\right)$$

Since this is a total differential (that is, it only depends on the final state, not how the particle got there), we can integrate it and call the result kinetic energy:

$$E_k = \int F \cdot dx = \int v \cdot dp = \frac{mv^2}{2}$$

This equation states that the kinetic energy (E_k) is equal to the integral of the dot product of the velocity (v) of a body and the infinitesimal change of the body's momentum (p). It is assumed that the body starts with no kinetic energy when it is at rest (motionless).

KINETIC ENERGY OF SYSTEMS

For a single point, or a rigid body that is not rotating, the kinetic energy goes to zero when the body stops.

However, for systems containing multiple independently moving bodies, which may exert forces between themselves, and may (or may not) be rotating; this is no longer true.

This energy is called 'internal energy'.

The kinetic energy of the system at any instant in time is simply the sum of the kinetic energies of the masses- including the kinetic energy due to the rotations.

An example would be the solar system. In the center of mass frame of the solar system, the Sun is (almost) stationary, but the planets and planetoids are in motion about it. Thus even in a stationary center of mass frame, there is still kinetic energy present.

However, recalculating the energy from different frames would be tedious, but there is a trick. The kinetic energy of the system from a different inertial frame can be calculated simply from the sum of the kinetic energy in the center of mass frame and adding on the energy that the total mass of bodies in the center of mass frame would have if it were moving at the relative speed between the two frames.

This may be simply shown: let V be the relative speed of the frame *k* from the center of mass frame *i* :

$$E_k = \int \frac{v_k^2 dm}{2} = \int \frac{(v_i + V)^2 dm}{2} = \int \frac{(v_i^2 + 2v_i V + V^2) dm}{2}$$

$$= \int \frac{v_i^2 dm}{2} + v \int v_i dm + \frac{V^2}{2} \int dm$$

However, let

$$\int \frac{v_i^2 dm}{2} = E_i$$

the kinetic energy in the center of mass frame, $\int v_i dm$ would be simply the total momentum which is by definition zero in the center of mass frame, and let the total mass:

$$\int dm = M.$$

Substituting, we get:

$$E_k = E_i + \frac{MV^2}{2}$$

The kinetic energy of a system thus depends on the inertial frame of reference and it is lowest with respect to the center of mass reference frame, *i.e..*, in a frame of reference in which the center of mass is stationary. In any other frame of reference there is an additional kinetic energy corresponding to the total mass moving at the speed of the center of mass.

Thus it sometimes is convenient to split the total kinetic energy of this body into the sum of the body's center-of-mass translational kinetic energy and the energy of rotation around the center of mass rotational energy:

$$E_k = E_t + E_r$$

where:

E_k is the total kinetic energy

E_t is the translational kinetic energy

E_r is the *rotational energy* or *angular kinetic energy* in the rest frame

Thus the kinetic energy of a tennis ball in flight has is the kinetic energy due to its rotation, plus the kinetic energy due to its translation.

Rotational Systems

If a rigid body is rotating, its *rotational kinetic energy* or angular kinetic energy about any line through the center of mass, and r is the distance of any mass dm from that line is simply sum of kinetic energies of its moving parts, and thus it is equal to:

$$E_r = \int \frac{v^2 dm}{2} = \int \frac{(rw)^2 dm}{2} = \frac{w^2}{2} \int r^2 dm = \frac{w^2}{2} I = \frac{1}{2} Iw^2$$

where:

I is the body's moment of inertia

ω is the body's angular velocity.

The moment of inertia must be taken about an axis through the center of mass and the rotation measured by ω must be around that axis.

Relativistic Kinetic Energy of Rigid Bodies

In special relativity, we must change the expression for linear momentum. Integrating by parts, we get:

$$E_r = \int v \cdot dp =$$

$$\int v \cdot d(m\gamma v) = m\gamma v \cdot v - \int m\gamma v \cdot dv = m\gamma v^2 - \frac{m}{2} = \int \gamma d(v^2)$$

Remembering that $\gamma = \left(1 - v^2 / c^2\right)^{-1/2}$, we get:

$$E_k = m\gamma v^2 - \frac{-mc^2}{2} \int \gamma d\left(1 - v^2 / c^2\right)$$
$$= m\gamma v^2 + mc^2 \left(1 - v^2 / c^2\right)^{1/2} + C$$

And thus:

$$E_k = m\gamma\left(v^2 + c^2\left(1 - v^2 / c^2\right)\right) + C$$
$$= m\gamma\left(v^2 + c^2 - v^2\right) + C = m\gamma c^2 + C$$

The constant of integration is found by observing that $\gamma = 1$ when v = 0, so we get the usual formula:

$$E_k = m\gamma c^2 - mc^2 = \frac{mc^2}{\sqrt{1 - v^2 / c^2}} - mc^2$$

If a body's speed is a significant fraction of the speed of light, it is necessary to use relativistic mechanics (the theory of relativity as expounded by Albert Einstein) to calculate its kinetic energy.

For a relativistic object the momentum p is equal to:

$$p = \frac{mc^2}{\sqrt{1 - (v / c)^2}},$$

where *m* is the rest mass, *v* is the object's speed, and *c* is the speed of light in vacuum.

Thus the work expended accelerating an object from rest to a relativistic speed is:

$$Ek = \frac{mc^2}{\sqrt{1 - (v / c)^2}} - mc^2.$$

The equation shows that the energy of an object approaches infinity as the velocity *v* approaches the speed of light *c*, thus it is impossible to accelerate an object across this boundary. The mathematical by-product of this calculation is the mass-energy equivalence formula—the body at rest must have energy content equal to:

$$E_{rest} = mc^2$$

At a low speed (v<<c), the relativistic kinetic energy may be approximated well by the classical kinetic energy. This is done by binomial approximation. Indeed, taking Taylor expansion for square root and keeping first two terms we get:

$$E_k \approx mc^2\left(1 + \frac{1}{2} v^2 / c^2\right) - mc^2 = \frac{1}{2} mv^2.$$

So, the total energy E can be partitioned into the energy of the rest mass plus the traditional Newtonian kinetic energy at low speed.

When objects move at a speed much slower than light (*e.g.* in everyday phenomena on Earth), the first two terms of the series predominate. The next term in the approximation is small for low speeds, and can be found by extending the expansion into a Taylor series by one more term:

$$E \approx mc^2\left(1+\frac{1}{2}v^2/c^2+\frac{3}{8}v^4/c^4\right)=mc^2+\frac{1}{2}mv^2+\frac{3}{8}mv^4/c^2.$$

For example, for a speed of 10 km/s the correction to the Newtonian kinetic energy is 0.07 J/kg (on a Newtonian kinetic energy of 50 MJ/kg) and for a speed of 100 km/s it is 710 J/kg (on a Newtonian kinetic energy of 5 GJ/kg), etc.

For higher speeds, the formula for the relativistic kinetic energy is derived by simply subtracting the rest mass energy from the total energy:

$$E = m\gamma c^2 - mc^2 = mc^2\left(\frac{1}{\sqrt{1-(v/c)}}-1\right)$$

The relation between kinetic energy and momentum is more complicated in this case, and is given by the equation:

$$E_k = \sqrt{p^2c^2+m^2c^4}-mc^2$$

This can also be expanded as a Taylor series, the first term of which is the simple expression from Newtonian mechanics.

What this suggests is that the formulas for energy and momentum are not special and axiomatic, but rather concepts which emerge from the equation of mass with energy and the principles of relativity.

Quantum Mechanical Kinetic Energy of Rigid Bodies

In the realm of quantum mechanics, the expectation value of the electron kinetic energy, $\langle\hat{T}\rangle$, for a system of electrons described by the wavefunction $|\psi\rangle$ is a sum of 1-electron operator expectation values:

$$\langle\hat{T}\rangle = -\frac{\hbar^2}{2m_e}\left\langle\psi\left|\sum_{i=1}^{N}\nabla_i^2\right|\psi\right\rangle$$

where m_e is the mass of the electron and ∇_i^2 is the Laplacian operator acting upon the coordinates of the ith electron and the summation runs over all electrons. Notice that this is the quantized version of the non-relativistic expression for kinetic energy in terms of momentum:

$$E_k = \frac{p^2}{2m}$$

The density functional formalism of quantum mechanics requires knowledge of the electron density *only*, *i.e.*, it formally does not require knowledge of the wavefunction. Given an electron density $\rho(\mathrm{r})$, the exact N-electron kinetic energy functional is unknown; however, for the specific case of a 1-electron system, the kinetic energy can be written as

$$T(\rho)=\frac{1}{8}\int\frac{\nabla\rho(\mathrm{r})\cdot\nabla\rho(\mathrm{r})}{\rho(\mathrm{r})}d^3r$$

where $T[\rho]$ is known as the Von Weizsacker kinetic energy functional.

KINETIC ENERGY CORRECTION

Before claiming that this formula explains the fine structure of the hydrogen atom, however, one needs to be careful. The correction is of the order a^4, which means it of the order v^4, where v is the electron speed. The kinetic energy used in the Hamiltonian when solving the Schrφdinger equation was just $p^2/2m$, which contributed to order a^2. However, the next term in the expansion of the true relativistic kinetic energy is of order $p^4$and hence will contribute to order a^4. So if one wishes to quote the energy splittings of the hydrogen atom accurate to order a^4, one had better include the contribution from this further correction. The relativistic kinetic energy of the electron can be expanded in terms of momentum as,

$$T=\frac{p^2}{2m}-\frac{p^4}{8m^3c^2}+\ldots$$

Therefore, the correction to the Hamiltonian is,

$$\Delta H_{rel}=-\frac{1}{8m^3c^2}p^4.$$

At first sight, this looks quite complicated, since it involves the operator $p^4=\hbar^4\nabla^4$. However, one can make use of the fact that,

$$\frac{p^2}{2m}=En-V$$

to get,

$$\Delta H_{rel}=-\frac{1}{2mc^2}\left(E_n^2-2E_nV+V^2\right).$$

With $V = -ze^2/r$, applying first-order perturbation theory to this Hamiltonian reduces to the problem of finding the expectation values of r^{-1} and r^{-2}. This can be done with some effort, and the result is,

$$\Delta E_{rel}=-(Za)^4mc^2\frac{1}{2n}\left[\frac{1}{\left(I+\frac{1}{2}\right)}-\frac{3}{4}\right].$$

Combining equations 57 and 52 and using the fact that j=l–1/2,l+1/2, the complete energy correction to order (Za)^{4}may be written

$$\Delta E_{fs} = \Delta E_{rel} + \Delta E_{so} = -(Za)^4 mc^2 \frac{1}{2n}\left[\frac{1}{\left(j+\frac{1}{2}\right)} - \frac{3}{4n}\right].$$

This energy correction depends only on j and is called the fine structure of the hydrogen atom, since it is of order a^2 ~ 10^{-4}times smaller than the principle energy splittings. This is why a is known as the fine-structure constant.

The fine structure of the hydrogen atom. Note that all levels are shifted down from the Bohr energies, and that for every n and l there are two states corresponding to j=l–1/2 and j=l+1/2, except for s states.

Also note that states with the same n and j but different l have the same energies, though this will be shown later not to be true due an effect know as the Lamb shift. As an aside, these fine structure splittings were derived by Sommerfeld by modifying the Bohr theory to allow elliptical orbits and then calculating the energy differences between the different states due to differences in the average velocity of the electron. By using the wrong method he got exactly the right answer, a coincidence which caused much confusion at the time.

Strictly speaking, equation 58 has only been shown to be correct for $I \neq 0$ states, although it turns out to be correct for all l. To do the calculation correctly for l=0, one needs to include the effect of an additional term in the Hamiltonian known as the Darwin term, which is purely an effect of relativistic quantum mechanics and can only be understood in the context of the Dirac theory.

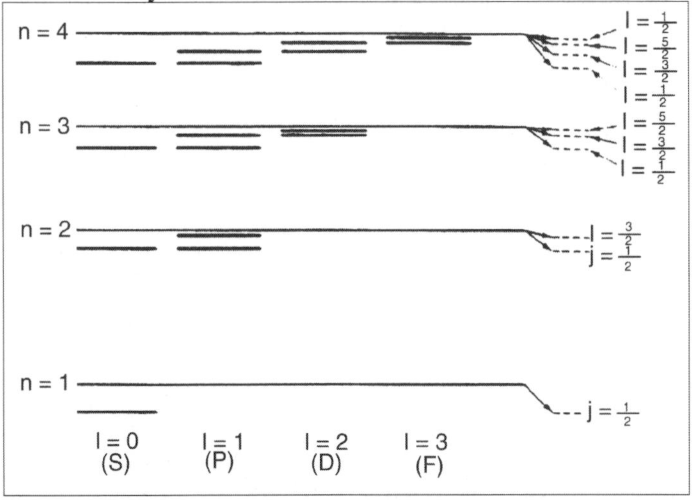

Fig. The Fine Structure of the Hydrogen Atom.
The Diagram is Not to Scale.

It is therefore appropriate to discuss the Dirac theory to achieve a more complete understanding of the fine structure of the hydrogen atom.

LAMINAR AND TURBULENT FLOW IN FLUID DYNAMICS

The nature of the flow of a fluid by injecting a fine filament of dye into the stream of flow and taking note of what happens to this filament. It was found in experiments that at low velocities the dye filament remained intact and that the filaments made parallel lines in the stream of flow. This is known as Laminar flow (or viscous or streamline). If the velocity of flow is gradually increased, the dye filament is eventually broken up and spread over the cross section of the pipe. This is turbulent flow, in which the particles of fluid are not moving in parallel lines but are moving across the general direction of flow.

If a fluid particle in a stream is disturbed, its inertia will tend to move it in a new direction, however the viscous forces from the surrounding fluid will tend to move it in the general direction of flow. If the shear forces are large enough to overcome any deviation, then we have viscous or laminar flow. However, if the shear forces are relatively weaker, and not sufficient to overcome the inertia of the particles, then we have turbulent flow.

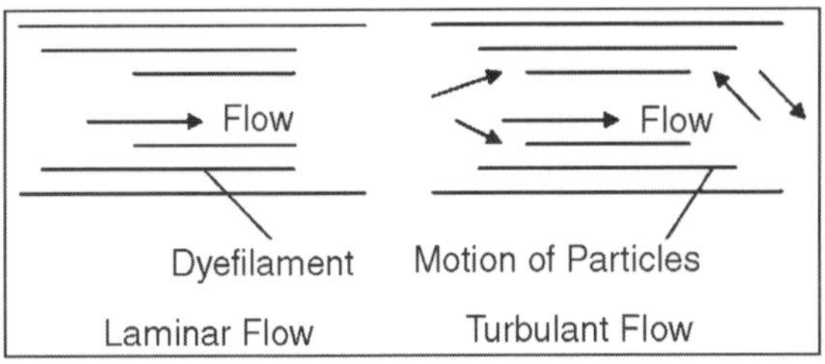

Therefore it is the ratio of the inertia to the viscous forces which determines whether flow will be laminar or turbulent.

The ratio of the inertia forces to the viscour forces is given by:

$$\rho\frac{cl}{\mu} \text{ (Reynolds Number)}$$

Therefore, it is the Reynolds number which determines whether a flow will be laminar or turbulent.

As Kinematic Viscosity (ν) $= \frac{\mu}{\rho}$,

Reynolds number can also be expressed as cl vl is the characteristic dimension in the system *e.g.* the diameter of the pipe in which the fluid is flowing.

Experiments have shown that if the Reynolds number is less than 2000, then the flow will be laminar. If it is above 2000 it is likely to be turbulent, however turbulent flow can exist at much higher Reynolds numbers.

Example: Water flows through a pipe 30mm diameter at a velocity of 10 m/s. If the dynamic viscosity of water is taken as 1.30×10^{-3} kg/ms and its density as 1000 kg/m^3, determine whether the flow will be laminar or turbulent.

Reynolds number,

$$= \rho \frac{c1}{\mu}$$

$$= \frac{1000 \times 10 \times 0.03}{1.3 \times 10^{-3}}$$

$$= 230769$$

This is significantly larger than 2000 and therefore the flow will be turbulent.

Problem If oil of specific gravity 0.85 and dynamic viscosity 14×10^{-2} kg/ms is pumped through the pipe at the same viscocity, what type of flow will occur. If we were to take a pipe of free flowing water and inject a dye into the middle of the stream, what would we expect to happen?

This,

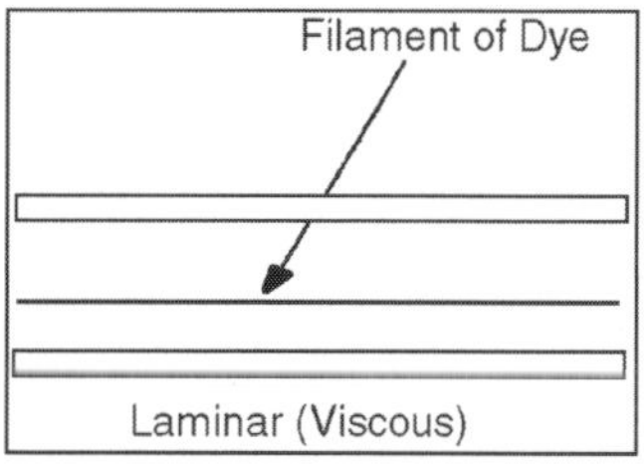

this,

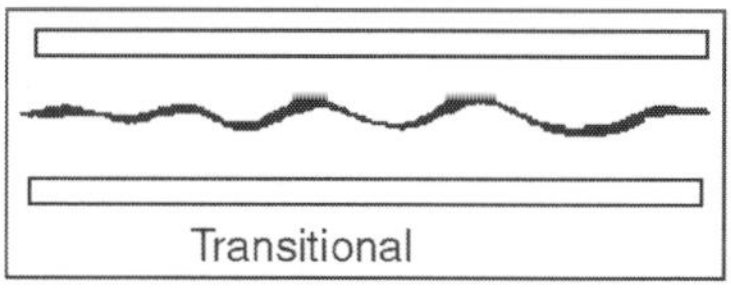

or this

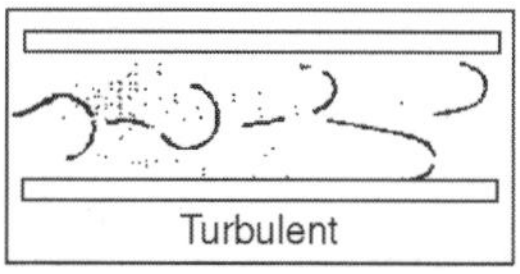

Actually both would happen - but for different flow rates. The top occurs when the fluid is flowing fast and the lower when it is flowing slowly.

The top situation is known as turbulent flow and the lower as laminar flow. In laminar flow the motion of the particles of fluid is very orderly with all particles moving in straight lines parallel to the pipe walls.But what is fast or slow? And at what speed does the flow pattern change? And why might we want to know this? The phenomenon was first investigated in the 1880s by Osbourne Reynolds in an experiment which has become a classic in fluid mechanics.

$$\frac{\rho u d}{\mu}$$

He used a tank arranged as above with a pipe taking water from the centre into which he injected a dye through a needle. After many experiments he saw that this expression

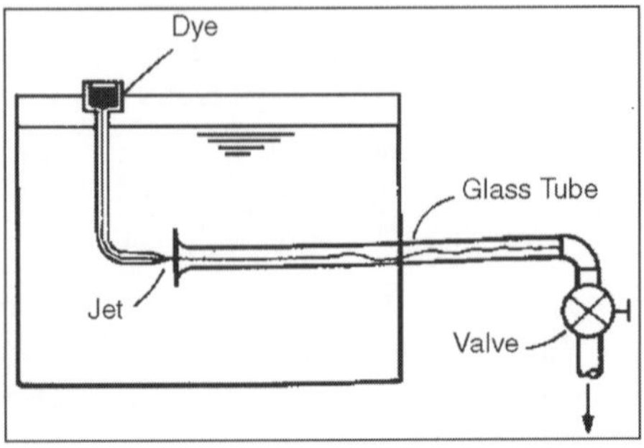

where ρ = density, u = mean velocity, d = diameter and μ = viscosity would help predict the change in flow type. If the value is less than about 2000 then flow is laminar, if greater than 4000 then turbulent and in between these then in the transition zone.

This value is known as the Reynolds number, Re:

$$\mathrm{Re} = \frac{\mathrm{pud}}{\mu}$$

Laminar flow: Re < 2000
Transitional flow: 2000 < Re < 4000
Turbulent flow: Re > 4000

Units of Reynolds number

We can fill in the equation with SI units:

$\rho = \mathrm{kg/m^3},\ u = \mathrm{m/s},\ d = \mathrm{m}$

$\mu = \mathrm{Ns/M^2} = \mathrm{kg/ms}$

$$\mathrm{Re} = \frac{\mathrm{pud}}{\mu} = \frac{\mathrm{kg}}{\mathrm{m^3}}\frac{\mathrm{m}}{\mathrm{s}}\frac{\mathrm{m}}{1}\frac{\mathrm{m}}{\mathrm{kg}} = 1$$

i.e. it has no units. A quantity that has no units is known as a non-dimensional (or dimensionless) quantity. Thus the Reynolds number, Re, is a non-dimensional number.

We can go through an example to discover at what velocity the flow in a pipe stops being laminar.

If the pipe and the fluid have the following properties:
water density ρ = 1000 kg/m^3
pipe diameter d = 0.5m
(dynamic) viscosity, μ = 0.55x10^3 Ns/m^2
We want to know the maximum velocity when the Re is 2000.

$$\mathrm{Re} = \frac{\mathrm{pud}}{\mu} = 2000$$

$$\mu = \frac{2000\mu}{pd} = \frac{2000 \times 0.55 \times 10^{-3}}{1000 \times 0.5}$$

$$m = 0.0022 \text{ m/s}$$

If this were a pipe in a house central heating system, where the pipe diameter is typically 0.015m, the limiting velocity for laminar flow would be, 0.0733 m/s. Both of these are very slow. In practice it very rarely occurs in a piped water system - the velocities of flow are much greater. Laminar flow does occur in situations with fluids of greater viscosity - *e.g.* in bearing with oil as the lubricant. At small values of Re above 2000 the flow exhibits small instabilities. At values of about 4000 we can say that the flow is truly turbulent. Over the past 100 years since this experiment, numerous more experiments have shown this phenomenon of limits of Re for many different Newtonian fluids - including gasses.

Abstract Number Mean

We can say that the number has a physical meaning, by doing so it helps to understand some of the reasons for the changes from laminar to turbulent flow.

$$Re = \frac{pud}{\mu} = \frac{\text{inertial forces}}{\text{viscous forces}}$$

It can be interpreted that when the inertial forces dominate over the viscous forces (when the fluid is flowing faster and Re is larger) then the flow is turbulent. When the viscous forces are dominant (slow flow, low Re) they are sufficient enough to keep all the fluid particles in line, then the flow is laminar.

LAMINAR FLOW

Laminar flow is also referred to as streamline or viscous flow. These terms are descriptive ofthe flow because, in laminar flow:

- Layers of water flowing over one another at differentspeeds with virtually no mixing between layers,
- Fluid particles move in definite andobservable paths or streamlines.
- The flow is characteristic of viscous (thick) fluid or isone in which viscosity of the fluid plays a significant part.
  - Re < 2000
  - 'low' velocity
  - Dye does not mix with water
  - Fluid particles move in straight lines
  - Simple mathematical analysis possible
  - Rare in practice in water systems.

TURBULENT FLOW

Turbulent flow is characterized by the irregular movement of particles of the fluid. There is nodefinite frequency as there is in wave motion. The particles travel in irregular paths with noobservable pattern and no definite layers.

- Re > 4000
- 'high' velocity
- Dye mixes rapidly and completely
- Particle paths completely irregular
- Average motion is in the direction of the flow
- Cannot be seen by the naked eye
- Changes/fluctuations are very difficult to detect. Must use laser.
- Mathematical analysis very difficult - so experimental measures are used
- Most common type of flow.

PRESSURE LOSS

Up to this point on the course we have considered ideal fluids where there have been no losses due to friction or any other factors. In reality, because fluids are viscous, energy is lost by flowing fluids due to friction which must be taken into account. The effect of the friction shows itself as a pressure (or head) loss.

In a pipe with a real fluid flowing, at the wall there is a shearing stress retarding the flow, as shown below.

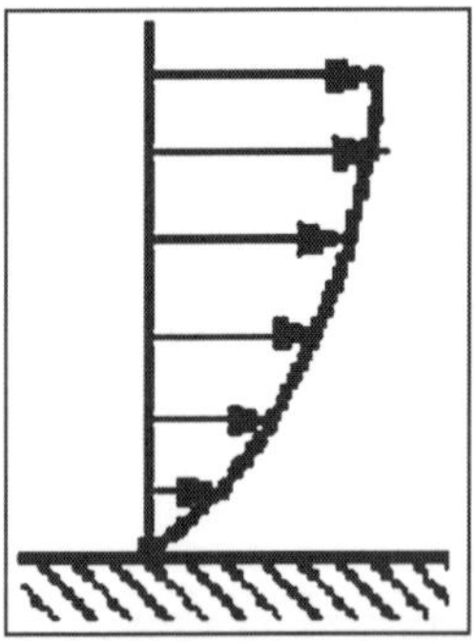

If a manometer is attached as the pressure (head) difference due to the energy lost by the fluid overcoming the shear stress can be easily seen. The pressure at 1 (upstream) is higher than the pressure at 2.

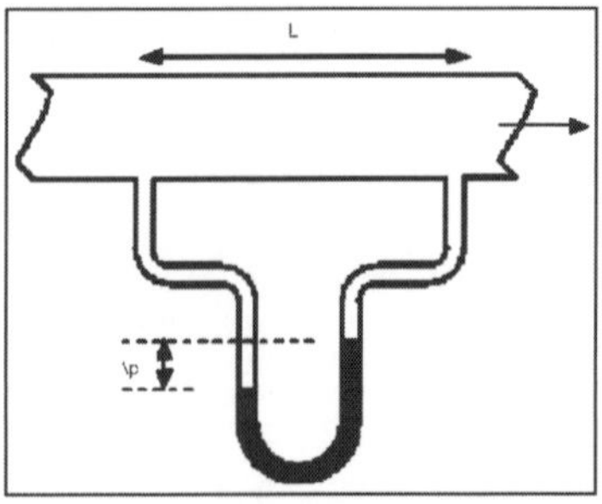

We can do some analysis to express this loss in pressure in terms of the forces acting on the fluid. Consider a cylindrical element of incompressible fluid flowing in the pipe, as shown,

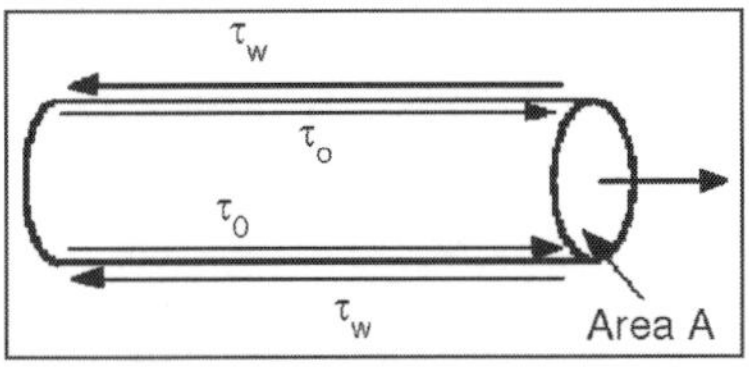

The pressure at the upstream end is p, and at the downstream end the pressure has fallen by Δp to (p-Δp). The driving force due to pressure (F = Pressure x Area) can then be written driving force = Pressure force at 1 - pressure force at 2,

$$pA-(p-\Delta p)A=\Delta p A=\Delta p\frac{\pi d^2}{4}$$

The retarding force is that due to the shear stress by the walls,

= shear stress × area over which it acts

= τ_w × area of pipe wall

= τ_w πdL

As the flow is in equilibrium, driving force = retarding force,

$$\Delta p\frac{\pi d^2}{4}=\tau_w \pi dL$$

$$\Delta p=\frac{\tau_{w4L}}{d}$$

Giving an expression for pressure loss in a pipe in terms of the pipe diameter and the shear stress at the wall on the pipe.

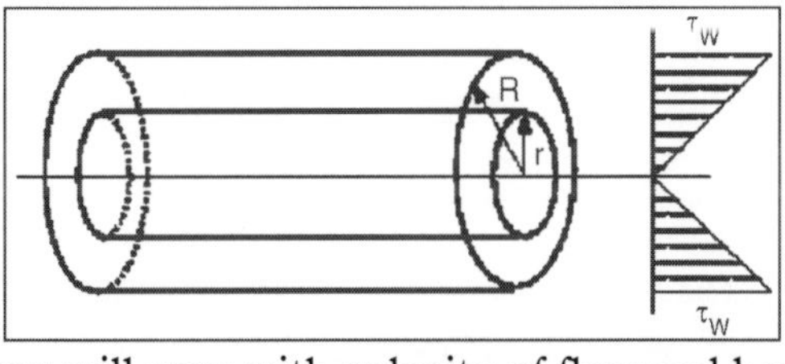

The shear stress will vary with velocity of flow and hence with Re. Many experiments have been done with various fluids measuring the pressure loss at various Reynolds numbers.

These results plotted to show a graph of the relationship between pressure loss and Re look similar to the figure below:

This graph shows that the relationship between pressure loss and Re can be expressed as

laminar Δp α μ

turbulent Δp α $\mu^{17(\text{or } 2.0)}$

As these are empirical relationships, they help in determining the pressure loss but not in finding the magnitude of the shear stress at the wall τ_w on a particular fluid. If we knew τ_w we could then use it to give a general equation to predict the pressure loss.

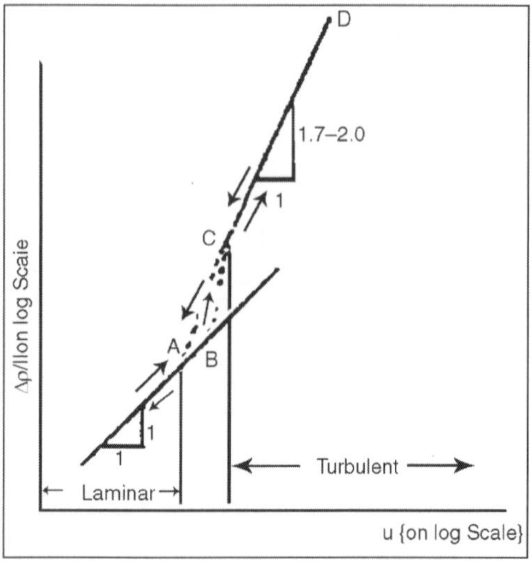

PRESSURE LOSS DURING LAMINAR FLOW

In general the shear stress τ_w. is almost impossible to measure. But for laminar flow it is possible to calculate a theoretical value for a given velocity, fluid and pipe dimension. In laminar flow the paths of individual particles of fluid do not cross, so the flow may be considered as a series of concentric cylinders sliding over each other - rather like the cylinders of a collapsible pocket telescope.

As before, consider a cylinder of fluid, length L, radius r, flowing steadily in the centre of a pipe.

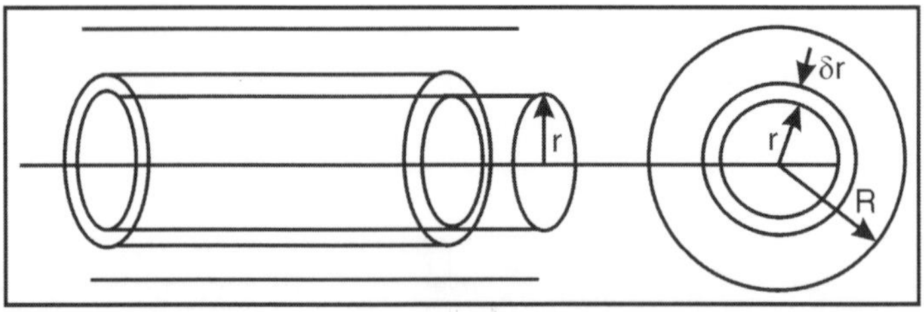

We are in equilibrium, so the shearing forces on the cylinder equal the pressure forces.

$$\tau 2\pi r L = \Delta p A = \pi r^2$$

$$\tau = \frac{\Delta p}{L}\frac{r}{2}$$

By Newtons law of viscosity we have,

$$\tau = \mu \frac{du}{dy},$$

where y is the distance from the wall. As we are measuring from the pipe centre then we change the sign and replace y with r distance from the centre, giving,

$$\tau = \mu \frac{du}{dr}$$

Which can be combined with the equation above to give,

$$\frac{\Delta p}{L}\frac{r}{2} = \mu \frac{du}{dr}$$

$$\frac{du}{dr} = -\frac{\Delta p}{L}\frac{r}{2\mu}$$

In an integral form this gives an expression for velocity,

$$u = -\frac{\Delta p}{L}\frac{1}{2\mu}\int r\,dr$$

Integrating gives the value of velocity at a point distance r from the centre,

$$u_r = -\frac{\Delta p}{L}\frac{r^2}{4\mu} + C$$

At $r = 0$, (the centre of the pipe), $u = u_{max}$, at $r = R$ (the pipe wall) $u = 0$, giving,

$$C = \frac{\Delta P}{L}\frac{R^2}{4\mu}$$

so, an expression for velocity at a point r from the pipe centre when the flow is laminar is,

$$u_r = \frac{\Delta p}{L}\frac{1}{4\mu}(R^2 - r^2)$$

Note how this is a parabolic profile (of the form $y = ax^2 + b$) so the velocity profile in the pipe looks similar to the figure below,

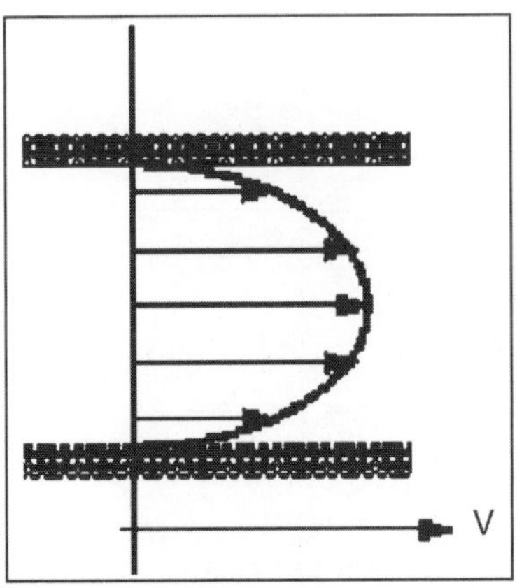

What is the discharge in the pipe?

$$Q = u_m A$$

$$u_m = \int_0^R u_r, dr$$

$$= \frac{\Delta p}{L} \frac{1}{4\mu} \int_0^R u_r \, dr$$

$$= \frac{\Delta p}{L} \frac{R^2}{8\mu} = \frac{\Delta p d^2}{32\mu L}$$

So the discharge can be written,

$$Q = \frac{\Delta p d^2}{32\mu L} \frac{\pi d^2}{4}$$

$$= \frac{\Delta p}{L} \frac{\pi d^2}{128\mu}$$

This is the Hagen-Poiseuille equation for laminar flow in a pipe. It expresses the discharge Q in terms of the pressure gradient,

$$\left(\frac{\partial p}{dx} = \frac{\Delta p}{L} \right),$$

diameter of the pipe and the viscosity of the fluid.

We are interested in the pressure loss (head loss) and want to relate this to the velocity of the flow. Writing pressure loss in terms of head loss h_f, *i.e.* $p = rgh_f$

$$u = \frac{pgh_f d^2}{32\mu L}$$

$$h_f = \frac{32\mu L u}{pgd^2}$$

This shows that pressure loss is directly proportional to the velocity when flow is laminar.

It has been validated many time by experiment.

It justifies two assumptions:

- Fluid does not slip past a solid boundary
- Newtons hypothesis.

DYNAMIC EQUATIONS OF MOTION

IMPORTANT DYNAMIC QUANTITIES

Some important dynamic quantities needed to describe forces are momentum p, angular momentum L and moment of inertia *I* are as follows:

$$p = mv, \quad L = r \times p, \quad I = \sum_i m_i r_i^2$$

Another quantity, which is useful for simplification of forces but not essential, is the instantaneous mass moment:

$$m = mR\hat{e}_R$$

where $R = R\ (t)$ = instantaneous radius of curvature at r on the curve, and $\hat{e}_R$ = unit vector directed to centre of circle of curvature.

Some useful relations for describing the motion of rotating rigid bodies, analogous to the above, are:

$$p = \omega \times m$$
$$L = I \cdot \omega$$

Newtonian Mechanics

It may be simple to write down the equations of motion in vector form using Newton's laws of motion, but the components may vary in complicated ways with spatial coordinates and time, and solving them is not easy. Often there is an excess of variables to solve for the problem completely, so Newton's laws are not the most efficient method for generally finding and solving for the motion of a particle. In simple cases of rectangular geometry, the use of Cartesian coordinates works fine, but other coordinate systems can become dramatically complex.

Newton's 2nd law for translation

The first developed and most famous is Newton's 2nd law of motion:

$$\mathbf{F} = \frac{\mathrm{d}\mathbf{p}}{\mathrm{d}t}$$

where $\mathbf{p} = \mathbf{p}(t)$ is the momentum of the particle and $\mathbf{F} = \mathbf{F}(t)$ is the resultant external force acting on the particle (not any force the particle exerts) - in each case at time t. This is a differential equation of momentum, so solving this equation obtains the momentum vector as a function of time. This version of the formula is actually not much use, since the momentum is simply

$$\mathbf{p} = m\mathbf{v} = m\frac{\mathrm{d}\mathbf{r}}{\mathrm{d}t}$$

for a particle at position $\mathbf{r} = \mathbf{r}(t)$, velocity $\mathbf{v} = \mathbf{v}(t)$, and of mass $m = m(t)$, again each at time t. In general mass can vary with time, if the system gains or loses mass in some way. Velocity $\mathbf{v}$ is usually known as a 'function of time'.
Hence in usual application, Newton's second law is a differential equation of position and time, rather than momentum and time, using the product rule for differentiation:

$$\begin{aligned}\mathbf{F} &= m\frac{d\mathbf{v}}{dt} + \mathbf{v}\frac{dm}{dt} \\ &= m\frac{d^2\mathbf{r}}{dt^2} + \frac{d\mathbf{r}}{dt}\frac{dm}{dt}\end{aligned}$$

where $\mathbf{a}$ = acceleration of particle. This is a differential equation in terms of position $\mathbf{r}$. The solution is $\mathbf{r}$ as a function of the spatial coordinates and time (in general), for example in cartesian coordinates $\mathbf{r} = (x, y, z, t)$, or in spherical polar coordinates $\mathrm{r} = (r, \theta, \phi, t)$. Time is usually a parameter of the spatial coordinates, so explicitly $\mathbf{r} = (x(t), y(t), z(t), \mathrm{t})$.

For a number of particles, the equation of motion for one particle i is:

$$\frac{d\mathbf{p}_i}{dt} = \mathbf{F}_E + \sum_{i \neq j} \mathbf{F}_{ij}$$

where $\mathbf{p}_i$ = momentum of particle i, $\mathbf{F}_{ij}$ = force **on** particle i **by** particle j, $\mathbf{F}_{ij}$ = force **on** body j **by** particle i, and $\mathbf{F}_E$ = resultant external force (due to any agent not part of system). Particle i does not exert a force on itself.

Newton's Second Law for Rotation

The analogue for rotation is:

$$\boldsymbol{\tau} = \frac{d\mathbf{L}}{dt}$$

where $\boldsymbol{\tau}$ denotes the torque acting on the body. Again this form isn't much use, so re-writing it:

$$\boldsymbol{\tau} = \frac{d\mathbf{r}}{dt} \times \mathbf{p} + \mathbf{r} \times \frac{d\mathbf{p}}{dt}$$

that is, the resultant torque acting on the system is the moment of the resultant force:

$$\boldsymbol{\tau} = \mathbf{r} \times \mathbf{F}$$

since momentum and velocity are always parallel by definition:

$$\mathbf{v} \times \mathbf{p} = \mathbf{0}$$

For rigid bodies, Newton's 2nd law for rotation takes the same form as for translation:

$$\mathbf{L} = I\boldsymbol{\omega}$$

again by the product rule:

$$\begin{aligned}\boldsymbol{\tau} &= I\frac{d\boldsymbol{\omega}}{dt} + \boldsymbol{\omega}\frac{dI}{dt} \\ &= \hat{\mathbf{n}}\left(I\frac{d^2\theta}{dt^2} + \frac{d\theta}{dt}\frac{dI}{dt}\right)\end{aligned}$$

Likewise, for a number of particles, the equation of motion for one particle i is:

$$\frac{d\mathbf{L}_i}{dt} = \boldsymbol{\tau}_E + \sum_{i \neq j} \boldsymbol{\tau}_{ij}$$

where $\mathbf{L}_i$ = angular momentum of particle i, $\boldsymbol{\tau}_{ij}$ = torque ON particle i BY particle j, $\boldsymbol{\tau}_{ji}$ = torque ON body j BY particle i, and $\boldsymbol{\tau}_E$ = resultant external torque (due to any agent not part of system).

Particle i does not exert a torque on itself.

Applications

Some examples of Newton's law include describing the motion of a pendulum:

$$mg\sin\theta = m\frac{d^2(\ell\theta)}{dt^2} + 0 \Rightarrow g\sin\theta = \ell\frac{d^2\theta}{dt^2}$$

a damped, driven harmonic oscillator:

$$\Gamma_0 \sin\omega_0 t = m\frac{d^2x}{d^2t} + bm\frac{dx}{dt} + \omega^2 x$$

or a ball thrown in the air, in air currents (such as wind) described by a vector field of resistive forces R = R(x, y, z, t):

$$-\frac{GmM}{|r|^2}\hat{e}_r + R = m\frac{d^2r}{dt^2} + 0 \Rightarrow -\frac{GM}{|r|^2}\ddot{e}_r + A = \frac{d^2r}{dt^2}$$

where G = gravitational constant, M = mass of the earth and A is the acceleration of the projectile due to the air currents at position r and time t. Newton's law of gravity has been used. The mass m of the ball cancels.

Euler's Equations for Rigid Body Dynamics

Euler also worked out analogous laws of motion to those of Newton. These extend the scope of Newton's laws to rigid bodies, but are essentially the same as above. A new equation Euler formulated is:

$$I\cdot\alpha + \omega \times (I\cdot\omega) = \tau$$

where I is the moment of inertia tensor.

General Planar Motion

The previous equations for planar motion can be used here: corollaries of momentum, angular momentum, *etc.* can immediately follow by applying the above definitions. For any object moving in any path in a plane,

$$r = r(t) = r\hat{e}_r$$

the following are general dynamic results.

The momentum and angular momenta are:

$$p = m\left(\hat{e}_r \frac{d^2 r}{dt^2} + r\omega \hat{e}_\theta\right)$$

$$L = mr \times \left(\hat{e}_r \frac{d^2 r}{dt^2} + r\omega \hat{e}_\theta\right)$$

The centripetal force is

$$F_\perp = -m\omega^2 R\hat{e}_r = -\omega^2 m$$

where again m is the mass moment, and the coriolis force is

$$F_C = 2\omega \frac{dr}{dt}\hat{e}_\theta = 2\omega v\hat{e}_\theta$$

IMPORTANCE OF COMPRESSIBILITY IN FLUID DYNAMICS

Compressibility is an important factor in aerodynamics. At low speeds, the compressibility of air is not significant in relation to aircraft design, but as the airflow nears and exceeds the speed of sound, a host of new aerodynamic effects become important in the design of aircraft.

These effects, often several of them at a time, made it very difficult for World War II era aircraft to reach speeds much beyond 800 km/h (500mph).

Some of the minor effects include changes to the airflow that lead to problems in control. For instance, the P-38 Lightning with its thick high-lift wing had a particular problem in high-speed dives that led to a nose-down condition. Pilots would enter dives, and then find that they could no longer control the plane, which continued to nose over until it crashed. Adding a "dive flap" beneath the wing altered the center of pressure distribution so that the wing would not lose its lift. This fixed the problem.

A similar problem affected some models of the Supermarine Spitfire. At high speeds the ailerons could apply more torque than the Spitfire's thin wings could handle, and the entire wing would twist in the opposite direction. This meant that the plane would roll in the direction opposite to that which the pilot intended, and led to a number of accidents.

Earlier models weren't fast enough for this to be a problem, and so it wasn't noticed until later model Spitfires like the Mk.IX started to appear. This was mitigated by adding considerable torsional rigidity to the wings, and was wholly cured when the Mk.XIV was introduced.

The Messerschmitt Bf 109 and Mitsubishi Zero had the exact opposite problem in which the controls became ineffective. At higher speeds the pilot simply couldn't move the controls because there was too much airflow over the control surfaces. The planes would become difficult to maneuver, and at high enough speeds less maneuverable aircraft could out-turn them.

Finally, another common problem that fits into this category is flutter. At some speeds the airflow over the control surfaces will become turbulent, and the controls will start to flutter. If the speed of the fluttering is close to a harmonic of the control's movement, the resonance could break the control off completely. This was a serious problem on the Zero. When problems with poor control at high speed were first encountered, they were addressed by designing a new style of control surface with more power. However this introduced a new resonant mode, and a number of planes were lost before this was discovered.

All of these effects are often mentioned in conjunction with the term "compressibility", but in a manner of speaking, they are incorrectly used. From a strictly aerodynamic point of view, the term should refer only to those side-effects arising as a result of the changes in airflow from an incompressible fluid (similar in effect to water) to a compressible fluid (acting as a gas) as the speed of sound is approached. There are two effects in particular, wave drag and critical mach.

Wave drag is a sudden rise in drag on the aircraft, caused by air building up in front of it. At lower speeds this air has time to "get out of the way", guided by the air in front of it that is in contact with the aircraft.

But at the speed of sound this can no longer happen, and the air which was previously following the streamline around the aircraft now hits it directly. The amount of power needed to overcome this effect is considerable. The critical mach is the speed at which some of the air passing over the aircraft's wing becomes supersonic.

At the speed of sound the way that lift is generated changes dramatically, from being dominated by Bernoulli's principle to forces generated by shock waves. Since the air on the top of the wing is traveling faster than on the bottom, due to Bernoulli effect, at speeds close to the speed of sound the air on the top of the wing will be accelerated to supersonic. When this happens the distribution of lift changes dramatically, typically causing a powerful nose-down trim. Since the aircraft normally approached these speeds only in a dive, pilots would report the aircraft attempting to nose over into the ground.

All of these effects have adverse effects on the control or performance of the plane. For this reason it's common to see references to aircraft that suffer from compressibility. The P-38 and Zero are particularly common examples, although in fact they are both bad ones.

THERMODYNAMICS

The term "compressibility" is also used in thermodynamics to describe the deviance in the thermodynamic properties of a real gas from those expected from an ideal gas.

The compressibility factor is defined as

$$Z = \frac{p\tilde{V}}{RT}$$

where p is the pressure of the gas, T is its temperature, and $\tilde{V}$ is its molar volume. In the case of an ideal gas, the compressibility factor Z is equal to unity, and the familiar ideal gas law is recovered:

$$p = \frac{RT}{\tilde{V}}$$

Z can, in general, be either greater or less than unity for a real gas. The deviation from ideal gas behaviour tends to become particularly significant (or, equivalently, the compressibility factor strays far from unity) near the critical point, or in the case of high pressure or low temperature. In these cases, a generalized Compressibility chart or an alternative equation of state better suited to the problem must be utilized to produce accurate results.

Earth Sciences

Table. Vertical, Drained Compressibilities

Material	β (m²/N)
Plastic clay	$2\times10^{-6} - 2.6\times10^{-7}$
Stiff clay	$2.6\times10^{-7} - 1.3\times10^{-7}$
Medium-hard clay	$1.3\times10^{-7} - 6.9\times10^{-8}$
Loose sand	$1\times10^{-7} - 5.2\times10^{-8}$
Dense sand	$2\times10^{-8} - 1.3\times10^{-8}$
Dense, sandy gravel	$1\times10^{-8} - 5.2\times10^{-9}$
Rock, fissured	$6.9\times10^{-10} - 3.3\times10^{-10}$
Rock, sound	$<3.3\times10^{-10}$
Water at 25°C (undrained)	4.6×10^{-10}

Compressibility is used in the Earth sciences to quantify the ability of a soil or rock to reduce in volume with applied pressure.

This concept is important for specific storage, when estimating groundwater reserves in confined aquifers. Geologic materials are made up of two portions: solids and voids (or same as porosity).

The void space can be full of liquid or gas. Geologic materials reduces in volume only when the void spaces are reduced, which expel the liquid or gas from the voids.

This can happen over a period of time, resulting in settlement. It is an important concept in geotechnical engineering in the design of certain structural foundations. For example, the construction of high-rise structures over underlying layers of highly compressible bay mud poses a considerable design constraint, and often leads to use of driven piles or other innovative techniques.

BULK MODULUS

The ratio of the compressive or tensile force applied to a substance per unit surface area to the change in volume of the substance per unit volume.

Also known as bulk modulus; compression modulus; hydrostatic modulus; modulus of compression; modulus of volume elasticity..

The bulk modulus (K) of a substance essentially measures the substance's resistance to uniform compression. It is defined as the pressure increase needed to effect a given relative decrease in volume. As an example, suppose an iron cannon ball with bulk modulus 160 GPa (gigapascal) is to be reduced in volume by 0.5 per cent . This requires a pressure increase of $0.005 \times 160\text{GPa} = 0.8$ GPa. If the cannon ball is subjected to a pressure increase of only 100 MPa, it will decrease in volume by a factor of 100MPa/160GPa = 0.000625, or 0.0625 per cent .

The bulk modulus K can be formally defined by the equation:

$$K = -V \frac{\partial p}{\partial V}$$

where p is pressure, V is volume, and $\partial p / \partial V$ denotes the partial derivative of pressure with respect to volume.

The inverse of the bulk modulus gives a substance's compressibility. Other moduli describe the material's response (strain) to other kinds of stress: the shear modulus describes the response to shear, and Young's modulus describes the response to linear strain. For a fluid, only the bulk modulus is meaningful.

For an anisotropic solid such as wood or paper, these three moduli do not contain enough information to describe its behaviour, and one must use the full generalized Hooke's law.

Strictly speaking, the bulk modulus is a thermodynamic quantity, and it is necessary to specify how the temperature varies in order to specify a bulk modulus: constant-temperature (K_T), constant-enthalpy (adiabatic K_S), and other variations are possible.

In practice, such distinctions are usually only relevant for gases.

Table. Bulk Modulus Values for Some Example Substances

Water	2.2×10^9 Pa (value increases at higher pressures)
Air	1.42×10^5 Pa (adiabatic bulk modulus)
Air	1.01×10^5 Pa (constant temperature bulk modulus)
Steel	1.6×10^{11} Pa
Solid helium	5×10^7 Pa (approximate)

For a gas, the adiabatic bulk modulus K_S is approximately given by,

$$Ks = kp$$

Where,

k is the adiabatic index, sometimes called Y.

p is the pressure.

In a fluid, the bulk modulus K and the density ρ determine the speed of sound c (pressure waves), according to the formula

$$C = \sqrt{\frac{K}{\rho}}.$$

Solids can also sustain transverse waves, for these one additional elastic modulus, for example the shear modulus, is needed to determine wave speeds.

3

Basic of Fluid Flow and Motion

INTRODUCTION

Thus Basic fluid motion can be described as some combination of

TRANSLATION

This is characterized by the velocity,

$$\vec{U} = u\vec{i} + v\vec{j} + w\vec{k}$$

$$\rho = \frac{P}{RT}$$

Dilatation

[volume change]

$$\vec{u}, \rho, p, e, h$$

Rotation

[About one, two or 3 axes].

$$\omega = \frac{1}{2}\left[\left(\frac{dw}{dy} - \frac{dv}{dz}\right)\vec{i} + \left(\frac{du}{dz} - \frac{dw}{dx}\right)j + \left(\frac{dv}{dx} - \frac{du}{dy}\right)k\right]$$

Note that rotation is a vector. We usually use a quantity called "vorticity" which is twice the rotation vector, to describe the amount of rotation in flows.

Shear Strain

The quantity "strain" thus has nine components.

They are:

- e_{xx}, e_{yy}, e_{zz} These are "normal strain" components.
- e_{xy}, e_{yz}, e_{zx} These are "shear strain" components.
- e_{yx}, e_{zy}, e_{xz} These are equal to the corresponding "shear strain" components above.

If this discussion brings back memories of your solid mechanics/ statics course, there is good reason for it.

The difference comes in the relation between strain and the stress: in the case of solids, the stress depends on the strain. In the case of fluids, the stress depends on the rate of strain.

Circulation

Defined as $d\vec{s}$ integrated around a closed contour. The negative sign is included such that positive circulation on a body corresponds to positive lift, and the integral is evaluated counter-clockwise.

From the preceding discussion, we see that r will be zero unless there is some vorticity, $\vec{\omega}$ contained within the contour. *r is an extremely useful quantity*: it helps us calculate lift, vortex strength, etc.

Important Points

$$d\vec{s} \times \vec{U} = 0$$

The circulation around a closed contour with net rotation and/or shear will be non-zero. It is, however, always possible to have a combination of rotation and/or shear that gives a zerocirculation.

CONCEPT OF FLUID

What is fluid mechanics? As its name suggests it is the branch of applied mechanics concerned with the statics and dynamics of fluids - both liquids and gases. The analysis of the behaviour of fluids is based on the fundamental laws of mechanics which relate continuity of mass and energy with force and momentum together with the familiar solid mechanics properties.

We normally recognise three states of matter: solid; liquid and gas. However, liquid and gas are both fluids: in contrast to solids they lack the ability to resist deformation. Because a fluid cannot resist the deformation force, it moves, it *flows* under the action of the force. Its shape will change continuously as long as the force is applied. A solid can resist a deformation force while at rest, this force may cause some displacement but the solid does not continue to move indefinitely.

The deformation is caused by *shearing* forces which act tangentially to a surface. Referring to the figure below, we see the force F acting tangentially on a rectangular (solid lined) element ABDC. This is a shearing force and produces the (dashed lined) rhombus element A'B'DC.

$$\sigma = \frac{\rho \text{subtance}}{\rho H_2O(\text{at } 4°C)}$$

Shearing force, F, acting on a fluid element.

We can then say:

A Fluid is a substance which deforms continuously, or flows, when subjected to shearing forces.

and conversely this definition implies the very important point that:

If a fluid is at rest there are no shearing forces acting.

All forces must be perpendicular to the planes which the are acting.

When a fluid is in motion shear stresses are developed if the particles of the fluid move relative to one another. When this happens adjacent particles have different velocities. If fluid velocity is the same at every point then there is no shear stress produced: the particles have zero *relative* velocity.

Consider the flow in a pipe in which water is flowing. At the pipe wall the velocity of the water will be zero. The velocity will increase as we move towards the centre of the pipe. This change in velocity across the direction of flow is known as velocity profile and shown graphically in the figure below:

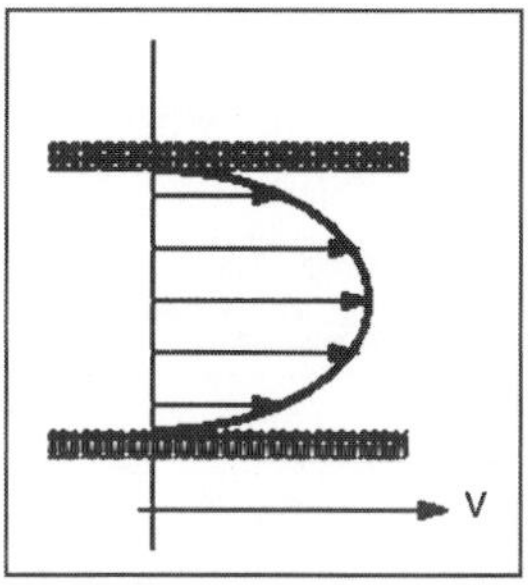

Fig. Velocity Profile in a Pipe.

Because particles of fluid next to each other are moving with different velocities there are shear forces in the moving fluid *i.e.* shear forces are normally present in a moving fluid. On the other hand, if a fluid is a long way from the boundary and all the particles are travelling with the same velocity, the velocity profile would look something like this:

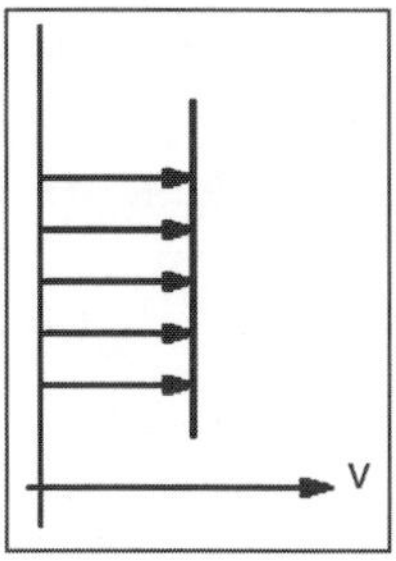

Fig. Velocity Profile in Uniform Flow

and there will be no shear forces present as all particles have zero relative velocity. In practice we are concerned with flow past solid boundaries; aeroplanes, cars, pipe walls, river channels etc. and shear forces will be present.

1. *Newton's law of viscosity*: How can we make use of these observations?

   We can start by considering a 3d rectangular element of fluid, like that in the figure below.

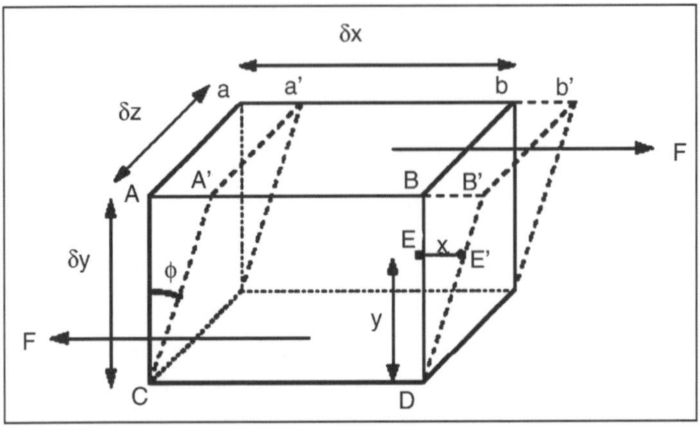

Fig. Fluid Element Under a Shear Force

The shearing force F acts on the area on the top of the element. This area is given by $A = \delta s \times \delta x$. We can thus calculate the *shear stress* which is equal to force per unit area *i.e.*

$$\text{shear stress}, \tau = \frac{F}{A}$$

The deformation which this shear stress causes is measured by the size of the angle f and is know as *shear strain*.

In a solid shear strain, f, is constant for a fixed shear stress t.

In a fluid f increases for as long as t is applied - the fluid flows.

It has been found experimentally that the *rate of shear stress* (shear stress per unit time, t/time) is directly proportional to the shear stress.

If the particle at point E (in the above figure) moves under the shear stress to point E' and it takes time *t* to get there, it has moved the distance *x*. For small deformations we can write shear strain,

$$\phi = \frac{x}{y}$$

$$\text{rate of shear strain} = \frac{\phi}{t}$$

$$= \frac{x}{ty} = \frac{x}{t}\frac{1}{y}$$

$$= \frac{u}{y}$$

where,

$$\frac{x}{t} = u$$

is the velocity of the particle at E.

Using the experimental result that shear stress is proportional to rate of shear strain then,

$$\tau = \text{constant} \times \frac{u}{y}$$

The term,

$$\frac{u}{y}$$

is the change in velocity with y, or the velocity gradient, and may be written in the differential form,

$$\frac{du}{dy}$$

The constant of proportionality is known as the dynamic viscosity, μ, of the fluid, giving

$$\tau = \mu \frac{du}{dy}$$

This is known as Newton's law of viscosity.

2. *Fluids vs. Solids*: Earlicr wc have discussed the differences between the behaviour of solids and fluids under an applied force. Summarising, we have;
    - For a solid the strain is a function of the applied stress (providing that the elastic limit has not been reached). For a fluid, the rate of strain is proportional to the applied stress.
    - The strain in a solid is independent of the time over which the force is applied and (if the elastic limit is not reached) the deformation disappears when the force is removed. A fluid continues to flow for as long as the force is applied and will not recover its original form when the force is removed.

    It is usually quite simple to classify substances as either solid or liquid. Some substances, however, (*e.g.* pitch or glass) appear solid under their own weight. Pitch will, although appearing solid at room temperature, deform and spread out over days - rather than the fraction of a second it would take water.

    Looking at properties of solids, when the elastic limit is reached they seem to flow. They become plastic. They still do not meet the definition of true fluids as they will only flow after a certain minimum shear stress is attained.

3. *Newtonian / Non-Newtonian Fluids*: Even among fluids which are accepted as fluids there can be wide differences in behaviour under stress. Fluids obeying Newton's law where the value of m is constant are known as Newtonian fluids. If m is constant the shear stress is linearly dependent on velocity gradient. This is true for most common fluids.
   Fluids in which the value of m is not constant are known as non-Newtonian fluids. There are several categories of these, and they are outlined briefly below.
   These categories are based on the relationship between shear stress and the velocity gradient (rate of shear strain) in the fluid. These relationships can be seen in the graph below for several categories

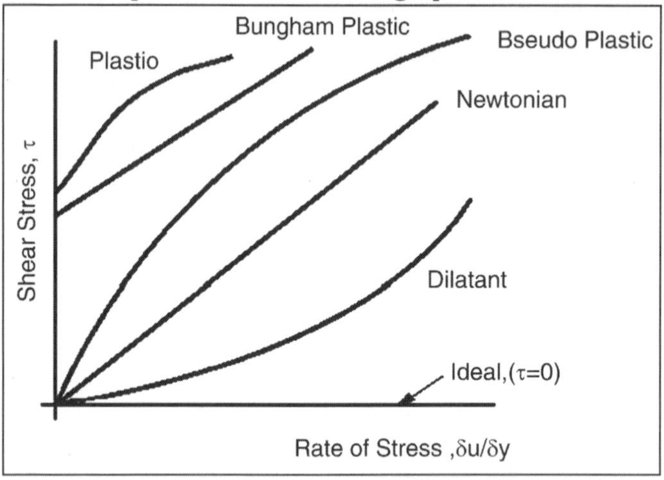

Fig. Shear Stress vs. Rate of Shear Strain du/dy

Each of these lines can be represented by the equation,

$$\tau = A + B\left(\frac{\delta u}{\delta y}\right)^n$$

where A, B and n are constants. For Newtonian fluids A = 0, B = m and n = 1.

Below are brief description of the physical properties of the several categories:

- *Plastic:* Shear stress must reach a certain minimum before flow commences.
- *Bingham plastic:* As with the plastic above a minimum shear stress must be achieved. With this classification n = 1. An example is sewage sludge.
- *Pseudo-plastic:* No minimum shear stress necessary and the viscosity decreases with rate of shear, *e.g.* colloidial substances like clay, milk and cement.

- *Dilatant substances;* Viscosity increases with rate of shear *e.g.* quicksand.
- *Thixotropic substances:* Viscosity decreases with length of time shear force is applied *e.g.* thixotropic jelly paints.
- *Rheopectic substances:* Viscosity increases with length of time shear force is applied
- *Viscoelastic materials:* Similar to Newtonian but if there is a sudden large change in shear they behave like plastic.

There is also one more - which is not real, it does not exist - known as the ideal fluid. This is a fluid which is assumed to have no viscosity. This is a useful concept when theoretical solutions are being considered - it does help achieve some practically useful solutions.

4. *Liquids vs. Gasses*: Although liquids and gasses behave in much the same way and share many similar characteristics, they also possess distinct characteristics of their own. Specifically,
   - A liquid is difficult to compress and often regarded as being incompressible. A gas is easily to compress and usually treated as such - it changes volume with pressure.
   - A given mass of liquid occupies a given volume and will occupy the container it is in and form a free surface (if the container is of a larger volume).

   A gas has no fixed volume, it changes volume to expand to fill the containing vessel. It will completely fill the vessel so no free surface is formed.
5. *Causes of Viscosity in Fluids*:
   - *Viscosity in Gasses*: The molecules of gasses are only weakly kept in position by molecular cohesion (as they are so far apart). As adjacent layers move by each other there is a continuous exchange of molecules. Molecules of a slower layer move to faster layers causing a drag, while molecules moving the other way exert an acceleration force. Mathematical considerations of this momentum exchange can lead to Newton law of viscosity.

   If temperature of a gas increases the momentum exchange between layers will increase thus increasing viscosity.

   Viscosity will also change with pressure - but under normal conditions this change is negligible in gasses.
   - *Viscosity in Liquids*: There is some molecular interchange between adjacent layers in liquids - but as the molecules are so much closer than in gasses the cohesive forces hold the molecules in place much more rigidly. This cohesion plays an important roll in the viscosity of liquids.

   Increasing the temperature of a fluid reduces the cohesive forces and increases the molecular interchange. Reducing cohesive forces reduces shear stress, while increasing molecular interchange increases shear stress. Because of this complex interrelation the effect of temperature on viscosity has something of the form:

$$\mu_T = \mu_0 (1 + AT + BT)$$

where μ_T is the viscosity at temperature TC, and μ_0 is the viscosity at temperature 0C. A and B are constants for a particular fluid.

High pressure can also change the viscosity of a liquid. As pressure increases the relative movement of molecules requires more energy hence viscosity increases.

PHASES OF MATTER

A phase diagram for a typical pure substance in the pressure (p) temperature (T) plane is sketched below.

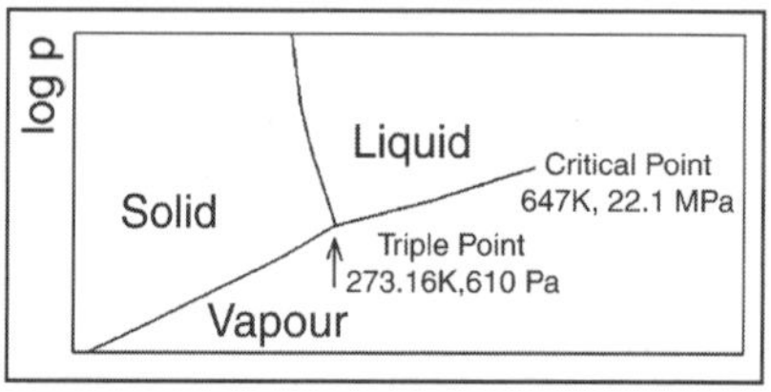

Fig. Pressure-temperature Phase Diagram of Water. Recall that 1.0135MPa = 10 ATM.

There are three equilibrium phases, the solid (s), liquid (l), and vapor (v) (or gas), with phase boundaries separating the phases. Crossing the phase boundaries results in a phase transitioni from one phase to the other.

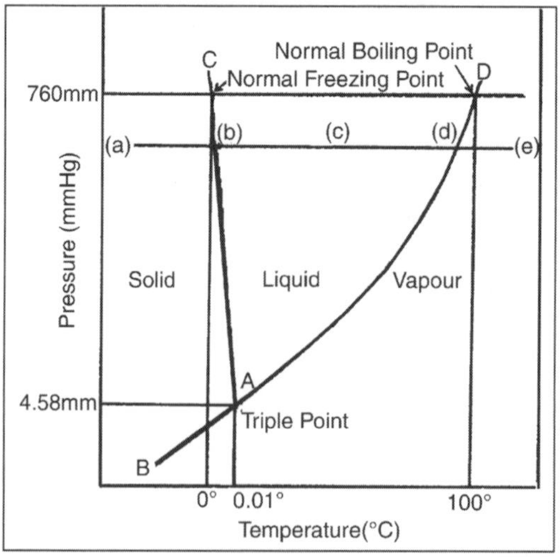

The liquid-vapor phase boundary (often called the vaporization curve), terminates in a critical point at sufficiently high temperatures and pressures. At temperatures above the critical temperature or pressures above the critical pressure there is no distinction between the vapor and liquid phases, and we have simply a single fluid phase. Indeed, at a microscopic level there is little difference between a liquid and a vapor; there are stronger interactions between the particles in the liquid, but there are no fundamental differences in symmetry between the two phases. Our description of fluid mechanics will apply equally well to liquids and vapors (gases).

HYDRODYNAMIC VARIABLES

How do we characterize fluid flow? We could track the positions of all of the particles in the fluid, by solving the equations of motion for the particles. While this might be feasible for a small number of particles, it is clearly absurd for any macroscopic system. Instead, we seek a reduced description in terms of a small number of macroscopic variables, which are functions of the fixed position r and the time t, such as the local density ρ(r,t), the local fluid velocity v(r,t), the local pressure p(r,t), and the local temperature T(r,t). We will stick with the more convenient, and conventional, Eulerian coordinate system, in which *r* is a coordinate fixed in space. One could also choose a coordinate system fixed to the particles, the Lagrangian system.

By throwing away most of the microscopic information, we hope to obtain a simpler, and in most cases, more useful, macroscopic description. However, such a simplification comes with a price: we will often need to introduce phenomenological parameters (such as the viscosity) whose values are not known a priori, but which must be taken from experiment. Exactly how we achieve this reduction from microscopic variables to macroscopic variables is the subject of kinetic theory; the reduced description is often called a continuum theory. Kinetic theory is extremely subtle (and interesting), and vexed the great physicist Boltzmann throughout his career. We won't have much to say about it in this course. However, I should point out that the macroscopic description must eventually break down when we begin probing length scales comparable to microscopic lengths. The relevant microscopic length is the mean free path of the particles in the fluid, which is loosely defined as the average distance a particle moves before colliding with another particle. For example, if we try to understand sound propagation starting from a macroscopic point of view, our analysis will certainly break down once the wavelength of sound is comparable to the mean free path. We see immediately that a continuum theory needs to be used with caution for a very tenuous, or rarefied, gas.

THE EQUATION OF CONTINUITY

Having introduced our continuum, or hydrodynamic, variables, we now need to construct equations of motion for these variables. We do this by appealing to conservation laws, such as conservation of mass, momentum, and energy. In this section we will focus on the conservation of mass and derive the *equation of continuity* for the mass density. Consider an arbitrary, fixed volume *V*, inside the fluid,

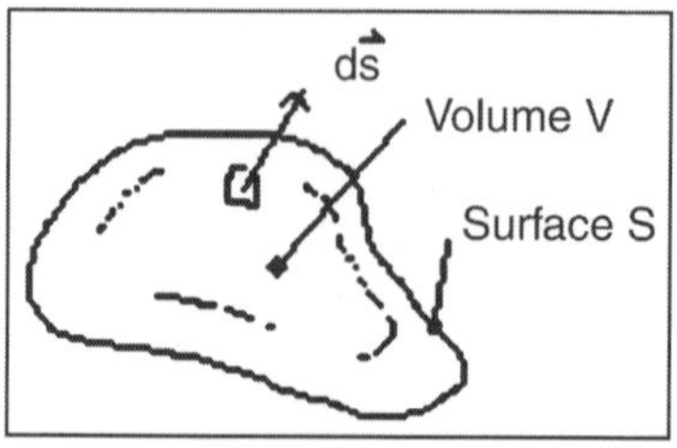

Fig. Fluid Volume Used for the Derivation of the Continuity Equation.

Fluid is entering and leaving *V* by crossing the surface *S*. If we let *dS* be an element of surface area which points outward, normal to the surface, and *v* be the fluid velocity at the surface, then it is the component of *v* in the direction of *dS* which is responsible for transporting mass out of *V*. Therefore, the mass flux (mass per unit time) transported through the infinitesimal surface area *dS* . Integrating this over the entire surface *S*, we have the net rate of change of mass in the volume *V* is

$$\text{rate of mass change in V} = -\int \rho \mathrm{v} \cdot \mathrm{dS} = -\int_{V} \nabla \cdot (\rho \mathrm{v}) \mathrm{dV},$$

where the minus sign has been added since the mass is decreasing, and the last equality follows from the divergence theorem. Now the total mass in the volume *V* is,

$$\text{Total Mass in V} = \int_{V} \rho \, \mathrm{dV},$$

and the rate at which this mass is changing with time is,

$$\text{rate of mass change in V} = \frac{\mathrm{d}}{\mathrm{dt}} \int_{V} \rho \, \mathrm{dV} = \int_{V} \frac{\partial \rho}{\partial \mathrm{t}} \mathrm{dV}.$$

Since mass is conserved in an arbitrarily small volume *V*, the integrands of Eqs. and we have,

$$\frac{\partial \rho}{\partial \mathrm{t}} + \nabla \cdot (\rho \mathrm{V}) = 0.$$

This is the equation of continuity, which is a mathematical statement of conservation of mass.

INCOMPRESSIBLE FLOW

Many liquids are hard to compress (*i.e.*, their volumes or densities don't change much when you squeeze them), so that the density ρ is essentially a constant. For such an incompressible fluid the equation of continuity simplifies to

$$\nabla \cdot v = 0 \,(\text{incompressible}),$$

so that the velocity field v is a solenoidal vector. This condition is analogous to the condition $\nabla \cdot B =$ which we encounter in electromagnetism—the magnetic field has zero divergence.

Incompressible flows are much easier to treat theoretically than compressible flows, so this is a very useful simplifying assumption. You might worry about applying it to gases, but it turns out that even there it often works well. In fact, if U is some characteristic flow velocity, and c is the speed of sound in the fluid, then the fluid can be treated as incompressible whenever $U << c$. This is often expressed in terms of the Mach number Ma, defined as

$$Ma = U / c;$$

then if $Ma << 1$ the fluid can be treated as incompressible. For instance, the speed of sound in air is 330 m/s (740 mi/h); therefore, fluid flow around people, cars, trains, as well as fluid flow in hurricanes and tornadoes, all satisfy the incompressibility condition. Air flow around commercial airliners ($Ma = 0.7$) violates this condition, as does supersonic flight.

STREAMLINES AND STREAMTUBES

In problems involving electric or magnetic fields we visualize the fields by introducing field lines (often called lines of force), which have the properties that they are everywhere tangent to the field vectors, and the density of the field lines is proportional to the magnitude (strength) of the field. We can use the same method to visualize the velocity field v by introducing streamlines, which have the following properties:

- A tangent to a streamline at a point is in the direction of the fluid velocity at that point;
- The density of streamlines in the vicinity of a point is proportional to the magnitude of the velocity at that point;
- The streamlines cannot intersect except at a point of zero velocity, otherwise the velocity would not be uniquely determined at that point.

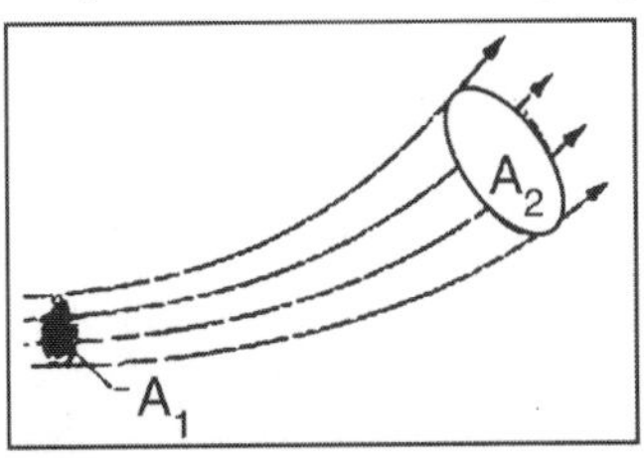

Fig. A Streamtube in a Fluid.

A *streamtube* is a tubular region of fluid surrounded by streamlines. Since streamlines don't intersect, the same streamlines pass through a streamtube at all points along its length. Let's take two cross-sections of a streamtube, with cross-sectional areas $A_1$and A_2

If these areas are made small enough then the fluid velocities across the cross-sections will be constant; let's call the speeds v_1 and v_2 (the directions of the velocities being perpendicular to the cross-sections). The rate at which mass is entering the streamtube is $\rho_1 v_1 A_1$; the rate at which it is leaving is $\rho_2 v_2 A_2$. If the mass inside the streamtube is not changing with time, so that the fluid is either incompressible or in steady-state, we have,

$$\rho_1 v_1 A_1 = \rho_2 v_2 A_2.$$

If, in addition, the fluid is incompressible, then $\rho_1 = \rho_2$, and we have

$$v_1 A_1 = v_2 A_2 A,$$

so that the fluid speed increases when the cross-sectional area of the streamtube decreases. If this seems counter-intuitive, the following analogy might help. You drive down a four lane highway, and due to construction one lane on your side has been closed. The traffic crawls towards the point where the lane has been closed, with cars haphazardly merging into your lane. You finally hit the single lane, and whiz through the construction area. The cross-sectional area has decreased, and your speed has increased.

CIRCULATION, VORTICITY, AND VORTICES

Fluids will invariably swirl around, and it is often useful to characterize the amount of swirling. This is done by introducing the *circulation* Γ, defined as,

$$\Gamma = \oint_C \mathrm{v} \cdot \mathrm{dl},$$

with the line integral taken is around a closed circuit *C* in a counter-clockwise sense. By using Stokes' theorem we may write this as a surface integral,

$$\oint_C \mathrm{v} \cdot \mathrm{dl} = \int_S (\nabla \times \mathrm{v}) \cdot \mathrm{dS}$$

$$= \int_S \mathrm{w} \cdot \mathrm{dS},$$

where the *vorticity* w of the fluid is defined as,

$$\mathrm{w} = \nabla \times \mathrm{v}.$$

If the vorticity of the fluid is zero, we say that the fluid is irrotational :

$$\nabla \times \mathrm{v} = 0 \qquad (\text{irrotational}).$$

Already mentioned that there are some useful analogies between magnetostatics and incompressible fluid flow. To transform the magnetostatics expressions into those for fluid flow, we make the substitutions $B/\mu_0 \rightarrow v$, $I \rightarrow \Gamma$, , and $\mu_0 \rightarrow \rho$, (where here ρ is the mass density of the fluid). If a fluid is both incompressible *and* irrotational, then $\nabla \times v = 0$ and $\nabla \cdot v = 0$, which are the same as the equations of electrostatics and magnetostatics in free space (*i.e.*, in the absence of charges and currents).

Magnetostatics

Incompressible fluid flow,

$$\nabla \cdot B = 0$$

$$\nabla \cdot v = 0$$

$$\int B \cdot dl = \mu_0 I \text{ (Ampére's Law)}$$

$$\int v \cdot dl = \Gamma \text{ (circulation)}$$

$$E_B = B^2 / 2_{\mu_0} \text{ (energy density of the magnetic field)}$$

$$E = \frac{1}{2}\rho v^2 \text{ (kinetic energy density of a fluid)}$$

Analogies between incompressible fluid flow and magnetostatics.

As an example of fluid flow with vorticity, we can consider a single *vortex* , which would be produced when draining a bathtub, and which is illustrated in Fig

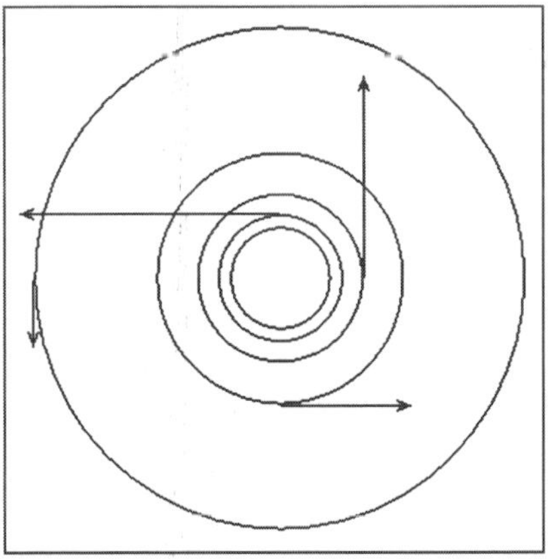

Fig. Velocity Field of a Vortex. Flow Lines are Drawn for Equal Increments of the Speed, as Shown by Arrows.

In cylindrical coordinates (r, θ, z) there is only one component of the velocity field, v_θ. In calculating the circulation, the line element $dl = rd\theta\hat{\theta}$, so that $v \cdot dl = v_\theta rd\theta$. If the circulation is independent of the integration path, then we must have $v_\theta = C/r$, with *C* a constant. The circulation is then,

$$\Gamma = \int_0^{2\pi}\left(\frac{C}{r}\right) r d\theta = 2\pi C,$$

so that $C = \Gamma / 2\pi$. Therefore, the velocity field of a vortex is,

$$v(r) = \frac{\Gamma}{2\pi r}\hat{\theta} \quad (\text{vortex}).$$

By making the substitutions suggested above, we can use this result to find the magnetic field produced by a wire carrying a current *I*:

$$B(r) = \frac{\mu_0 I}{2\pi r}\hat{\theta} \quad (\text{current} - \text{carrying wire}).$$

CAPILLARITY

WHAT IS CAPILLARY ACTION?

Capillary action is a physical effect caused by the interactions of a liquid with the walls of a thin tube. The capillary effect is a function of the ability of the liquid to wet a particular material.

The liquid for which this effect is most commonly seen is water, because water is capable of strong surface interactions and because water is ubiquitous.

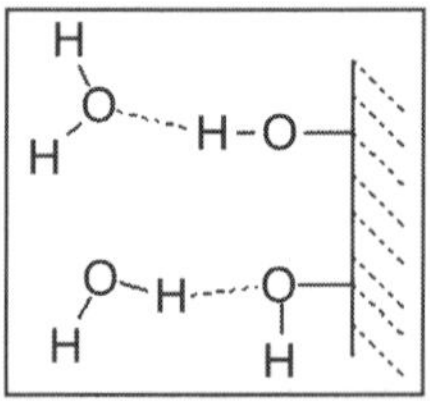

Water climbs up a thin glass tube because of the strong hydrogen-bonding interactions between the water and the oxygens (and terminal hydrogens) at the surface of the glass (SiO_2; surface oxygens are typically bonded to hydrogen). The energetic gain from the new intermolecular interactions must be balanced against gravity, which attempts to pull the liquid back down. Therefore, the narrower the tube, the higher the liquid will climb, because a narrow column of liquid weighs less than a thick one.

One gets the same effect for pores and narrow spaces in any hydrophilic object. While I have not been able to find any information on the Web, it is my suspicion that a narrow tube coated with a hydrophobic substance (such as Teflon) should not show a capillary effect.

Capillary Action

Capillary action is the result of adhesion and surface tension. Adhesion of water to the walls of a vessel will cause an upward force on the liquid at the edges and result in a meniscus which turns upward. The surface tension acts to hold the surface intact, so instead of just the edges moving upward, the whole liquid surface is dragged upward.

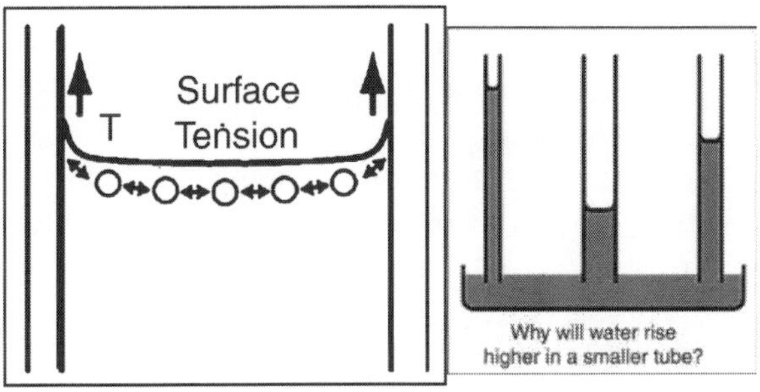

Capillary action, capillarity, capillary motion, or wicking is the ability of a substance to draw another substance into it. The standard reference is to a tube in plants but can be seen readily with porous paper. It occurs when the adhesive intermolecular forces between the liquid and a substance are stronger than the cohesive intermolecular forces inside the liquid. The effect causes a concave meniscus to form where the substance is touching a vertical surface. The same effect is what causes porous materials such as sponges to soak up liquids.

A common apparatus used to demonstrate capillary action is the *capillary tube*. When the lower end of a vertical glass tube is placed in a liquid such as water, a concave meniscus forms. Surface tension pulls the liquid column up until there is a sufficient weight of liquid for gravitational forces to overcome the intermolecular forces. The weight of the liquid column is proportional to the square of the tube's diameter, but the contact length (around the edge) between the liquid and the tube is proportional only to the diameter of the tube, so a narrow tube will draw a liquid column higher than a wide tube. For example, a glass capillary tube 0.5 mm in diameter will lift a theoretical 2.8 cm column of water. Actual observations show shorter total distances.

With some pairs of materials, such as mercury and glass, the interatomic forces within the liquid exceed those between the solid and the liquid, so a convex meniscus forms and capillary action works in reverse.

Examples

In hydrology, capillary action describes the attraction of water molecules to soil particles. Capillary action is responsible for moving groundwater from wet areas of the soil to dry areas. Differences in soil matric potential (ψ_m) drive capillary action in soil.

Capillary action is also essential for the drainage of constantly produced tear fluid from the eye, two canalicula of tiny diameter are present in the inner corner of the eyelid, also called the lacrimal ducts; their openings can be seen with the naked eye within the lacrimal sacs when the eyelids are everted.

Paper towels adsorb liquid through capillary action, allowing a fluid to be transferred from a surface to the towel. The small pores of a sponge act as small capillaries, causing it to adsorb a comparatively large amount of fluid.

Some modern sport and exercise fabrics, such as Coolmax, use capillary action to "wick" sweat away from the skin. These are often referred to as wicking fabrics, presumably after the capillary properties of a candle wick.

Chemists utilize capillary action in thin layer chromatography, in which a solvent moves vertically up a plate via capillary action. Dissolved solutes travel with the solvent at various speeds depending on their polarity.

Capillary Action Calculation

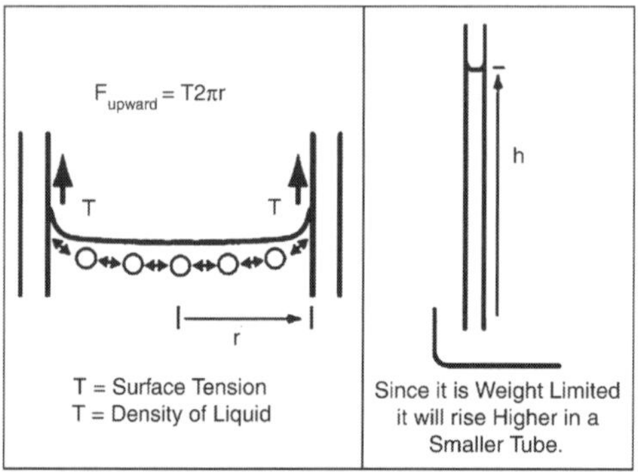

The height h to which Capillary action will lift water depends upon the weight of water which the surface tension will lift:

$$\mathrm{T}2\pi \mathrm{r} = \rho \mathrm{g}\left(\mathrm{h}\pi \mathrm{r}^2\right)$$

The height to which the liquid can be lifted is given by,

$$\mathrm{h} = \frac{2\mathrm{T}}{\rho \mathrm{rg}}$$

Discussion: With notes on the dimension in SI units, the height h of a liquid column (m) is given by,

$$\mathrm{h} = \frac{2\gamma \cos\theta}{\rho \mathrm{rg}}$$

...where:

γ = The liquid-air surface tension (J/m^2 or N/m)
θ = Contact angle
ρ = Density of liquid (kg/m^3)
g = Acceleration due to gravity (m/s^2)
r = Radius of tube (m)

For a water-filled glass tube in air at sea level,

γ = 0.0728 J/m^2 at 20 °C
θ = 20° (0.35 rad)

$\nabla = 1000 \text{ kg/m}^3$
$g = 9.8 \text{ m/s}^2$
...and so the height of the water column is given by:

$$h \approx \frac{1.4 \times 10^{-5}}{R}.$$

Thus for a 2 m wide (1 m radius) tube, the water would rise an unnoticeable 0.014 mm. However, for a 2 cm wide tube, the water would rise 1.4 mm and for tube with radius 0.1 mm, the water would rise 14 cm (about 6 inches).

Capillary Wave

A capillary wave is a wave travelling along the interface between two fluids, whose dynamics are dominated by the effects of surface tension. Capillary waves are common in nature and the home and are often referred to as ripples. The wavelength of capillary waves is typically less than about a centimeter.

The dispersion relation for capillary waves is,

$$w^2 = \frac{\sigma}{\rho + \rho'} k^3$$

where ω is the frequency, σ the surface tension, ρ the density of the heavier fluid, ρ' the density of the lighter fluid and k the wavenumber. The wavelength is,

$$\lambda = \frac{2\pi}{k}$$

The waves with large wavelengths are generally also affected by gravity and are then called gravity-capillary waves. Their dispersion relation reads, for infinite depth of the two fluids,

$$w^2 = \frac{g(\rho - \rho')}{\rho + \rho'} k + \frac{\sigma}{\rho - \rho'} k^3$$

where ω is the frequency, g the acceleration due to gravity, σ the surface tension, ρ the density and k the wavenumber. This class of waves involves ocean surface waves.

In water on earth this is observed: Gravity waves have a group velocity half the phase velocity.

Following a single wave in a group one can see the wave appearing at the back of the group, growing and finally disappearing at the front of the group.

Therefore an interesting and common situation occurs when the dispersion caused by gravity cancels out the dispersion due to the capillary effect. At wavelength around 2 cm the capillary effect causes group velocity to equal phase velocity. The dispersion is zero, and a wave ridge can travel for long distances.

Shorter (*i.e.* 2 mm) ripples do the opposite: the wave appears at the front of the group, growing and finally disappearing at the back of the group.

Capillary Pressure

In fluid statics, capillary pressure is the difference in pressure across the interface between two immiscible fluids. The pressure difference is proportional to the surface tension, ã, and inversely proportional to the effective radius, *r*, of the interface, it also depends on the wetting angle, θ, of the liquid on the surface of the capillary.

$$\mathrm{Pc} = \frac{2\gamma\cos\theta}{\mathrm{r}}$$

The capillary pressure p_c is defined as,

$$p_c = p_g - p_w$$

for the case of one phase being gas with pressure p_g and the other phase being water with pressure p_w.

The equation for capillary pressure is only valid under capillary equilibrium, which means that there can not be any flowing phases.

Capillary Number

In fluid dynamics, the capillary number represents the relative effect of viscous forces versus surface tension acting across an interface between a liquid and a gas, or between two immiscible liquids. It is defined as

$$\mathrm{Ca} \overset{\mathrm{def}}{=} \frac{\mu \mathrm{V}}{\sigma}$$

where μ is the viscosity of the liquid, *v* is a characteristic velocity and σ is the surface or interfacial tension between the two fluid phases.

For low capillary numbers (a rule of thumb says less than 10^{-5}), flow in porous media is dominated by capillary forces.

VAPOUR PRESSURE

Vapour pressure is the pressure of a vapor in equilibrium with its non-vapor phases. All solids and liquids have a tendency to evaporate to a gaseous form, and all gases have a tendency to condense back. At any given temperature, for a particular substance, there is a partial pressure at which the gas of that substance is in dynamic equilibrium with its liquid or solid forms. This is the vapor pressure of that substance at that temperature.

In meteorology, the term *vapor pressure* is used to mean the partial pressure of water vapor in the atmosphere, even if it is not equilibrium, and the equilibrium vapor pressure is specified as such. Meteorologists also use the term saturation vapor pressure to refer to the equilibrium vapor pressure of water or brine above a flat surface, to distinguish it from equilibrium vapor pressure which takes into account the shape and size of water droplets and particulates in the atmosphere. This article is about our first definition of vapor pressure, or what meteorologists would call equilibrium vapor pressure.

Equilibrium vapor pressure is an indication of a liquid's evaporation rate. It relates to the tendency of molecules and atoms to escape from a liquid or a solid. A substance with a high vapor pressure at normal temperatures is often referred to as *volatile*. The higher the vapor pressure of a material at a given temperature, the lower the boiling point.

The vapor pressure of any substance increases non-linearly with temperature according to the Clausius-Clapeyron relation. The boiling point of a liquid is the temperature where the vapor pressure equals the ambient atmospheric pressure. At the boiling temperature, the vapor pressure becomes sufficient to overcome atmospheric pressure and lift the liquid to form bubbles inside the bulk of the substance.

Units of Vapor Pressure

The most common unit for vapor pressure is the torr. 1 torr = 1 mm Hg (one millimeter of mercury). The international unit for pressure is: 1 pascal = a force of 1 newton per square meter = 10 dyn/cm^2 = 0.01 mbar= 0.0075 mmHg = 0.00000969 atm= 0.0003545414 psi .

Equilibrium Vapor Pressure of Solids

Equilibrium vapor pressure can be defined as the pressure reached when a condensed phase is in equilibrium with its own vapor. In the case of an equilibrium solid, such as a crystal, this can be defined as the pressure when the rate of sublimation of a solid matches the rate of deposition of its vapor phase. For most solids this pressure is very low, but some notable exceptions are naphthalene, dry ice (the vapor pressure of dry ice is 5.73 MPa (831 psi, 56.5 atm) at 20 degrees Celsius, meaning it will cause most non-ventilated containers to explode if sealed inside), and ice. Ice will still continue to disappear even though the ambient temperature is below the freezing point of water. All solid materials have a vapor pressure. However, due to their often extremely low values, measurement can be rather difficult. Typical techniques include the use of thermogravimetry and gas transpiration.

Relation between Solid and Liquid Vapor Pressures

It may be noted that the vapor pressure of a substance in liquid form is usually different from the vapor pressure of the same substance in solid form. If the temperature is such that the vapor pressure of the liquid is higher than that of the solid, liquid will vaporize but vapor will condense to a solid, *i.e.* the liquid is freezing. If the temperature is such that the vapor pressure of the liquid is lower than that of the solid, solid will vaporize but vapor will condense to a liquid, *i.e.* the solid is melting. At the temperature that equalizes the two vapor pressures, an equilibrium exists between solid and liquid phases. This temperature is referred to as the melting point.

Water Vapor Pressure

Graph of water vapor pressure versus temperature. Note that at the boiling point of 100°C, the vapor pressure equals the standard atmospheric pressure of 760 Torr.

Water, like all liquids, starts to boil when its vapor pressure reaches its surrounding pressure. At higher elevations the atmospheric pressure is lower and water will boil at a lower temperature. The boiling temperature of water for pressures around 100 kPa can be approximated by,

$$T_v = 100 + 0.0002772 \cdot (p - 101000) - 1.24 \cdot 10^{-9} \cdot (p - 101000)^2$$

where the temperature T_v is in degrees Celsius and the pressure p is in pascals. One gets the vapor pressure by solving this equation for p.

In meteorology, the international standard for the vapour pressure of water over a flat surface is given by the Goff-Gratch equation.

Partial Pressures

Raoult's law gives an approximation to the vapor pressure of mixtures of liquids. It states that the activity (pressure or fugacity) of a single-phase mixture is equal to the mole-fraction-weighted sum of the components' vapor pressures:

$$P_{tot} = \sum_i P_i X_i$$

where p is vapor pressure, i is a component index, and X is a mole fraction. The term $p_i X_i$ is the vapor pressure of component i in the mixture. Raoult's Law is applicable only to non-electrolytes (uncharged species); it is most appropriate for non-polar molecules with only weak intermolecular attractions (such as London forces).

Systems that have vapor pressures higher than indicated by the above formula are said to have positive deviations. Such a deviation suggests weaker intermolecular attraction than in the pure components, so that the molecules can be thought of as being "held in" the liquid phase less strongly than in the pure liquid. An example is the azeotrope of approximately 95% ethanol and water. Because the azeotrope's vapor pressure is higher than predicted by Raoult's law, it boils at a temperature below that of either pure component.

There are also systems with negative deviations that have vapor pressures that are lower than expected. Such a deviation is evidence for stronger intermolecular attraction between the constituents of the mixture than exists in the pure components.

Thus, the molecules are "held in" the liquid more strongly when a second molecule is present. An example is a mixture of trichloromethane (chloroform) and 2-propanone (acetone), which boils above the boiling point of either pure component.

EXAMPLES OF VAPOR PRESSURES

Gas	Vapor pressure (bar)	Vapor pressure (mmhg)	Temperature
Helium	1	750	@ -269.15 °C
Propane	22	16500	@ 55 °C
Butane	2.2	1650	@ 20 °C
Carbonyl sulfide	12.55	9412	@ 25 °C
Acetaldehyde	0.987	740	@ 20 °C
Freon 113	0.379	284	@ 20 °C
Methyl isobutyl ketone	0.02648	19.86	@ 25 °C
Tungsten	0.001	0.75	@ 3203 °C

COMPRESSIBILITY

This article is about thermodynamics and fluid mechanics. For other uses, see Compression.

Table. Material Properties

- *Specific heat*:

$$c = \frac{T}{N}\left(\frac{\partial S}{\partial T}\right)$$

- *Compressibility*:

$$\beta = -\frac{1}{V}\left(\frac{\partial V}{\partial p}\right)$$

- *Thermal expansion*:

$$\alpha = \frac{1}{V}\left(\frac{\partial V}{\partial T}\right)$$

In thermodynamics and fluid mechanics, compressibility is a measure of the relative volume change of a fluid or solid as a response to a pressure (or mean stress) change.

$$\beta = -\frac{1}{V}\frac{\partial V}{\partial p}$$

where *V* is volume and *p* is pressure. The above statement is incomplete, because for any object or system the magnitude of the compressibility depends strongly on whether the process is adiabatic or isothermal. Accordingly we define the isothermal compressibility as:

$$\beta_T = -\frac{1}{V}\left(\frac{\partial V}{\partial p}\right)_T$$

where the subscript T indicates that the partial differential is to be taken at constant temperature.

The adiabatic compressibility as:

$$\beta_s = -\frac{1}{V}\left(\frac{\partial V}{\partial p}\right)_S$$

where S is entropy. For a solid, the distinction between the two is usually negligible.

The inverse of the compressibility is called the bulk modulus, often denoted K (sometimes B). That page also contains some examples for different materials.

Importance of Compressibility in Fluid Dynamics

Compressibility is an important factor in aerodynamics. At low speeds, the compressibility of air is not significant in relation to aircraft design, but as the airflow nears and exceeds the speed of sound, a host of new aerodynamic effects become important in the design of aircraft. These effects, often several of them at a time, made it very difficult for World War II era aircraft to reach speeds much beyond 800 km/h (500mph).

Some of the minor effects include changes to the airflow that lead to problems in control. For instance, the P-38 Lightning with its thick high-lift wing had a particular problem in high-speed dives that led to a nose-down condition. Pilots would enter dives, and then find that they could no longer control the plane, which continued to nose over until it crashed. Adding a "dive flap" beneath the wing altered the center of pressure distribution so that the wing would not lose its lift. This fixed the problem.

A similar problem affected some models of the Supermarine Spitfire. At high speeds the ailerons could apply more torque than the Spitfire's thin wings could handle, and the entire wing would twist in the opposite direction. This meant that the plane would roll in the direction opposite to that which the pilot intended, and led to a number of accidents. Earlier models weren't fast enough for this to be a problem, and so it wasn't noticed until later model Spitfires like the Mk.IX started to appear. This was mitigated by adding considerable torsional rigidity to the wings, and was wholly cured when the Mk.XIV was introduced.

The Messerschmitt Bf 109 and Mitsubishi Zero had the exact opposite problem in which the controls became ineffective. At higher speeds the pilot simply couldn't move the controls because there was too much airflow over the control surfaces. The planes would become difficult to maneuver, and at high enough speeds less maneuverable aircraft could out-turn them.

Finally, another common problem that fits into this category is flutter. At some speeds the airflow over the control surfaces will become turbulent, and the controls will start to flutter. If the speed of the fluttering is close to a harmonic of the control's movement, the resonance could break the control off completely. This was a serious problem on the Zero. When problems with poor control at high speed were first encountered, they were addressed by designing a new style of control surface with more power. However this introduced a new resonant mode, and a number of planes were lost before this was discovered.

All of these effects are often mentioned in conjunction with the term "compressibility", but in a manner of speaking, they are incorrectly used. From a strictly aerodynamic point of view, the term should refer only to those side-effects arising as a result of the changes in airflow from an incompressible fluid (similar in effect to water) to a compressible fluid (acting as a gas) as the speed of sound is approached. There are two effects in particular, wave drag and critical mach.

Wave drag is a sudden rise in drag on the aircraft, caused by air building up in front of it. At lower speeds this air has time to "get out of the way", guided by the air in front of it that is in contact with the aircraft. But at the speed of sound this can no longer happen, and the air which was previously following the streamline around the aircraft now hits it directly. The amount of power needed to overcome this effect is considerable. The critical mach is the speed at which some of the air passing over the aircraft's wing becomes supersonic.

At the speed of sound the way that lift is generated changes dramatically, from being dominated by Bernoulli's principle to forces generated by shock waves. Since the air on the top of the wing is traveling faster than on the bottom, due to Bernoulli effect, at speeds close to the speed of sound the air on the top of the wing will be accelerated to supersonic. When this happens the distribution of lift changes dramatically, typically causing a powerful nose-down trim. Since the aircraft normally approached these speeds only in a dive, pilots would report the aircraft attempting to nose over into the ground.

All of these effects have adverse effects on the control or performance of the plane. For this reason it's common to see references to aircraft that suffer from compressibility. The P-38 and Zero are particularly common examples, although in fact they are both bad ones.

CLASSIFICATION OF FLUID FLOW

Fluid flow can be very smooth, calm, and regular, but generally the flow of a fluid is not so disciplined- it becomes a vagabond and starts flowing in random patterns. What are different flow patterns, how are flows classified, and what are the fluid paths in those patterns?

While studying the motion of a rigid body we do not have to bother about the relative motion of the particles of the rigid body as they are very firmly fixed to each other and move as a whole.

But for the study of the motion of fluids, things are not so simple because the fluid particles are attached with each other with very weak forces. There are various relative motions and a lot of possibilities for relative motion between the fluid particles.

To make things somewhat simple or for making the flow analysis feasible, fluid flow is visualized as a composition of fluid elements. These elements are defined by using certain similarities or patterns and mathematics is applied to them to study fluid flow comprehensively.

Rotational or Irrotational Flow

To classify any flow as Rotational or Irrotational the angular motion of the fluid elements is analysed. If the angle between the two intersecting lines of the boundary of the fluid element changes while moving in the flow, then the flow is a Rotational Flow. But if the fluid element rotates as a whole and there is no change in angles between the boundary lines then the flow cannot be Rotational Flow, so it is Irrotational Flow. This means that there should be some deformation in the fluid element in a Rotational Flow. Such deformation of the fluid element or the shear strain is necessarily caused by tangential forces or shear stresses. Shear stresses are caused by viscosity, thus the flow of viscous fluids is rotational. But this does not mean that the flow of non-viscous or ideal fluid is always irrotational. The flow of ideal fluids can be rotational by external work or heat interaction.

LAMINAR OR TURBULENT FLOW

Laminar Flow

The flow of a fluid moving with a moderate speed has fluid layers moving past other layers as if some sheets are moving over other layers. Such flow of fluids is called Laminar Flow. In Laminar Flow viscous shear stresses act between these layers of the fluid which defines the velociity distribution among these layers of flow. In Laminar Flows the shear stresses are defined by Newton's equation for shear stress.

Turbulent Flow

As the flow speed of the otherwise calm layers increases, these smoothly moving layers start moving randomly, and with further increase in flow velocity, the flow of fluid particles becomes completely random and no such laminar layers exist any more. Shear stresses in the Turbulent Flow are more than those in Laminar Flow. A dimensionless parameter, Reynolds Number, is defined as the ratio of inertial and viscous force to characterize these two types of flow patterns.

With increase in flow velocity the intial forces increase so the Reynonlds Number. For moderate flows the Reynolds Number is below 2000 and for Turbulent Flows it is well above 2300. For the transition region between the two types the Reynolds Number varies between 2000-4000.

Fundamentals of Fluid Flow

How does a flowing fluid behaves,and how do the different parameters vary in a flowing fluid? Fluid flow analysis involves theoretical and experimental study of the flow of fluid by defining flow models and observation of variation of parameters. This series will discuss Fundamentals of Fluid Flow.

FLOW MEASUREMENT

MANOMETER

Pressure (P) is the ratio of the force (F) applied to a surface to the surface area (A).

$$P = F / A$$

A manometer is an device employed to measure pressure. There are a variety of manometer designs. A simple, common design is to seal a length of glass tubing and then bend the glass tube into a U-shape. The glass tube is then filled with a liquid, typically mercury, so that all trapped air is removed from the sealed end of the tube. The glass tube is then positioned with the curved region at the bottom.

After the mercury settles to the bottom of the manometer, a vacuum is produced in the sealed tube. The open tube is connected to the system whose pressure is being measured. In the sealed tube, there is no gas to exert a force on the mercury. In the tube connected to the system, the gas in the system exerts a force on the mercury. The net result is that the column of mercury in the left (sealed) tube is higher than that in the right (unsealed) tube. The difference in the heights of the columns of mercury is a measure of the pressure of gas in the system.

In the example at the left, the top of the left column of mercury corresponds to 875 mm on the scale. The top of the right column of mercury corresponds to 115 mm. The difference in heights is 875 mm - 115 mm = 760. mm, which indicates that the pressure is 760. mm Hg or 760. torr.

This method for measuring pressure led to the use of millimeters of mercury (mm Hg) as a unit of pressure. Today 1 mm Hg is called 1 torr. A pressure of 1 torr or 1 mm Hg is literally the pressure that produces a 1 mm difference in the heights of the two columns of mercury in a manometer. To understand how the height of a column of mercury can be used as a unit of pressure and how the unit of torr is related to the SI unit of pascal (1 Pa = 1 N/ m^2), consider the following mathematical analysis of the behaviour of the manometer. The force exerted by the column of mercury in a tube arises from the gravitational acceleration of the column of mercury. Newton's Second Law provides an expression for this force:

$$F = m\,g$$

In this equation, m is the mass of mercury in the column and g = 9.80665 m/sec^2 is the gravitational acceleration. This force is distributed over the cross-sectional area of the column (A). The pressure resulting from the column of mercury is thus,

$$P = \frac{mg}{A}$$

The mass of mercury is given by the product of the density of mercury (d_{Hg}) and the volume of mercury (V). For a cylindrical column of mercury, the volume of mercury is the product of the cross-sectional area and the height of the column (h). These relationships product the following equation.

$$P = \frac{mg}{A} = \frac{d_{Hg} V_g}{A} = \frac{d_{Hg} AV_g}{A} = d_{Hg} Ahg$$

This equation clearly shows that the height of a column of mercury is directly proportional to the pressure exerted by that column of mercury. The difference in heights of the two columns of mercury in a manometer can thus be used to measure the difference in pressures between the two sides of the manometer.

The relation between torr and Pa is also clearly evident. Using d_{Hg} = 13.5951 g cm^{-3}, one finds that 1 torr = 133 Pa or 1 atm = 760 torr = 101 kPa.

Analog Manometers

One of the earliest pressure measuring instruments is still in wide use today because of its inherent accuracy and simplicity of operation. It's the U-tube manometer, which is a U-shaped glass tube partially filled with liquid. This manometer has no moving parts and requires no calibration. Manometry measurements are functions of gravity and the liquid's density, both physical properties that make the U-tube manometer a NIST standard for accuracy.

Digital Manometers

A digital manometer is a microprocessor based instrument that can be stationary or portable. It has communication output capabilities that can be used for process control or transferring measurement data. Due to the network communication capabilities, digital or "smart" manometers are well suited for field measurement and process control.

Pitot tube

A Pitot tube is a pressure measuring instrument used to measure fluid flow velocity, and more specifically, used to determine the airspeed of an aircraft. The Pitot tube was invented by Italian-born French engineer Henri Pitot in the early 1700s, and was modified to its modern form in the mid 1800s by French scientist Henry Darcy.

The basic Pitot tube simply consists of a tube pointing directly into the fluid flow. As this tube contains air, a pressure can be measured as the moving air is brought to rest. This pressure is the stagnation pressure of the air, also known as the total pressure, or sometimes (particularly in aviation circles) the pitot pressure.

The measured stagnation pressure cannot of itself be used to determine the airspeed. However, since Bernoulli's equation states that stagnation pressure = static pressure + dynamic pressure, then the dynamic pressure is simply the difference between the static pressure and the stagnation pressure. The static pressure is generally measured using the static ports on the side of the fuselage. The dynamic pressure is then determined using a diaphragm inside an enclosed container. If the air on one side of the diaphragm is at the static pressure, and the other at the stagnation pressure, then the deflection of the diaphragm is proportional to the dynamic pressure, which can then be used to determine the indicated airspeed of the aircraft. The diaphragm arrangement is typically contained within the airspeed indicator, which converts the dynamic pressure to an airspeed reading by means of mechanical levers.

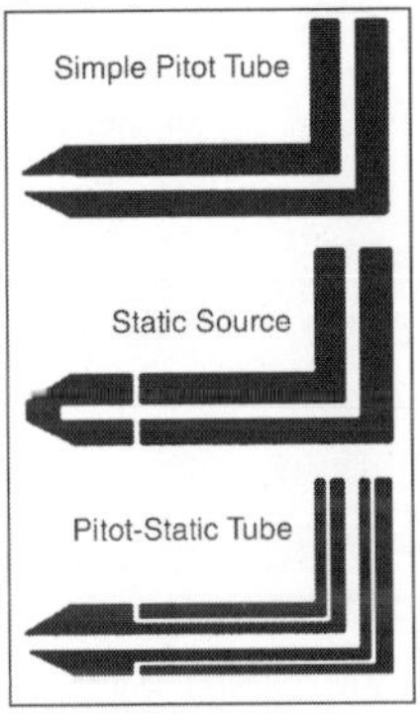

Fig. Types of Pitot Tube

Fig. Airliners have Pitot Probes Protruding from the Cockpit for Measuring Airspeed

Instead of static ports, a Pitot-static tube (also called a Prandtl tube) may be employed, which has a second tube coaxial with the Pitot tube with holes on the sides, outside the direct airflow, to measure the static pressure.

Pitot tubes on aircraft commonly have heating elements to prevent the tube from becoming clogged with ice. The failure of these systems has catastrophic consequences, like in the case of Austral Líneas Aéreas Flight 2553.

Applications in Mining

In mining applications the velocities being measured are those flowing in ducts or tubing where measurements by an anemometer would be difficult to obtain. In these kinds of measurements, the most practical instrument to use is the Pitot tube. The Pitot tube can be inserted through a small hole in the duct with the Pitot connected to a U-tube water gauge or some other differential pressure gauge for determining the velocity inside the duct.

The quantity of air flowing through the duct each minute can then be estimated from:

Volume Flow Rate (cubic feet per minute) = Duct Area (square feet) × Velocity (feet per minute)

Velocity is typically measured in feet per minute in such applications.

Venturi Meter

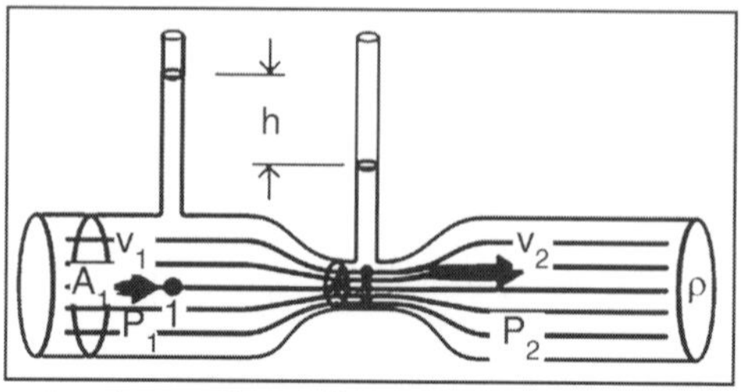

A Venturi meter is shown in a diagram, the pressure in "1" conditions is higher than "2", and the relationship between the fluid speed in "2" and "1" respectively, is the same as for pressure.

The Venturi effect is an example of Bernoulli's principle, in the case of fluid flow through a tube or pipe with a constriction in it. The fluid velocity must increase through the constriction to satisfy the equation of continuity, while its pressure must decrease due to conservation of energy: the gain in kinetic energy is supplied by a drop in pressure or a pressure gradient force. The effect is named after Giovanniee Battista Venturi, (1746–1822), an Italian physicist. The limiting case of the Venturi effect is choked flow, in which a constriction in a pipe or channel limits the total flow rate through the channel, because the pressure cannot drop below zero in the constriction. Choked flow is used to control the delivery rate of water and other fluids through spigots and other valves.

Referring to the diagram to the right, using Bernoulli's equation in the special case of incompressible fluids (such as the approximation of a water jet), the theoretical pressure drop at the constriction would be given by $(\rho/2)(v_2^2 - v_1^2)$.

Experimental Apparatus

This is a Venturi tube demonstration apparatus built out of PVC pipe and operated with a vacuum pump.

- *Venturi Tubes*: The simplest apparatus, as shown in the photograph and diagram, is a tubular setup known as a Venturi tube or simply a venturi. Fluid flows through a length of pipe of varying diameter. To avoid undue drag, a venturi tube typically has an entry cone of 30 degrees and an exit cone of 5 degrees.

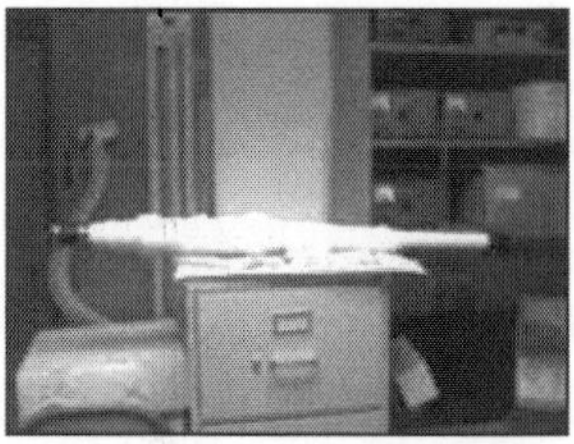

A venturi can also be used to mix a fluid with air. If a pump forces the fluid through a tube connected to a system consisting of a venturi to increase the water speed (the diameter decreases), a short piece of tube with a small hole in it, and last a venturi that decreases speed (so the pipe gets wider again), air will be sucked in through the small hole because of changes in pressure. At the end of the system, a mixture of fluid and air will appear.

- *Orifice plate*: Venturi tubes are more expensive to construct than a simple orifice plate which uses the same principle as a tubular scheme, but the orifice plate causes significantly more permanent energy loss and is less accurate.

In Chronic Aortic Regurgitation, after the initial large stroke volume is released, the Venturi effect draws walls together, transiently obstructing flow causing a Pulsus Bisferiens.

Practical Uses

The Venturi effect is visible in:

- The capillaries of the human circulatory system, where it indicates aortic regurgitation
- Large cities where wind is forced between buildings.
- Inspirators that mix air and flammable gas in barbecues, gas stoves, bunsen burners and airbrushes.
- Water aspirators that produce a partial vacuum using the kinetic energy from the faucet water pressure
- Steam siphon using the kinetic energy from the steam pressure to create a partial vacuum
- Atomizers that disperse perfume or spray paint (*i.e.* From a spray gun)
- Foam firefighting nozzles and extinguishers
- Carburetors that use the effect to suck gasoline into an engine's intake air stream.
- Protein skimmers, a filtration device for saltwater aquaria.
- In automated pool cleaners that use pressure-side water flow to collect sediment and debris.

- The modern day barrel of the clarinet, which uses a reverse taper to speed the air down the tube, enabling better tone, response and intonation.
- Compressed air operated industrial vacuum cleaners
- Venturi scrubbers used to clean flue gas emissions
- Injectors (or sometimes called ejectors) used to add chlorine gas in water treatment chlorination systems

A simple way to demonstrate the Venturi effect is to squeeze and release a flexible hose that is normal shape: the partial vacuum produced in the constriction is sufficient to keep the hose collapsed.

Venturi tubes are also used to measure the speed of a fluid, by measuring pressure changes at different segments of the device. Placing a liquid in a U-shaped tube and connecting the ends of the tubes to both ends of a venturi is all that is needed. When the fluid flows though the venturi the pressure in the two ends of the tube will differ, forcing the liquid to the "low pressure" side. The amount of that move can be calibrated to the speed of the fluid flow.

Orifice Meters

Good water management requires that flow rates be accurately measured. A simple but economical and accurate device for the measurement of flow rates from pipes discharging to the atmosphere is the orifice meter. The orifice meter is sometimes called a pipe orifice, an end-cap orifice, or more descriptively, a circular orifice weir.

Orifice meters are usually circular orifices placed at the end of a horizontal discharge pipe. Flow rates are calculated from the orifice characteristics and a measure of the pressure behind the orifice. The head (or pressure) on the orifice is measured with a water manometer to obtain a high degree of accuracy (Typically within 2% of true values).

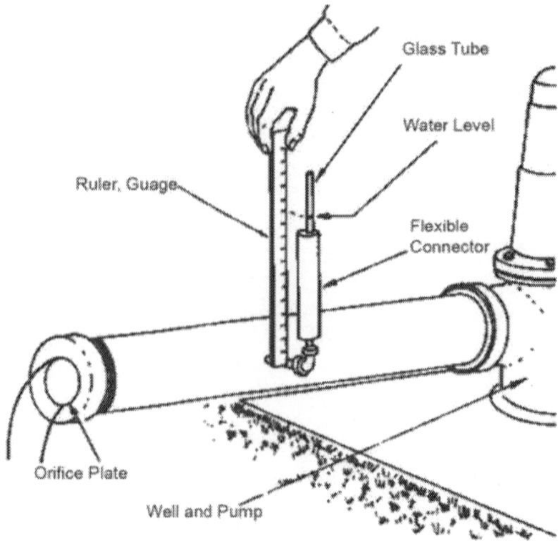

Because of the simplicity of design and the few components of the orifice meter, it can readily be constructed by individuals with only average mechanical skills. An orifice meter can be constructed from a pipe end-cap which has been properly drilled, if the discharge pipe is threaded to allow the end cap to be added for flow measurements or removed after its use.

The cost of the orifice meter is minimal when self-constructed. The orifice meter is very accurate because of the sensitivity of the water manometer, and, because of the absence of moving parts or the need for calibration, it is often used as a means of calibrating other flow-measuring devices.

Details of Construction

The orifice consists of a perfectly round hole in the center of a circular steel plate. The orifice must be cut with clean, square edges. The plate must be 1/16-inch thick around the circumference of the hole. The plate is fastened against the outer end of a level discharge pipe so that the orifice is centered on the pipe. The end of the pipe must be cut square so that the plate will be vertical. A properly machined pipe end-cap may be used as the orifice plate.

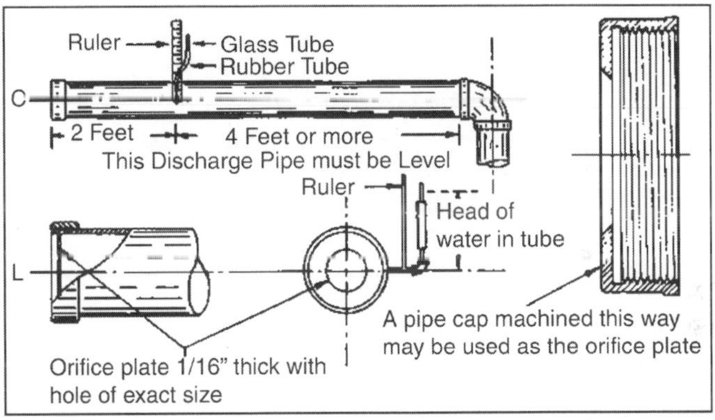

The bore of the pipe should be smooth and free of any obstruction that might cause excessive turbulence. This includes ensuring that there are no elbows, valves or other fittings closer than 6 feet upstream from the orifice. This approach must also be straight and level.

To measure pressure head on the orifice with sufficient accuracy, a glass tube manometer should be used. Exactly 24 inches from the orifice plate, the pipe wall is tapped mid-way between the top and bottom with a 1/8-1/4 inch diameter hole. Burrs inside the pipe caused by the drilling or tapping should be filed to ensure a smooth surface. The glass tube is connected with a short piece of flexible tubing to a pipe nipple threaded into the tapped hole. The glass tube is mounted vertically with a scale set to measure the height of water in the tube above the centertine of the pipe. When water is pumped through the orifice so that it is flowing full, the height of the water in the tube is the pressure head on the orifice.

Orifice Equations

For any given size of orifice discharge pipe, the rate of flow varies in a known manner with the pressure head as measured with the glass tube manometer. Standard tables, such as Table , give the flow in gallons per minute (gpm) for various combinations of orifice diameter and pipe diameter

Discharge through the orifice is computed from equation.

$$Q = 8.02c/\sqrt{h}$$

equation.

where,

Q = Orifice discharge (gpm),

C = Coefficient which varies with the ratio of the orifice Diameter to the pipe diameter, as well as with all of the other factors affecting flows in orifices of this construction,

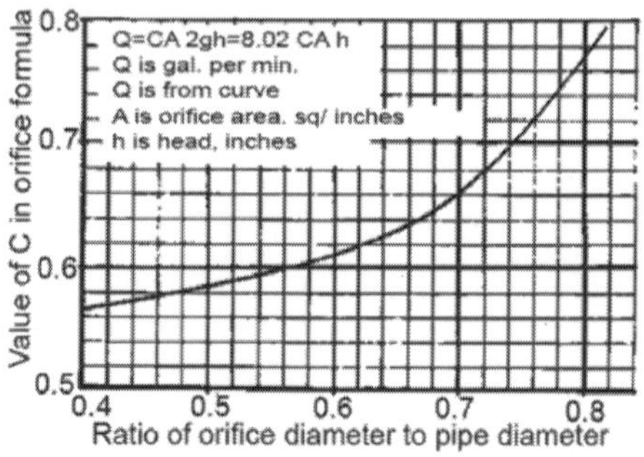

A = cross sectional area of the orifice (in ^{2}), and

h = head on the orifice measured above its centerline (in).

The value of its coefficient, C, varies with the ratio of the orifice diameter to pipe diameter because of changes in streamlines is flow converges through orifices of various sizes. The value of C is also a function of friction losses through the orifice. Therefore, the values of C given in Figure are only applicable when the orifice meter is constructed as previously described.

Example Calculations

The use of Equation 1 is illustrated by an example.

The following data apply:

- Orifice constructed as previously described
- Orifice diameter = 4 in
- Discharge pipe inside diameter = 8 in
- Pressure head, h = 30 in

Calculations are performed in 4 steps:

1. Ratio of orifice diameter to pipe diameter = 4 in / 8 in = 0.5
2. From Figure 3, C = 0. 587

3. Cross sectional area of orifice = pi d ^{2}/4 = (3.14)(4 in) ^{2}/4 = 12.56 in ^{2}
4. Q = 8.02 CA h
   Q = 8.02 (0.587) (12.56 in2) 30in
   Q = 324 gpm

Comparing this result with that shown in Table (325 gpm) shows the utility of Equation . The discrepancies between values in Table and those calculated result from inaccuracies in interpolating C values from Figure.

Ensuring Accuracies

The orifice meter provides the capability of measuring flow rates with accuracies within 2% of the actual rates if it is properly constructed and operated. Besides constructing the parts accurately and setting up the device correctly in the field, additional precautions must be taken to obtain accurate results. First, the diameter of the orifice should be less than 0.8 and greater than 0.4 of the inside diameter of the discharge pipe. The value of C changes rapidly as the ratio of diameters increases. Therefore, to ensure greatest accuracy, the ratio should be less than 0.7.

Second, the manometer tube must be completely free of obstruction and air bubbles when measurements of pressure head are made. Air bubbles can be eliminated by lowering the manometer tube to allow water to flow through it.

MOUTHPIECE

Mouthpiece usually refers to the part of an object which comes near or in contact with one's mouth during use, such as the mouthpiece of a smoking pipe, telephone or musical instrument.

It is also used figuratively and with negative connotations to indicate the role of a spokesperson or mass media venue that is used to perpetuate the views or agenda of another, as in "the newspaper became a *mouthpiece* for its owner's political views."

Fig. MouthPieces for Brass Instruments.

ROTAMETER

A Rotameter is a device that measures the flow rate of liquid or gas in a closed tube. It is occasionally misspelled as 'rotometer'.

It belongs to a class of meters called variable area meters, which measure flow rate by allowing the cross sectional area the fluid travels through to vary, causing some measurable effect.

A rotameter consists of a tapered tube, typically made of glass, with a float inside that is pushed up by flow and pulled down by gravity. At a higher flow rate more area (between the float and the tube) is needed to accommodate the flow, so the float rises.

Floats are made in many different shapes, with spheres and spherical ellipses being the most common. The float is shaped so that it rotates axially as the fluid passes. This allows you to tell if the float is stuck since it will only rotate if it is not. Readings are usually taken from the top of the float. Some manufacturers may use a different standard, so it is always best to check the documentation provided with the device.

Note that the 'float' does not actually float in the fluid: it has to have a higher density than the fluid otherwise it will float to the top even if there is no flow.

Advantages:

- A rotameter requires no external power or fuel, it uses only the inherent properties of the fluid, along with gravity, to measure flow rate.
- A rotameter is also a relatively simple device that can be mass manufactured out of cheap materials, allowing for widespread use in places such as third world countries.

Disavantages:

- Due to its use of gravity, a rotameter must always be vertically oriented and right way up, with the fluid flowing upwards.
- Due to its reliance on the ability of the fluid or gas to displace the float, the graduations on a given rotameter will only be accurate for a given substance. The main property of importance is the density of the fluid, however viscosity may also be significant. Floats are ideally designed to be insensitive to viscosity, however this is seldom verifiable from manufacturers specs. Either separate rotameters for different densities and viscosities may be used, or multiple scales on the same rotameter can be utilized.
- Rotameters normally require the use of glass (or other transparent material), otherwise the user cannot see the float. This limits their use in many industries to benign fluids, such as water.
- Rotameters are not easily adapted for reading by machine: although magnetic floats that drive a follower outside the tube are available.

FLOW MEASUREMENT AND FLUID VISCOSITY

Viscosity is the internal resistance of a fluid to flow. Water has low viscosity, while syrup has high viscosity. When syrup is cold it is very much more viscousthan when it is hot. Viscosity is highly temperature dependent.

ABSOLUTE AND KINEMATIC VISCOSITY

There are two ways to measure the viscosity of a fluid. One is to rotate a cup ordisk in the fluid and measure the rotational speed along with the torque required tokeep it rotating.

The density of the fluid doesn't play any part in the measurement.Viscosity is the torque divided by the speed and divided bythe contacting surfacearea. When measured this way, the viscosity is called "absolute viscosity." Itsunits are centipoise, pascal-seconds, or lb/ft-sec.

Another way to measure viscosity is to allow a certain volume of fluid to drain by gravity out of a container through a capillary tube or some other restriction.The time to drain is directly related to viscosity and is often reported in "seconds."

CONTINUITY EQUATION

Even though a detailed analysis of fluid flow can be extremely difficult, the basic concepts involved in fluid flow problems are fairly straightforward. These basic concepts can be applied in solving fluid flow problems through the use of simplifying assumptions and average values, where appropriate.

Even though this type of analysis would not be sufficient in the engineering design of systems, it is very useful in understanding the operation of systems and predicting the approximate response of fluid systems to changes in operating parameters. The basic principles of fluid flow include three concepts or principles; the first two of which the student has been exposed to in previous manuals. The first is the principle of momentum(leading to equations of fluid forces) which was covered in the manual on Classical Physics.

The second is the conservation of energy (leading to the First Law of Thermodynamics) which was studied in thermodynamics. The third is the conservation of mass (leading to the continuity equation) which will be explained in this module.

PROPERTIES OF FLUIDS

A *fluid* is any substance which flows because its particles are not rigidly attached to one another. This includes liquids, gases and even some materials which are normally considered solids, such as glass. Essentially, fluids are materials which have no repeating crystalline structure. Several properties of fluids were discussed in the Thermodynamics section of this text.

These included temperature, pressure, mass, specific volume and density. *Temperature* was defined as the relative measure of how hot or cold a material is. It can be used to predict the direction that heat will be transferred. *Pressure* was defined as the force per unit area. Common units for pressure are pounds force per square inch (psi).

Mass was defined as the quantity of matter contained in a body and is to be distinguished from weight, which is measured by the pull of gravity on a body. The *specific volume* of a substance is the volume per unit mass of the substance. Typical units are ft^3/lbm. *Density*, on the other hand, is the mass of a substance per unit volume.

Typical units are lbm/ft^3. Density and specific volume are the inverse of one another. Both density and specific volume are dependant on the temperature and somewhat on the pressure of the fluid. As the temperature of the fluid increases, the density decreases and the specific volume increases.

Since liquids are considered incompressible, an increase in pressure will result in no change in density or specific volume of the liquid. In actuality, liquids can be slightly compressed at high pressures, resulting in a slight increase in density and a slight decrease in specific volume of the liquid.

The properties outlines below are general properties of fluids which are of interest in engineering. The symbol usually used to represent the property is specified together with some typical values in SI units for common fluids. Values under specific conditions (temperature, pressure etc.) can be readily found in many reference books. The dimensions of each unit is also give in the MLT system (see later in the section on dimensional analysis for more details about dimensions.)

Density

The density of a substance is the quantity of matter contained in a unit volume of the substance. It can be expressed in three different ways.

Mass Density

Mass Density, ρ, is defined as the mass of substance per unit volume.

Units: Kilograms per cubic metre, kg/m^3 (or $^{kgm-3}$)

Dimensions: ML^{-3}

Typical values: Water = 1000^{kgm-3}, Mercury = 13546^{kgm-3} Air = 1.23^{kgm-3}, Paraffin Oil = 800^{kgm-3}. (at pressure =1.013 = 1.013×10^{-5} Nm^{-2} and Temperature = 288.15 K.)

Specific Weight

Specific Weight ω, (sometimes, and sometimes known as *specific gravity*) is defined as the weight per unit volume.

or

The force exerted by gravity, g, upon a unit volume of the substance. The Relationship between g and ω can be determined by Newton's 2^{nd}Law, since weight per unit volume = mass per unit volume g,

$$\omega = \rho g$$

Units: Newton's per cubic metre, N/M^3 (or Nm^{-3})

Dimensions: ML^{-2}T^{-2}.

Typical values:

Water = 9814 Nm^{-3}, Mercury = 132943 Nm^{-3}, Air = 12.07 Nm^{-3}, Paraffin Oil =7851 Nm^{-3}

Relative Density

Relative Density, σ, is defined as the ratio of mass density of a substance to some standard mass density. For solids and liquids this standard mass density is the maximum mass density for water (which occurs at 4°c) at atmospheric pressure.

$$s = \frac{\rho_{\text{sibstamce}}}{\rho_{H_2O(\omega 4°c)}}$$

Units: None, since a ratio is a pure number.

Dimensions: 1.

Typical values: Water = 1, Mercury = 13.5, Paraffin Oil =0.8.

Viscosity

Viscosity, m, is the property of a fluid, due to cohesion and interaction between molecules, which offers resistance to sheer deformation. Different fluids deform at different rates under the same shear stress. Fluid with a high viscosity such as syrup, deforms more slowly than fluid with a low viscosity such as water. All fluids are viscous, "Newtonian Fluids" obey the linear relationship given by Newton's law of viscosity which we saw earlier,

$$\tau = \mu \frac{du}{dy},$$

where τ is the shear stress,

Units: Nm^{-2}; kg m^{-1}s^{-2}

Dimensions ML^{-1}T^{-2}.

$$\frac{du}{dy}$$

is the velocity gradient or rate of shear strain, and has

Units: radianss^{-1}

Dimensions t^{-1}

μ is the "coefficient of dynamic viscosity"-see below.

Coefficient of Dynamic Viscosity

The Coefficient of Dynamic Viscosity, μ, is defined as the shear force, per unit area, (or shear stress τ), required to drag one layer of fluid with unit velocity past another layer a unit distance away.

$$\mu = \tau \frac{du}{dy} = \frac{\text{Force}}{\text{Area}} \frac{\text{Velocity}}{\text{Distance}} = \frac{\text{Force} \times \text{Time}}{\text{Area}} = \frac{\text{Mass}}{\text{Length} \times \text{Area}}$$

Units: Newton seconds per square metre, Nsm^{-2} or Kilograms per meter per second, $kgm^{-1}\,s^{-1}$. (Although note that μ is often expressed in Poise, P, where 10 P = 1 $kgm^{-1}\,s^{-1}$.)

Typical values: Water =1.14 × 10^{-3} Air =1.78 × 10^{-5} $kgm^{-1}s^{-1}$, Mercury = 1.552 $kgm^{-1}\,s^{-1}$,

Paraffin Oil =1.9 $kgm^{-1}\,s^{-1}$.

Kinematic Viscosity

Kinematic Viscosity, ν, is defined as the ratio of dynamic viscosity to mass density.

$$\nu \frac{\mu}{\rho}$$

Units: square metres per second, m^2s^{-1} (Although note that n is often expressed in Stokes, St, where,

$$10^4 \text{ St} = 1\,m^2s^{-1}), \text{ St} = 1\,m^2s^{-1})$$

Dimensions: $L^2\,T^{-1}$.

Typical values:

Water =1.14 × 10^{-6} m^2s^{-1}, Air = 1.46 × 10^{-5} m^2s^{-1}, Mercury =1.145 × 10^{-4} m^2s^{-1},

Paraffin Oil = 2.375 × 10^{-3} m^2s^{-1}

BUOYANCY

Buoyancy is defined as the tendency of a body to float or rise when submerged in a fluid. Weall have had numerous opportunities of observing the buoyant effects of a liquid. When we goswimming, our bodies are held up almost entirely by the water. Wood, ice, and cork float onwater. When we lift a rock from a stream bed, it suddenly seems heavier on emerging from thewater. Boats rely on this buoyant force to stay afloat. The amount of this buoyant effect wasfirst computed and stated by the Greek philosopher Archimedes. When a body is placed in afluid, it is buoyed up by a force equal to the weight of the water that it displaces.

4

Kinematics of Fluid in Motion

KINEMATIC CLASSIFICATION OF FLUID MOTION

One way of classifying different types of motion is according to their kinematic properties. For example, there is a large class of motions dealing with divergenceless fluids (commonly referred to as motion of *incompressible fluids*. Another class is motion wherein the vorticity is everywhere zero. This is referred to as *irrotational motion*. Motions of solid bodies, representing perhaps the most widely studied class in applied mechanics, falls under the heading of *deformationless motion*.

STREAMFUNCTION FOR TWO-DIMENSIONAL FLOW

If a particular two-dimensional motion is divergenceless, implying that,

$$\frac{\partial u}{\partial x}+\frac{\partial v}{\partial y}=0,$$

then the velocity field can be represented in terms of a single scalar function commonly referred to as the streamfunction. If we let ψ represent the streamfunction, then

$$u=-\frac{\partial \psi}{\partial y}$$

and,

$$v=-\frac{\partial \psi}{\partial y}$$

Equation is a necessary condition that such a function as implied by Eq. exist. It is readily verified that equations satisfy the condition, provided that

$$\frac{\partial}{\partial x}\frac{\partial \psi}{\partial y}=\text{ppxy}\left(\frac{\partial \psi}{\partial x}\right),$$

which holds as long as ψ possesses first and second derivatives.

VELOCITY POTENTIAL

If a particular fluid motion is irrotational, meaning that ,

$$\nabla \times \vec{V} = 0,$$

it is then possible to represent the field of flow in terms of a single scalar function ϕ as,

$$\vec{V} = -\nabla \phi$$

or, componentwise,

$$u = -\frac{\partial \phi}{\partial x},$$

$$v = -\frac{\partial \phi}{\partial y}, \text{and}$$

$$w = -\frac{\partial \phi}{\partial z}.$$

The quantity ϕ, referred to as the velocity potential, is a function of x,y,z,t in the most general case. Unlike the streamfunction, which is applicable for divergenceless, two-dimensional flow, the velocity potential may exist even for three-dimensional flow. However, the necessary condition for the existence of ϕ as defined by Eq. is somewhat more restrictive since, in effect, Eq. requires that

$$\frac{\partial w}{\partial y} = \frac{\partial v}{\partial z}, \frac{\partial u}{\partial z} = \frac{\partial w}{\partial x}, \frac{\partial v}{\partial x} = \frac{\partial u}{\partial y}.$$

LAPLACIAN FLOW

A two-dimensional field of flow which is both divergenceless and irrotational can be characterized by either one of the scalar functions ψ or ϕ. This type of flow, which bears the name of Laplace, possesses a number of interesting features which have been extensively investigated, and forms the bulk of much of the classical literature on the subject of hydrodynamics. Only some of the properties of this type of flow will be dealt with here. For a more exhaustive treatment, the reader is referred to the incomparable work of Lamb.

CONFORMAL MAPPING

It is possible at once to write down the most general solution for ψ and ϕ as governed by the equations of Art.. This can be done by introducing the concept of functions of a complex variable. If we let,

$$W = \phi + i\psi$$

and,

$$Z = x + iy$$

then,

$$W = f(Z)$$

represents the most general solution for ϕ and ψ, where f is any differentiable function at all. In any particular case, the problem is reduced to that of finding a function f which will satisfy certain boundary conditions.

The proof that Eq. satisfies the basic conditions on ϕ and ψ expressed by Eqs. is straightforward. If we differentiate W with respect to x we obtain,

$$\frac{\partial \phi}{\partial x} + i\frac{\partial \psi}{\partial x} = \frac{df}{dz}\frac{\partial z}{\partial x} = \frac{df}{dz}.$$

Differentiating W *with respect to* y, *on the other hand, gives,*

$$\frac{\partial \phi}{\partial y} + i\frac{\partial \psi}{\partial y} = \frac{df}{dz}\frac{\partial z}{\partial y} = i\frac{df}{dz}.$$

Eliminating df/dZ *between (214) and (215) yields,*

$$\frac{\partial \phi}{\partial y} + i\frac{\partial \psi}{\partial y} = i\frac{\partial \phi}{\partial x} - \frac{\partial \psi}{\partial x}.$$

This equation, however, can only be satisfied if,

$$\frac{\partial \phi}{\partial y} = -\frac{\partial \psi}{\partial x},$$

and ,

$$\frac{\partial \psi}{\partial y} = -\frac{\partial \phi}{\partial x},$$

where it is considered that the quantities ϕ and ψ are real. These equations are in accord with euations, as can be seen by equating the two expressions for u and the two expressions for v. The conditions in fact represent the Cauchy-Riemann conditions which are required if the function f(Z) is to possess a unique derivative.

The relation between W and Z stipulated by implies that for each point x, y in the Z-plane there corresponds at least one value of ϕ and ψ in the W-plane. One can map out the isolines of ϕ and ψ in the x, y plane or isolines of x and y in the ϕ,ψ plane. In either case the sets of isolines are orthogonal.

Consider an arbitrary curve C on the Z-plane. This can be mapped on the W-plane through a given transform w = F(z). Suppose the curve C and its mapped counterpart. Any point on the curve in the Z-plane can be represented by x, y or by the polar coordinates r, ϑ. The complex quantity Z can be written in terms of r and ϑ as,

$$Z = x + iy = r\cos\vartheta + i(r\sin\vartheta) = re^{i\vartheta}.$$

The corresponding point ϕ,ψ in the W-plane can be expressed by the polar coordinates Rθ. In terms of these quantities the complex potential is,

$$W = Re^{i\theta}$$

The solution of a problem of Laplacian flow in the vicinity of a fixed boundary is tantamount to finding that transform w = F(z) which maps the boundary as the straight line ψ = constant in the W-plane. That is, for a fixed boundary the flow must be along the boundary and consequently the latter must be a surface of constant ψ.

Consider the problem of ascertaining the two-dimensional, irrotational, divergenceless flow in the neighbourhood of the solid boundary formed by two planes intersecting at an arbitrary angle α.

The boundary, c – c', can be mapped on the W-plane as a straight line by use of the following transform:

$$W = z^N.$$

If we introduce the complex polar form of W and Z given by Eqs., we have from Eq.

$$Re^{i\theta} = r^n\, e^{in\vartheta},$$

which implies that,

$$R = z^N.$$

and,

$$\theta = n\vartheta$$

If we then choose the value of n *as,*

$$n = \frac{\pi}{\text{alpha}}$$

then the curve c - c' will represent the axis of ϕ on the W plane, *i.e.* the line $\phi = 0$, which satisfies the condition that the boundary is a streamline.

The streamlines for other positions in the fluid can now be mapped from the W-plane back onto the Z-plane by use of the relations. As a specific example suppose $a = 2\pi / 3$ radians, then,

$$n = 3/2$$

and,

$$r = R^{2/3},$$
$$\vartheta = 2/3\theta.$$

For any particular isoline ψ = constant, the value of R consistent with a selected ϕ can be found from the isoline of ψ in the W-plane and mapped on the Z-plane using the transformed coordinates r, ϑ from equations. The isolines of ϕ can be mapped on the Z-plane in a similar manner. The resulting orthogonal fields of ψ and ϕ. Either set of isolines determines the flow uniquely in the neighbourhood of the boundary. The isolines of ψ obtained in this illustration represent a family of hyperbolas whose major axis is the isoline $\phi = 0$ and whose asymptotes are the boundary lines c – c'. If we express W as the complex potential $\phi + i\psi$ in Eq. then we find,

$$\phi + i\psi = r^n \cos n\vartheta + i r^n \sin n\vartheta,$$

whence,

$$\phi = r^{3/2} \cos 3/2\vartheta$$
$$\psi = r^{3/2} \sin 3/2\vartheta$$

for this example. The velocity components V_τ and V_ϑ are given by,

$$V_\tau = -\frac{\partial \phi}{\partial r}, \qquad V_\vartheta = -\frac{1}{r}\frac{\partial \phi}{\partial \vartheta},$$

for Laplacian flow in general.

Hence at point 0 the velocity is zero, *i.e.* the point of intersection of the straight portions of the boundary represents a stagnation point in the flow. Eqs. indicate that the velocity increases with distance from the boundary, the magnitude simply being,

$$V\sqrt{V_\tau^2 + V_\vartheta^2} = \frac{3}{2} r^{1/2}.$$

Consequently, the solution can only apply over a finite distance from the boundary if the motion is to be physically real (*i.e.* have a finite velocity at infinity).

A thorough discussion of conformal transformations is beyond the scope of this text. For a more thorough treatment including a wide variety of solved problems of this type, reference is made to Milne-Thompson (1938).

FLUID KINEMATICS

GENERAL CONSIDERATIONS

The the most important basic knowledge of mathematics and physics with respect to fluid mechanics. This knowledge is needed to describe fluid flows or derive and construct basic equations of fluid mechanics in order to solve flow problems. Here it is important to know that fluid mechanics is primarily interested in the velocity field $U_j(x_i, t)$ at initial and boundary conditions, and in the accompanying pressure field $P(x_i, t)$, *i.e.* fluid mechanics tries to describe flow processes in field variables. This representation results in "Eulerian presentation" of fluid flows. This is best suited for the solution of flow problems and is thus applied in experimental, analytical and numerical fluid mechanics. Thanks to the introduction of field quantities for the thermodynamic properties of a fluid, like *e.g.* the pressure $P(x_i, t)$, the temperature $T(x_i, t)$, the density

$\rho(x_i, t)$, the internal energy $e(x_i, t)$ as well as for the molecular transport quantities, as the dynamic vis-cosity $\mu(x_i, t)$, the heat conductivity $\lambda(x_i, t)$ and the diffusion coefficients $D(x_i, t)$ so that a complete presentation of fluid mechanics is possible. With the inclusion of diffusive transport quantities, *i.e.* the molecular heat trans-port $q_i(x_i, t)$, the molecular mass transport $m_i(x_i, t)$ the molecular momen-tum transport $t_{ij}(x_i, t)$, it is possible to formulate the conservation laws for mass, momentum and energy for general application. The basic equations of fluid mechanics can thus be formulated locally, and hold for all flow problems in the same form. The differences in the solutions result from the different initial and boundary conditions that define the actual flow problems which enter into the solutions by the integration of the locally formulated basic quations.

Experience shows that the derivation of basic equations of fluid mechanics can be achieved in the easiest way if considerations are carried for fluid elements, *i.e.* by employing the "Lagrangian consideration" for the derivation of equations. The "Lagrange considerations" assumes that a fluid can be split up at a fixed time $t = 0$, in "marked elements" with the mass $\delta m_{\Re}$, the pres-sure $P_{\Re}$, the temperature $T_{\Re}$, the density $\rho_{\Re}$, the internal energy $e_{\Re}$ etc. The element with the index $\Re$ possesses also a velocity $(U_j)_{\Re}$, which is defined as Lagrangian velocity and which is always linked to the fluid element marked Basics of Fluid Kinema-tics once with as well as to all other quantities labelled with $\Re$. In fluid mechan-ics, these are also designated as substantial quantities and always employed to derive the basic laws of fluid mechanics in an easily comprehensible way. As the following considerations will show, the basic knowledge of mechanics gained in physics can be transferred in the most simple way into fluid mechanics by deriving the basic equations for fluid elements by introducing the Lagrange considerations.

SUBSTANTIAL DERIVATIVES

While one defines a generally as a substantial quantity, the derivation of fluid mechanics equations often require the total differential da to be employed.

$$d\alpha_{\Re} = \frac{\partial\alpha}{\partial t}dt + \frac{\partial\alpha}{\partial x_1}(dx_1)_{\Re} + \frac{\partial\alpha}{\partial x_2}(dx_2)_{\Re} + \frac{\partial\alpha}{\partial x_2}(dx_3)_{\Re}$$

The fluid element motion in space can be described as follows:

$(dx_1) = (U_1)\, dt = U_1 dt$

$(dx_2) = (U_2)\, dt = U_2 dt$

$(dx_3) = (U_3)\, dt = U_3 dt$

The transition of substational velocities $(U_i)_{\Re}$ to the field quantities U_i in equation is permissible, as at the time $t(x_i) = x_i(t)$ and thus $(U_i)(t) = U_i(x_i, t)$. Therefore generally it holds:

$$d\alpha_{\Re} = \frac{\partial\alpha}{\partial t}dt + \frac{\partial\alpha}{\partial x_1}U_1 dt \frac{\partial\alpha}{\partial x_2}U_2 dt + \frac{\partial\alpha}{\partial x_3}U_3 dt$$

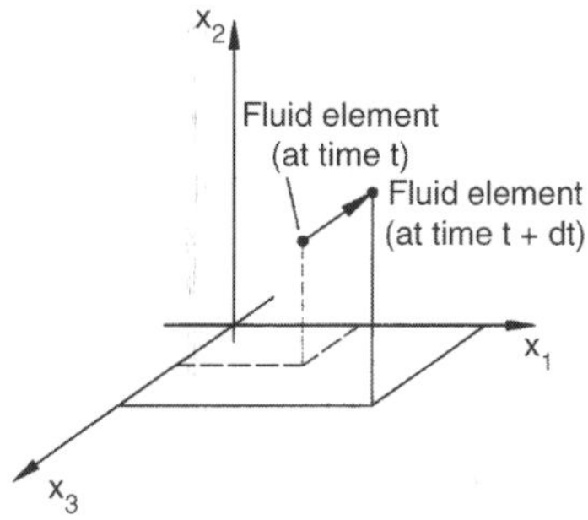

Fig. Motion of a fluid element in space

and for the substantial time derivative with $\alpha_{\Re}(t) = \alpha(x_i, t)$ when $x_{\Re} = x_i$ at the time t it holds:

$$\frac{d\alpha_{\Re}}{dt} = \frac{\partial \alpha}{\partial t} + U_i \frac{\partial \alpha}{\partial x_i} =: \frac{D\alpha}{Dt}$$

where $(D\alpha/Dt)$ is the substantial time derivative of the field quantity $\alpha(x_i, t)$ with respect to time and the operator:

$$\frac{D}{Dt} := \frac{\partial}{\partial t} + U_i \frac{\partial}{\partial x_i}$$

indicates how the substantial derivative of a field variable is to be calculated. The operator D/Dt may only be applied to field quantities. When one applies D/Dt to the velocity field $U_j(x_i, t)$, the substantial acceleration results, *i.e.* the local acceleration which a fluid element experiences in a flow field at a point x_i at time t where $U_j(x_i, t)$ exists.

$$\frac{DU_j}{Dt} = \frac{\partial U_j}{\partial t} + U_i \frac{\partial U_j}{\partial x_i}$$

The substantial derivative plays an important role in the derivation of the momentum equation of fluid mechanics in Euler variables. In the acceleration term in Euler variables, four partial derivatives occur per momentum direction $j = 1, 2, 3$, one time derivative and three derivatives with respect to the space coordinates x_1, x_2 and x_3 *i.e.* the spatial derivatives $(\partial U_j / \partial x_i)$ multiplied by U_i occur in the substantial acceleration. These nonlinear terms in the resulting momentum equations lead to mathematical complications when flow problems are to be solved. They prevent the application of the superposition principle of solutions, result in solution bifurcations, *i.e.* in multiple solutions for equal initial and boundary conditions and in coupled velocity fluctuations, *e.g.* in turbulent flows. The treatment of these nonlinear terms is considered in the presentations of this book. It is important that their significance is understood in detail as part of the acceleration term of fluid elements. It is important to realize that not only the temporal changes of the velocity field that lead to accelerations of fluid elements, but also the motion of a fluid element in a non-uniform velocity field.

RELATIVE MOTIONS

Considerations of the velocities at two points separated by (dx_i) result in the following relation:

$$U_j(x_i + \delta x_i, t) = U_j(x_i,t) + \frac{\partial U_j}{\partial x_i}\delta x_i + \ldots$$

or transcribed:

$$U_j(x_i + \delta x_i, t) = \underbrace{U_j(x_i,t)}_{\text{Translation}} + \overbrace{\frac{1}{2}\left(\frac{\partial U_j}{\partial x_i} - \frac{\partial U_i}{\partial x_j}\right)\delta x_i}^{\text{Rotation}} + \underbrace{\frac{1}{2}\left(\frac{\partial U_j}{\partial x_i} - \frac{\partial U_i}{\partial x_j}\right)\delta x_i}_{\text{Deformation}}$$

This relationship makes it clear that the velocity at the adjacent point $x_i + \delta x_i$ is composed of the translation by velocity at point $p(x_i)$, a rotational velocity around this point and a deformation velocity in this point. The components for $j = 1, 2, 3$ read:

$$j = 1 : U_1\,(x_i + \delta x_i, t) = \left[U_1 + \frac{\partial U_1}{\partial x_1}\delta x_1 + \frac{1}{2}\left(\frac{\partial U_1}{\partial x_2} - \frac{\partial U_2}{\partial x_1}\right)\delta x_2\right.$$

$$\left.+\frac{1}{2}\left(\frac{\partial U_1}{\partial x_3} - \frac{\partial U_3}{\partial x_1}\right)\delta x_3 + \frac{1}{2}\left(\frac{\partial U_1}{\partial x_2} + \frac{\partial U_2}{\partial x_1}\right)\delta x_2 + \frac{1}{2}\left(\frac{\partial U_1}{\partial x_3} + \frac{\partial U_3}{\partial x_1}\right)\delta x_3\right]$$

$$j = 2 : U_2\,(x_i + \delta x_i, t) = \left[U_2 + \frac{\partial U_2}{\partial x_2}\delta x_2 + \frac{1}{2}\left(\frac{\partial U_2}{\partial x_3} - \frac{\partial U_3}{\partial x_2}\right)\delta x_3\right.$$

$$\left.+\frac{1}{2}\left(\frac{\partial U_2}{\partial x_1} - \frac{\partial U_1}{\partial x_2}\right)\delta x_1 + \frac{1}{2}\left(\frac{\partial U_2}{\partial x_3} + \frac{\partial U_3}{\partial x_2}\right)\delta x_3 + \frac{1}{2}\left(\frac{\partial U_2}{\partial x_1} + \frac{\partial U_1}{\partial x_2}\right)\delta x_1\right]$$

$$j = 3 : U_3\,(x_i + \delta x_i, t) = \left[U_3 + \frac{\partial U_3}{\partial x_3}\delta x_3 + \frac{1}{2}\left(\frac{\partial U_3}{\partial x_1} - \frac{\partial U_1}{\partial x_3}\right)\delta x_1\right.$$

$$\left.+\frac{1}{2}\left(\frac{\partial U_3}{\partial x_2} - \frac{\partial U_2}{\partial x_3}\right)\delta x_2 + \frac{1}{2}\left(\frac{\partial U_3}{\partial x_1} + \frac{\partial U_1}{\partial x_3}\right)\delta x_1 + \frac{1}{2}\left(\frac{\partial U_3}{\partial x_1} + \frac{\partial U_2}{\partial x_3}\right)\delta x_2\right]$$

With the above equations the most general motions of fluid elements can now be described, *i.e.* the velocity in any point of a fluid element can be stated

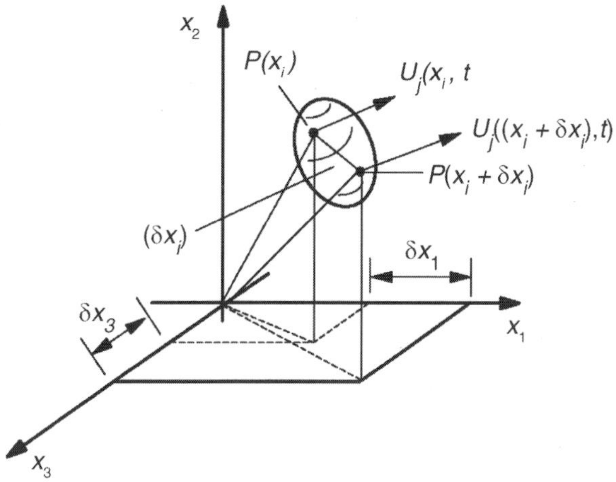

Fig. Relative motions in a fluid element

as the sum from the translation velocity of a reference point, a rotational motion around this point, *i.e.* the resulting speed of the rotation and a superimposed deformation velocity.

The different addends of the equations can be regrouped as follows

$$j = 1 : U_1(x_i) + \left[\frac{\partial U_1}{\partial x_1}\delta x_1 + \frac{1}{2}\left(\frac{\partial U_1}{\partial x_2} + \frac{\partial U_2}{\partial x_1}\right)\delta x_1 + \frac{1}{2}\left(\frac{\partial U_1}{\partial x_3} + \frac{\partial U_3}{\partial x_1}\right)\delta x_3\right]$$

$$+ \left\{\frac{1}{2}\left(\frac{\partial U_1}{\partial x_2} - \frac{\partial U_2}{\partial x_1}\right)\delta x_1 + \frac{1}{2}\left(\frac{\partial U_1}{\partial x_3} - \frac{\partial U_3}{\partial x_1}\right)\delta x_1\right\}$$

$$j = 2 : U_2(x_2) + \left[\frac{\partial U_2}{\partial x_2}\delta x_2 + \frac{1}{2}\left(\frac{\partial U_2}{\partial x_1} + \frac{\partial U_1}{\partial x_2}\right)\delta x_1 + \frac{1}{2}\left(\frac{\partial U_2}{\partial x_3} + \frac{\partial U_3}{\partial x_2}\right)\delta x_3\right]$$

$$+ \left\{\frac{1}{2}\left(\frac{\partial U_2}{\partial x_1} - \frac{\partial U_1}{\partial x_2}\right)\delta x_1 + \frac{1}{2}\left(\frac{\partial U_2}{\partial x_3} - \frac{\partial U_3}{\partial x_2}\right)\delta x_3\right\}$$

$$j = 2 : U_3(x_3) + \left[\frac{\partial U_3}{\partial x_3}\delta x_3 + \frac{1}{2}\left(\frac{\partial U_3}{\partial x_2} + \frac{\partial U_2}{\partial x_3}\right)\delta x_2 + \frac{1}{2}\left(\frac{\partial U_3}{\partial x_1} + \frac{\partial U_1}{\partial x_3}\right)\delta x_1\right]$$

$$+ \left\{\frac{1}{2}\left(\frac{\partial U_3}{\partial x_2} - \frac{\partial U_2}{\partial x_3}\right)\delta x_2 + \frac{1}{2}\left(\frac{\partial U_3}{\partial x_1} - \frac{\partial U_1}{\partial x_3}\right)\delta x_1\right\}$$

The expressions in front of the rectangular brackets represent the translation which is given by the following velocity vector:

$$U_j(x_i, t) = \{U_1, U_2, U_3\}$$

In the square brackets the product of the deformation tensor:

$$D_{ij}(x_i, t) = \begin{Bmatrix} \frac{\partial U_1}{\partial x_1} & \frac{1}{2}\left(\frac{\partial U_1}{\partial x_2}+\frac{\partial U_2}{\partial x_1}\right) & \frac{1}{2}\left(\frac{\partial U_1}{\partial x_3}+\frac{\partial U_3}{\partial x_1}\right) \\ \frac{1}{2}\left(\frac{\partial U_2}{\partial x_1}+\frac{\partial U_1}{\partial x_2}\right) & \frac{\partial U_2}{\partial x_2} & \frac{1}{2}\left(\frac{\partial U_2}{\partial x_3}+\frac{\partial U_3}{\partial x_2}\right) \\ \frac{1}{2}\left(\frac{\partial U_3}{\partial x_1}+\frac{\partial U_1}{\partial x_3}\right) & \frac{1}{2}\left(\frac{\partial U_3}{\partial x_2}+\frac{\partial U_2}{\partial x_3}\right) & \frac{\partial U_3}{\partial x_3} \end{Bmatrix}$$

and of the "'distance vector"'
$\{\delta x_i\} = \{\delta x_1, \delta x_2, \delta x_3\}$
is shown, while in the curved brackets the vector product of
$\{\delta x_i\} = \{\delta x_1, \delta x_2, \delta x_3\}$
and

$$2\omega_k = \left\{\left(\frac{\partial U_3}{\partial x_2}-\frac{\partial U_2}{\partial x_3}\right);\left(\frac{\partial U_1}{\partial x_3}-\frac{\partial U_3}{\partial x_1}\right);\left(\frac{\partial U_2}{\partial x_1}-\frac{\partial U_1}{\partial x_2}\right)\right\}$$

is shown. Thus it can be written:
$U_j(x + dx_i, t) = U_j(x_i, t) + D_{ij}(x_i, t)\,\delta x_i + \varepsilon_{ijk\omega k}\,(x_i, t)\,\delta x_i$

This relation again expresses the facts of the case that the total motion of a point $P'(x_i + dx_i)$ can be understood from the translation motion of the

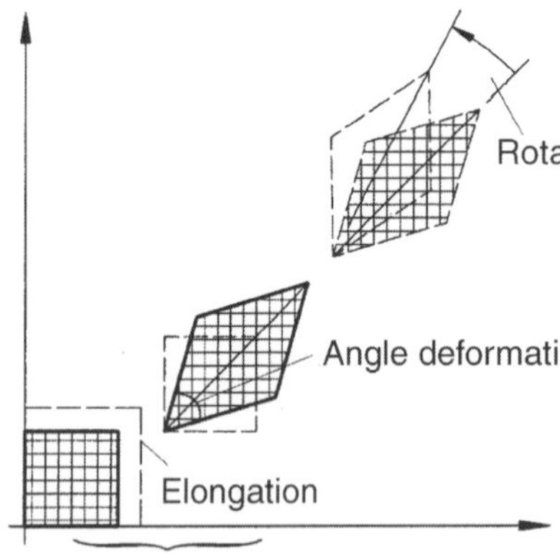

Fig. Illustration of the translation motion, the deformation and the rotation of a fluid element

point $P(x_i)$., superimposed by deformation motions and rotational motions around P (x_i).

The different parts translation, deformation and rotation can be taken from the sequence of a fluid element. For the two-dimensional case with $U_1 = u$ and $U_2 = v$ as well as $x_1 = x$ and $x_2 = y$.

The last named manner of motion requires:

$$\frac{\partial U2}{\partial x_1} = \frac{\partial U_1}{\partial x_2},$$

so that $\omega_3 = 0$.

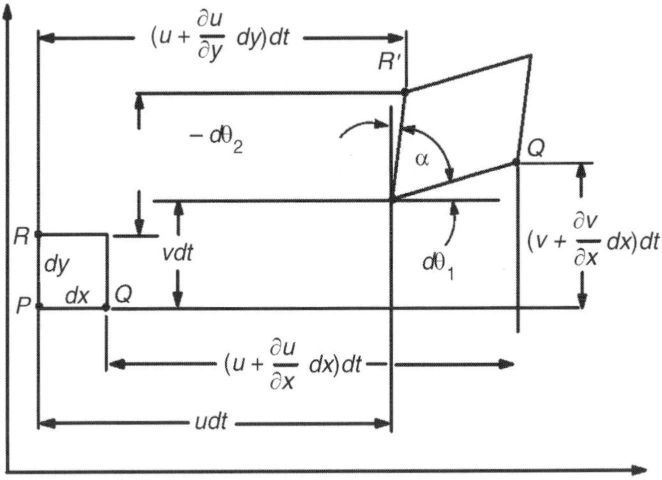

Fig. Translation, deformation and rotation of a fluid element due to the velocity components u and v

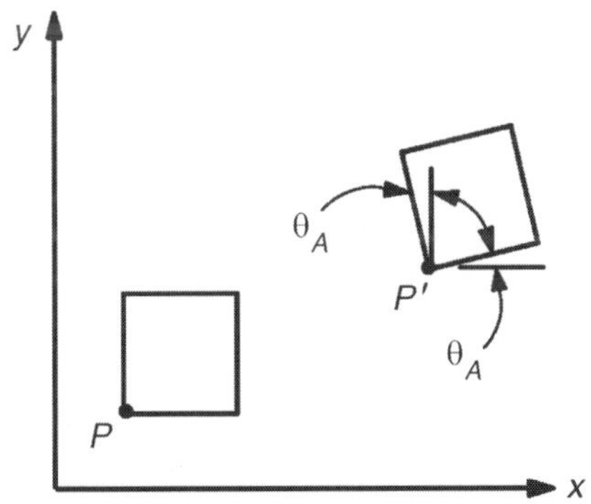

Fig. Translation motion of a fluid element with rotation mit Rotation

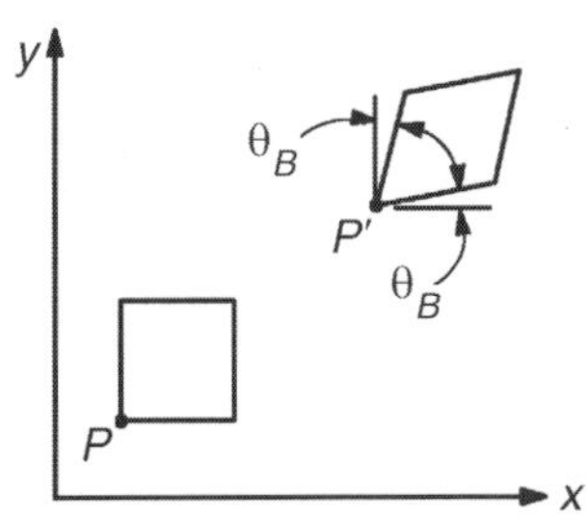

Fig. Translation motion of a fluid element with deformation

MOTION OF FLUID ELEMENTS

Flow kinematics is a vast field and a comprehensive treatment is beyond the scope of this book, which is meant to give only an introduction into sub-domains of fluid mechanics, among others also into fluid kinematics. To such an introduction belongs the treatment of path lines of particles, *i.e.* the computation of space curves along which marked fluid elements move in a fluid. Further, the computation of sweeping paths shall be treated, *i.e.* the "'marked path"' is to be computed which leaves a fixed injected tracer in a flow field. Both, the computation of path lines and of sweeping paths is of importance for the entire experimental flow mechanics, where it is often tried to get an insight in the process of flows by observations or also by quantitative measurements of the temporal changes of position of "'flow markers"'.

PATH LINES OF FLUID ELEMENTS

When one subdivides at the time $t = 0$ the entire domain of a flow field that is of interest into fluid elements and when one states the space coordinates of the mass centers of gravity of the each element in a coordinate system at the time $t = 0$, one achieves a marked fluid domain such that the position vector $\mathfrak{R}$:

$$\{ x_i \}_{\mathfrak{R},0} = \{ x_i \,(\mathrm{t} = 0)\}\}_{\mathfrak{R}}$$

is assigned to a fluid particle. Each of the moving fluid particles, resulting from the subdivision, marked and moving for $-\infty < t < +\infty$ is defined as a fluid element, that keeps its identity $0 \leq t < \infty$ forever.

When kinematic considerations are carried out, only the motions of the in-dividual fluid elements are of interest. These considerations result for each fluid element in a separate and for the marked element characteristic path line. The computation of these path lines will be explained in the following. Here it is assumed that the flow field determining the fluid element motions is known.

As the velocity of a fluid element is only time-dependent, it follows from $\mathrm{d}\{ x_i \}_{\mathfrak{R}}/\mathrm{d}t = \{U_i\}_{\mathfrak{R}}$, that as a path line of a fluid element the frequency locus: t

$$\{x_i \,(\mathrm{t})\}_{\mathfrak{R}} = \{x_i\}_{\mathfrak{R},0} + \int_0^t \{U_i(t')\}_{\mathfrak{R}} \mathrm{d}t'$$

is to be understood. The position vector defined in this way for each moment in time t contains as a parameter the position vector of the particle defined at the time $t = 0$ *i.e.* $\mathfrak{R}$, *i.e.* $\{ x_1 \}_{\mathfrak{R},0}$. Now the identity $\{U_i\} = \{U_i\}$ can be introduced into the considerations, *i.e.* at a certain moment in time t holds:

$$\frac{\mathrm{d}\{x_i\}_{\mathfrak{R}}}{\mathrm{d}t} = \{U_i\}_{\mathfrak{R}} = \{U_i\}$$

The equals sign between the substantial velocity $\{U_i\}_{\mathfrak{R}}$ and $\{U_i\}$ existing at the moment in time t indicates that the identity $\{x_i\}_{\mathfrak{R}} = \{x_i\}$ which exists at time t justifies equating the substantial velocity $\{U_i\}_{\Re}$ with the field size $\{U_i\}$. For the components $\{x_i\}_{\Re}$ of the particle motion it holds therefore:

$$\frac{d\{x_i\}_{\Re}}{dt} = \{U_i\} \text{ or } \frac{d\{x_i\}_{\Re}}{dt} = U_i$$

These differential equations have to be solved for i = 1, 2, 3 in order to determine the path lines of fluid elements. The differential quotient in the relation states that as a solution of the above differential equation the path line of a fluid element is obtained whose position was defined at the moment in time $t = 0$ with $\{x_i\}_{\Re, 0}$.

The general way of proceeding when defining path lines will be explained and made clear by the example stated below. The components of the flow velocity field shall be given:

$$U_1 = x_1 (1 + t), U_2 = -x_2 \text{ und } U_3 = -x_3 t$$

When one inserts these statements on the velocity field in the above indicated differential equations for the path lines of a fluid element, one obtains:

$$\frac{d\{x_1\}_{\Re}}{dt} = x_1(1+t),$$

$$\frac{d\{x_2\}_{\Re}}{dt} - -x_2,$$

and

$$\frac{d\{x_3\}_{\Re}}{dt} = -x_3 t.$$

This law of differential equations can be solved now and results in the following solutions holding for the path lines of all fluid elements:

$$(x_1(t))_{\Re} = C_1 \exp\left[t + \frac{t^2}{2}\right]$$

$$(x_2(t))_{\Re} = C_2 \exp[-t]$$

$$(x_3(t))_{\Re} = C_3 \exp\left[-\frac{t^2}{2}\right].$$

When one considers a fluid element of interest which had the position coordinates (1, 1, 1)) at the time $t = 0$, then from the initial conditions for each of the equations in the introduced constants C_α result uniformly as:

$$C_1 = C_2 = C_3 = 1,$$

i.e. for the case considered here all integration constants are equal. For the path lines of this fluid element yields:

$$x_1 (t)_{\Re} = \exp\left[t + \frac{t^2}{2}\right],$$

$$x_2\,(t)_{\Re} = \exp[-t]$$

$$x_3\,(t) = \exp\left[-\frac{t^2}{2}\right].$$

This path line is presented spatially in Fig. 4.2

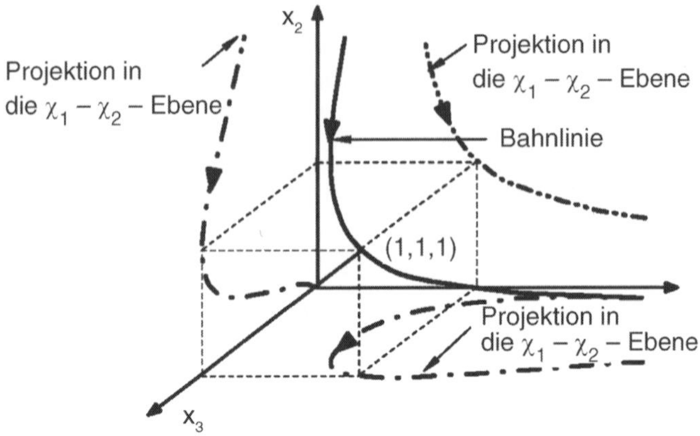

Fig. Spatial path line of the considered fluid element

When one selects a particle whose position at time $t = 0$ showed different position coordinates, the integration constants C_1 , C_2 , C_3 change accordingly and a different path line results. Thus the path line is an "'individual'" property of a fluid element which in the resulting equations is determined by the flow field and the position of the fluid element at the time $t = 0$.

The general solution for the position coordinates (x_i) ,0, which a fluid element takes at the time $t = 0$ is obtained when one inserts these coordinates in the general solutions for the path line coordinates for the determination of the integration constants $C_\alpha(\alpha = 1, 2, 3)$ This results in:

$$C_\alpha = (x_i),\,0$$

and thus in the general solutions for the path line coordinates:

$$(x_1\,(t)) = (x_{\Re,0})_1 \exp\left[t + \frac{t^2}{2}\right]$$

$$(x_2\,(t)) = (x_{\Re,0})_2 \exp[-t]$$

$$(x_3\,(t)) = (x_{\Re,0})_1 \exp\left[-\frac{t^2}{2}\right].$$

These yield the space curves, with the time t as parameter, which represent each curve points of the path lines of fluid elements.

For further explanation, the following two-dimensional velocity field is also considered:[1]

$$U_1 = x_1\,,\ U_2 = x_2\,(1 + 2t) \text{ und } U_3 = 0.$$

With these data the following law of differential equations for the coordinates of the path lines of fluid elements can be formulated:

$$\frac{d(x_1)_{\Re}}{dt} = x_1, \frac{d(x_2)_{\Re}}{dt} = x_2(1+2t), \frac{d(x_3)_{\Re}}{dt} = 0$$

The solution of the third differential equation results in a constant that states in which plane the two-dimensional flow considerations are carried out. For the path line coordinates x_1 (t) and x_2 (t) it is computed:

$$(x_1)_{\Re} = C_1 \exp[t],\ (x_2)_{\Re} = C_2 \exp[t + t^2],\ (x_3)_{\Re} = C_3$$

When one computes the path line of the fluid element which at the time t = 0 took the coordinates (1,1,0), the result is:

$$(x_1)_{\Re} = \exp[t],\ (x_2)_{\Re} = \exp[t + 1],\ (x_3)_{\Re} = 0$$

When one resolves the equation obtained for (x_1) with respect to time, it yields:

$$t = 1\text{n}(x_1)_{\Re}$$

When inserted in the solution for $(x_2\,)_{\Re}$ for two-dimensional path lines in the plane $x_1 - x_2$ – the following functional relation between $(x_1)_{\Re}$ and $(x_2)_{\Re}$ yields:

$$(x_2)_{\Re} = (x_1)_{\Re}^{(1+1\text{n}(x_1)\Re)}$$

Sweeping paths of locally injected tracer materials

As mentioned before, it is usual in experimental fluid mechanics to gain qualitative insight in a flow process by injecting a continuous fluid tracer at a fixed position. This leads to a marked "fluid thread" which is carried with the flow and thus marks/traces the course of the flow. When the exact course of the flow is of interest, quantitative evaluations of the location coordinates of sweeping paths of locally installed tracer materials are required. These evaluations can, based on the derivations stated below, be carried out with methods of flow kinematics.

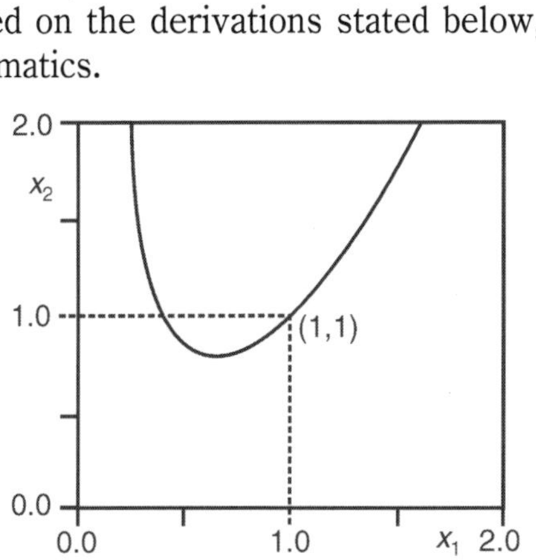

Fig. Path line of the flow running parallel to the plane $x_1 - x_2$

Attention is drawn to the fact that this flow field is not source-free , thus it violates the requirements/demands of the continuity equation. This is, however, insignificant for the purely kinematic considerations mentioned here.

A fluid particle marked with a tracer, *e.g.* an air particle or any other gas particle marked with smoke, or a water or fluid particle marked with colour, which at the time t is located at the position $\{x_i\} = \{x_i(t)\}_{\Re}$ must have passed the injection point for the tracer at a moment in time $(t - \tau)$, in order to be present as a marked particle at the point $\{x_i\}$ *i.e.* it holds :

$$\{x_i(t)\}_{\Re} = \{x_i(t - \tau)\}_S$$

Hence the way covered by a marked fluid element up to the time t can be computed as path line of the element that fulfills the condition *i.e.* a path line with the initial condition that for $t = \tau$ the fluid element held the position of the location coordinates of the injection point. The sweeping path thus is composed of the sum of the path lines of individual particles. For each individual marked particle of a sweeping path a parameter τ is introduced, which for $0 \leq \tau \leq t$ covers all parts of a sweeping path. It is therefore important to vary the parameter τ in the solution equations in order obtain the entire sweeping path.

The above short explanations shall be made clear again by way of an example, which is handled on the basis of the three-dimensional velocity field used above:

$$U_1 = x_1\,(1 + t),\ U_2 = -x_2 \text{ and } U_3 = -x_3 t$$

This velocity field yields the law of differential equations for the motion of a fluid element in space :

$$\frac{d(x_1)_S}{dt} = x_1(1+t)\,,\ \frac{d(x_2)_S}{dt} = x_1\,,\ \frac{d(x_3)_S}{dt} = -x_3 t.$$

As a solution one obtains for the components$(x_1)_S$, $(x_2)_S$ and $(x_3)_S$ according to equation:

[1] The index s signifies that the location coordinate of the sweeping path is meant.

$$(x_1)_S = C_1 \exp\left[t(1+\frac{t}{2})\right],\ (x_2)_S = C_2 \exp\,[-t],\ (x_3)_S = C_3 \exp\left[-\frac{t^2}{2}\right].$$

When one inserts now the initial conditions, that $(x_1)_S = (x_1)_{t=\tau} = 1$, $(x_2)_S = (x_2)_{t=\tau} = 1$, $(x_3)_S = (x_3)_{t=\tau} = 1$ was present for $t = \tau$, *i.e.* that the position (1, 1, 1) serves as an injection point of the tracer, one obtains:

$$C_1 = \exp\left[-\tau(1+\frac{\tau}{2})\right];$$

$$C_2 = \exp\,[\tau] \text{ and } C_3 = \exp\left[\frac{\tau^2}{2}\right].$$

Inserted in the solutions for $(x_1)_S$, $(x_2)_S$ and $(x_3)_S$ the equation of the frequency locus defined as sweeping path yields for all times:[2] :

$$(x_1)_S = \exp\left[t(1+\frac{t}{2}) - \tau(1+\frac{\tau}{2})\right],$$

$$(x_1)_S = \exp\left[-(t-\tau)\right],$$

$$(x_1)_S = \exp\left[1\frac{1}{2}(t^2 - \tau^2)\right]$$

When one wants to make visible the course of a sweeping path at a moment in time t (partly), one has to insert the value of t in the above equation in order to obtain in this way the equation of a space curve, with τ as a parameter. Here t is determined by the period of time $[\tau_1, \tau_2]$ of the tracer injection in (1, 1, 1) with $-\infty < \tau_1 < \tau_2 < t$. For $\tau_1 \rightarrow -\infty$, $\tau_2 = t$ and $t = 0$ yields:

$$(x_1)_S = \exp\left[-\tau(1+\frac{\tau}{2})\right],$$

$$(x_2)_S = \exp[\tau], \; -\infty < \tau < 0$$

$$(x_3)_S = \exp\left[\frac{\tau^2}{2}\right]$$

The course of this space curve is shown in 4.4. It indicates the sweeping path existing at the moment in time $\tau = 0$ (made visible from $\tau = -\infty$ to $\tau = 0$, the projections of the sweeping path into the main level of the Cartesian coordinate system are also introduced.

When one compares the equation for the sweeping path fixed by the space point (1, 1, 1) with the equations for the path line of a fluid element, stated for the same flow field, one realizes that path lines and sweeping paths are not identical for non-stationary flows. Only in the case of a stationary flow field path lines and sweeping paths are identical, as can be shown easily by the following considerations.

As a space curve is concerned here, the statement in x_1, x_2, x_3-coordinates is appropriate. The definition x_S indicates that the location coordinates of a sweeping path are meant.

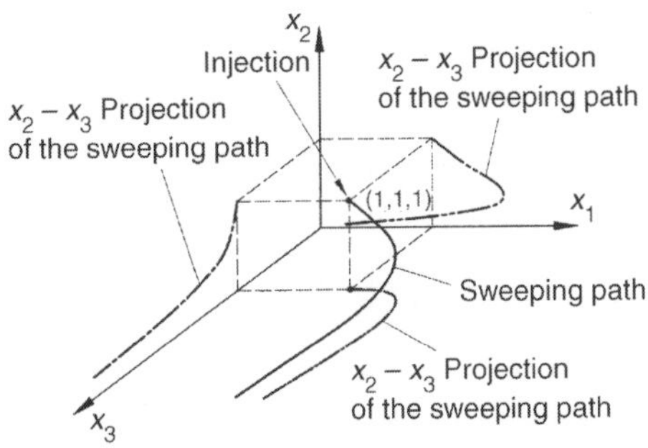

Fig. Sweeping path for the moment in time t = 0. with fluid tracer injec-tions between $\tau = -\infty$ and $\tau = 0$ at the position (1, 1, 1)

Considering the stationary velocity field:

$$U_1 = 2x_1 , U_2 = -x_2, U_3 = -x_3$$

one obtains for the path line of a fluid element the following differential equation:

$$\frac{d(x_1)_\Re}{dt} = 2x_1, \frac{d(x_2)_\Re}{dt} = -x_2, \frac{d(x_3)_\Re}{dt} = x_3$$

for $t = 0$ it shall be assumed that $(x_1)_\Re = (x_2)_\Re = (x_3)_\Re = 1$ so that in the solution

$$(x_1) = C_1 \exp [2t] , (x_2)_\Re = C_2 \exp [- t] , (x_3) = C_3 \exp [-t]$$

holds and thus the path line is stated as follows:

$$(x_1)_\Re = \exp [2t] , (x_2 \quad)_\Re = \exp [-t], (x_3)_\Re = \exp [-t]$$

with $-\infty < t < \infty$. For the computation of the sweeping paths the solution can be employed again and C_1 , C_2, C_3 can be computed such that it is claimed that at the time $t = \tau$ holds:

$$(x_1(t = \tau))_S = 1, (x_2(\mathrm{t} = \tau))_S = 1, (x_3(t = \tau))_S = 1$$

Therefore it holds:

$$C_1 = \exp [-2\tau] , C_2 = \exp [\tau] , C_3 = \exp [\tau]$$

or as an equation for the sweeping path:

$$(x_1)_S = \exp[2(t - \tau)], (x_2)_S = \exp [-(t - \tau)], (x_3)_S = \exp [-(t - \tau)]$$

thus t being defined, and the range of values of t is defined by the period of time of the tracer injection. In the case that tracer substance is injected at all times, *i.e.* $-\infty < \tau < \infty$, equation yield the same curve. When the tracer injection is limited in time, one obtains as a visible sweeping path a corresponding part of the path line.

$$U_1 = x_1, U_2 = x_2(1 + 2t) \text{ and } U_3 = 0$$

which leads to the differential equations:

$$\frac{d(x_1)_\Re}{dt} = x_\Re, \frac{d(x_2)_\Re}{dt} = x_\Re(1+2t), \frac{d(x_3)_\Re}{dt} = 0$$

with the solutions:

$$(x_1)_\Re = C_1 \exp [t], (x_2)_\Re = C_2 \exp [t(t + 1)], (x_3)_\Re = C_3$$

On the other hand, when one demands that the particle located at the time t in the point x_1, x_2, x_3 passes the injection point (1, 1, 0) of a tracer at the moment in time τ the integration constants C_1, C_2 and C_3 can be obtained from the following conditional equations:

$$C_1 \exp [\tau] = 1, C_2 \exp [\tau(\tau + 1)] = 1, C_3 = 0$$

These “'constants"' can now be inserted in equation again, where $C_3 = 0$ signifies that the sweeping path runs in the plane $x_1 - x_2$, *i.e.* in the plane $x_3 = 0$, and is described there by the following equations for the position coordinates $(x_1)_S$ and $(x_2)_S$

$$(x_1)_S = \exp [t - \tau] ,$$

$$(x_2)_S = \exp [t(t + 1) - \tau(\tau + 1)]$$

For the moment in time $t = 0$ the course of the sweeping path is:

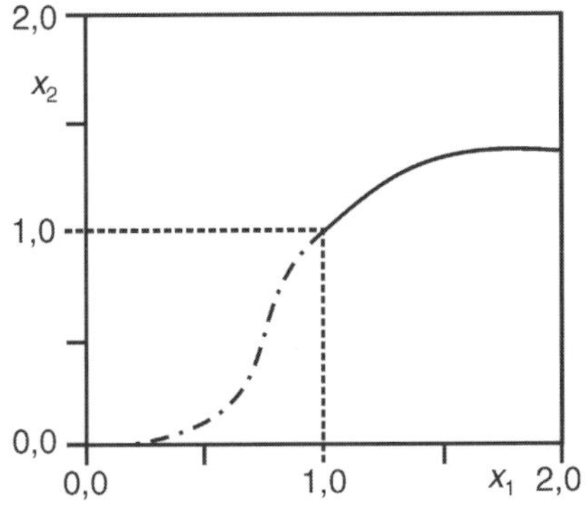

Fig. Sweeping path in the plane $x_1 - x_2$ (fully drawn line corresponds to $-\infty \le \tau \le 0$, broken line to $0 \le \tau \le \infty$)

$$(x_1)_S = \exp[-\tau],$$
$$(x_2)_S = \exp[-\tau(\tau + 1)]$$

in the domain $-\infty < \tau < \infty$.

From the equation for $(x_1)_S$ follows:

$$t = -\ln(x_1)_S$$

Inserted in the equation for $(x_2)_s$ at the time $t = 0$ the existing course of the sweeping path made visible by the injection of tracer material in $-\infty < \tau < \infty$ results in the plane $x_1 - x_2$

$$(x_2)_S = (x_1)_S^{(1-1n(x_1)_S)} \text{ in the domain } 0 < (x1)_S < \infty$$

$$(x_2)_S = [(x_1)_S]^{(1-1n(x_1)_S)}$$

KINEMATIC BOUNDARY CONDITION

At an impermeable boundary, the flow of fluid *relative to the boundary* must be tangential to it. If the boundary is fixed in space, this means that the component of fluid velocity normal to the boundary must be zero. However, if the boundary is moving, then the normal component of the fluid must be equal to the velocity of the boundary normal to itself. This applies whether there is a solid on one side of the boundary or not. If not, then we have the case of a free surface of the fluid, such as the sea surface.

Let $\vec{U}$ represent the velocity of the fluid boundary and $\vec{V}$ be the fluid velocity of the boundary. Then the kinematic boundary condition is simply,

$$\left(\vec{U} - \vec{V}\right) \cdot \vec{n} = 0$$

where $\vec{n}$ is the unit normal to the boundary.

The kinematic boundary condition for an impermeable boundary only, *i.e.* a boundary *across* which there is no flux of matter of the macroscopic scale. In those cases where a change of state of the fluid occurs at the boundary or where the boundary surface displays a selective permeability phenomenon, the condition is no longer rigourously applicable.

A specific case in point is the effect of a flux of evaporation at the sea surface. This is certainly important in thermodynamic problems although in nearly all kinematic or dynamical problems the effect of evaporation can be ignored so far as the kinematic boundary condition is concerned.

The boundary condition can be written in a more convenient form for a free surface of the fluid. In any physically real situation a free surface simply represents the boundary between two different fluids, *e.g.* water and air, water and steam, oil and water, etc.

That is, the free surface exists by virtue of a difference in the neighbouring fluids. The difference can be a chemical difference or merely a physical difference (difference in phase of the same substance, *e.g.* water/water vapor). The important thing in connection with our present discussion is that there exist certain physical or thermal properties of the two fluids which are continuous across the free surface. An example is temperature.

Pressure is another example in those cases where the effect of surface tension can be ignored.

Suppose $f(x,y,z,t)$ represents the field of some property in the fluid concerned and $f'(x,y,z,t)$ the field of the same property in the adjacent fluid. If the property f is one which is continuous across the free surface (which delineates the boundary between these two fluids) then,

$$f(x,y,z,t) = f'(x,y,z,t)$$

it at the boundary. If the functions f and f' are known, then Eq.actually defines the boundary, *i.e.*

$$F'(x,y,z,t) = \mathrm{f} - \mathrm{f}' = 0$$

is the implicit equation of the free surface. Usually this is written in the explicit form

$$\mathrm{z} = \eta(\mathrm{x}, \mathrm{y}, \mathrm{t}),$$

where η denotes the function representing the elevation of the free surface above some reference plane at position x, y and time t.

The time rate of change of the property f following the motion of the fluid at the boundary is $(\mathrm{Df}/\mathrm{Dt})_z = \eta$. The condition that the fluid at the boundary flows along the boundary and never leaves the boundary is equivalent to the condition that the individual rate of change of f at the boundary is equivalent to the rate of change of f' on the other side of the boundary as one follows with the particle motion of the first fluid. Thus,

$$\frac{\mathrm{Df}}{\mathrm{Dt}} = \left(\frac{\mathrm{Df}}{\mathrm{Dt}} - \frac{\mathrm{Df}'}{\mathrm{Dt}}\right)_{\mathrm{z=n}} = 0$$

represents the kinematic boundary condition. In view of Eq., it will be noted that

$$F = \eta(x, y, t) - z$$

so that Eq. can be written as,

$$\frac{\partial \eta}{\partial t} + u_\eta \frac{\partial \eta}{\partial x} + v_\eta \frac{\partial \eta}{\partial y} - w_\eta = 0$$

or,

$$w_\eta = \frac{\partial \eta}{\partial t} + u_\eta \frac{\partial \eta}{\partial x} + v_\eta \frac{\partial \eta}{\partial y}$$

where u_η, v_η, w apply at the boundary.

If the free surface is *stationary* (*i.e.* $\partial \eta / \partial t = 0$) and slopes only in the x-direction, then,

$$w_\eta = u_\eta \frac{\partial \eta}{\partial x}.$$

This expression can be verified by reference to Fig. below. If a particle starting at position A is to remain on the surface as itmoves over the differential distance *ds*, then we must have

$$w_\eta dt = \frac{\partial \eta}{\partial x} u_\eta dt$$

which is equivalent.

If the surface has no slope but is ascending at the rate of $\partial \eta / \partial t$ then it is obvious that in order that the fluid remain at the surface, then w_η must satisfy the relation

$$w_\eta = \frac{\partial \eta}{\partial t}.$$

Equation is applicable when both effects are present, including the effect of surface slope in the y-direction as well as the x-direction.

In many problems the nonlinear terms in equation are of second order compared with $\partial \eta / \partial t$ and equation represents a good approximation. However, one should always investigate the order of magnitude of these terms before making such an approximation.

APPLICATION OF THE BOUNDARY CONDITION FOR LAPLACIAN FLOW ADJACENT TO A SOLID CYLINDER

Consider a solid cylinder of circular cross-section which travels at constant velocity *U* normal to its axis through a fluid of very vast extent. That is, it is presumed that the linear dimensions of the fluid are many orders of magnitude greater than the diameter of the cylinder. The flow of fluid induced by the motion of the cylinder can be expected to be essentially two-dimensional, considering that the cylinder is very long compared with its diameter. This implies that the picturre of flow is the same in any plane which is normal to the axis of the cylinder.

If the flow is presumed to be Laplacian (*i.e.* divergenceless and irrotational) then we can evaluate the pattern of flow by the methods outlined in Art. 1.11, making use of the appropriate boundary conditions. It is sufficient to state at this point that Laplacian flow adjacent to the cylinder is a rather idealized type of motion, however it does possess certain characteristics which warrant its investigation. Practically speaking, the flow would exist immediately after an impulse is imparted to a cylinder which was originally at rest in a static fluid. The boundary condition to be satisfied is equation. We will take the origin of the reference frame at the center of the cylinder, and thus refer all quantities relative to the moving cylinder. The normal component of the fluid velocity to the surface is,

$$\left(V_r\right)_{r=a} = U\cos\theta.$$

Since the flow is presumed to be Laplacian, it follows that V_r can be expressed as,

$$\left(V_r\right) = -\frac{\partial\phi}{\partial_r}.$$

Consequently, equation can be written as,

$$\left(\frac{\partial\phi}{\partial r}\right)_{r=a} = -U\cos\theta$$

The transform $W = f(Z)$, which is capable of satisfying condition, is,

$$W = AZ^n, \qquad A = \text{const.}$$

which implies that,

$$\phi = Ar^n \cos n\theta$$

$$\phi = Ar^n \sin n\theta.$$

Applying condition (234a) gives,

$$An\, a^{(n-1)} \cos n\theta = -U \cos\theta$$

which can be satisfied only if,

$$n = \pm 1$$

$$A = -\frac{U}{na^{n-1}}.$$

The velocity at distance *r* is,

$$V = \sqrt{\left(\frac{\partial \phi}{\partial r}\right)^2 + \left(\frac{1}{r}\frac{\partial \phi}{\partial \theta}\right)^2} = Anr^{n-1}$$

or,

$$V = U\left(\frac{r}{a}\right)^{n-1}.$$

In order to select the proper sign of *n* in equation we introduce a second condition which is simply that the motion is to be confined to the neighbourhood of the cylinder; that is, the fluid is not significantly disturbed at a great distance from the cylinder. This requires that we take n = –1 and, consequently,

$$\phi = U\frac{a^2}{r}\cos\theta,$$

$$\psi = -U\frac{a^2}{r}\sin\theta,$$

$$V = U\left(\frac{a}{r}\right)^2.$$

The streamlines and isolines of velocity potential repesented by equations. The families of isolines are circles which pass through the point 0 if extended inside the cylinder.

It is of interest to compute the kinetic energy of the fluid motion induced by the passage of the cylinder through the fluid. This is given by the integral,

$$E_k = \frac{1}{2}\rho\int_a^\infty\int_0^{2\pi} V^2\, r\, dr\, d\theta,$$

where ρ is the mass density of the fluid, presumed constant, and E_k is the total kinetic energy of that fluid contained between two planes of unit distance apart and normal to the axis of the cylinder and extending indefinitely in the *x* and *y* directions. If we carry out the integration, making use of equation, we find that

$$E_k = \frac{1}{2}\pi a^2 \rho u^2$$

or,

$$E_k = \frac{1}{2}M'U^2$$

where $M' = \rho\pi a^2$ represents the mass of fluid displaced by the cylinder per unit length of cylinder.

If the cylinder is accelerated, then a force commensurate with the mass of the cylinder *plus* the mass *M'* is required to produce the given acceleration.

That is, of the kinetic energy being supplied to the cylinder, some is used to accelerate the cylinder and some is used to accelerate the water in the neighbourhood of the cylinder. The effective mass of the cylinder in the fluid therefore includes the mass *M'*.

KELVIN'S THEOREM OF MINIMUM KINETIC ENERGY

Kelvin has established the remarkable theorem that Laplacian flow has less kinetic energy than any other motion consistent with the same normal motion of the boundaries of the fluid. A practical criterion can be immediately inferred from this theorem in connection with the foregoing example the effective added mass, M_1, is a minimum in the case of Laplacian flow around a cylinder. In the more usual case where turbulent flow exists in the neighbourhood of the cylinder, we should expected to find that,

$$C_M \equiv \frac{M_\theta}{M} > 1$$

for a circular cylinder.

ADDITION OF VELOCITY POTENTIALS

If two velocity fields $\vec{V}_1$ and $\vec{V}_2$ can be represented by means of the velocity potentials ϕ_1 and ϕ_2, respectively, then their superposition can be represented by

$$\vec{V}_1 + \vec{V}_2 = -\nabla(\phi_1 + \phi_2).$$

In other words, the algebraic sum $(\phi_1 + \phi_2)$ defines the superimposed fields of $\vec{V}_1$ and $\vec{V}_2$.

As an illustration, suppose that it is desired to determine the irrotational and divergenceless flow of fluid past a circular cylinder which is held fixed. Since the flow is presumed to be Laplacian, the pattern can be established merely by superimposing a velocity *-U* on the pattern of flow. This will effectively stop the cylinder, and at the same time alter the pattern of the streamlines and velocity potential. The velocity *-U* can be represented by the potential

$$\phi_2 = U_x = U_r \cos\theta.$$

Thus if we let,

$$\phi_1 = U\frac{a^2}{r}\cos\theta$$

from equation then the potential applicable to flow past a stationary cylinder is,

$$\phi = \phi_1 + \phi_2 = U\left(r + \frac{a^2}{r}\right)\cos\theta.$$

We could also take,

$$\psi_1 = -U\frac{a^2}{r}\sin\theta,$$
$$\psi_2 = U_y = U_r \sin\theta$$

and the superposition of these streamfunctions gives,

$$\psi = \psi_1 + \psi_2 = U\left(r - \frac{a^2}{r}\right)\sin\theta.$$

Continuityequation

The continuity equation is simply a mathematical expression of the principle of conservation of mass. For a control volume that has a single inlet and a single outlet, the principle of conservation of mass states that, for steady-state flow, the mass flow rate into the volume must equal the mass flow rate out. The continuity equation for this situation is expressed by Equation.

$$\dot{m}_{inlet} = \dot{m}_{outlet}$$
$$(\rho Av)_{inlet} = (\rho Av)_{outlet}$$

For a control volume with multiple inlets and outlets, the principle of conservation of mass requires that the sum of the mass flow rates into the control volume equal the sum of the mass flow rates out of the control volume. The continuity equation for this more general situation is expressed by Eq.

$$\sum \dot{m}_{inlet} = \sum \dot{m}_{outlet}$$

One of the simplest applications of the continuity equation is determining the change in fluid velocity due to an expansion or contraction in the diameter of a pipe.

Example: Continuity Equation - Piping Expansion

Steady-state flow exists in a pipe that undergoes a gradual expansion from a diameter of 6 in. to a diameter of 8 in. The density of the fluid in the pipe is constant at 60.8 lbm/ft3.

If the flow velocity is 22.4 ft/sec in the 6 in. section, what is the flow velocity in the 8 in. section?

Solution: From the continuity equation we know that the mass flow rate in the 6 in. section must equal the mass flow rate in the 8 in. section. Letting the subscript 1 represent the 6 in. section and 2 represent the 8 in. section we have the following.

$$\dot{m}_1 = \dot{m}_2$$
$$\rho_1 A_1 v_1 = \rho_2 A_2 v_2$$

$$v_2 = v_1 \frac{\rho_1}{\rho_2}\frac{A_1}{A_2}$$

$$= V_t \frac{\pi r_1^2}{\pi r_2^2}$$

$$= \left(22.4\frac{ft}{sec}\right)\frac{(3\,in)^2}{(4\,in)^2}$$

$$v_2 = 12.6\frac{ft}{sec}$$

So by using the continuity equation, we find that the increase in pipe diameter from 6 to 8 inches caused a decrease in flow velocity from 22.4 to 12.6 ft/sec.

The continuity equation can also be used to show that a decrease in pipe diameter will cause an increase in flow velocity.

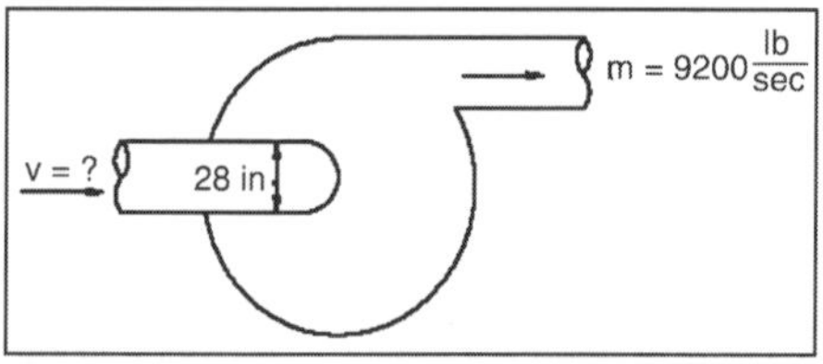

Fig. Continuity Equation

Example: Continuity Equation - Centrifugal Pump The inlet diameter of the reactor coolant pump shown in Figure is 28 in. while the outlet flow through the pump is 9200 lbm/sec. The density of the water is 49 lbm/ft3. What is the velocity at the pump inlet?

Solution:

$$A_{inlet} = \pi r^2 = (3.14)\left(14\text{ in}\left(\frac{1\,ft}{12\,in}\right)\right)$$

$$= 4.28\,ft^2$$

$$\dot{m}_{inlet} = \dot{m}_{outlet} = 9200\frac{lbm}{sec}$$

$$(\rho A v)_{inlet} = 9200\frac{lbm}{sec}$$

$$v_{inlet} = \frac{9200\frac{lbm}{sec}}{Ap}$$

$$= \frac{9200\dfrac{\text{lbm}}{\text{sec}}}{(4.28\,\text{ft}^2)\left(49\dfrac{\text{lbm}}{\text{ft}^2}\right)}$$

$$v_{\text{inlet}} = 43.9\frac{\text{ft}}{\text{sec}}$$

The above example indicates that the flow rate into the system is the same as that out of the system. The same concept is true even though more than one flow path may enter or leave the system at the same time.

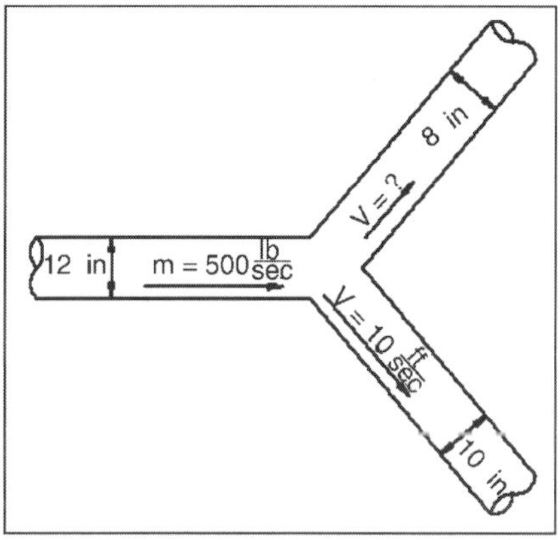

Fig. "Y" Continuity for Example Problem

The mass balance is simply adjusted to state that the sum of all flows entering the system is equal to the sum of all the flows leaving the system if steady-state conditions exist. An example of this physical case is included in the following example.

Example: Continuity Equation - Multiple Outlets

A piping system has a "Y" configuration for separating the flow as shown in Figure. The diameter of the inlet leg is 12 in., and the diameters of the outlet legs are 8 and 10 in. The velocity in the 10 in. leg is 10 ft/sec. The flow through the main portion is 500 lbm/sec. The density of water is 62.4 lbm/ft3. What is the velocity out of the 8 in. pipe section?

Solution:

$$A_8 = \pi\left(4\text{ in.}\frac{1\,\text{ft}}{12\,\text{in.}}\right)^2 - 0.349\text{ ft}^2$$

$$A_{10} = \pi\left(5\text{ in.}\frac{1\,\text{ft}}{12\,\text{in.}}\right)^2 - 0.545\text{ ft}^2$$

$$\sum \dot{m}_{\text{inlets}} = \sum \dot{m}_{\text{outlets}}$$

$$\dot{m}_{12} = \dot{m}_{10} + \dot{m}_8$$

$$\dot{m}_8 = \dot{m}_{12} - \dot{m}_{10}$$

$$(\rho A v)_8 = \dot{m}_{12} - (\rho A v)_{10}$$

$$v_8 = \frac{\dot{m}_{12} - (\rho A v)_{10}}{(\rho A)_8}$$

$$= \frac{500\dfrac{\text{lbm}}{\text{sec}} \left(62.4\dfrac{\text{lbm}}{\text{ft}^2}\right)(0.545\,\text{ft}^2)\left(10\dfrac{\text{ft}}{\text{sec}}\right)}{\left(62.4\dfrac{\text{lbm}}{\text{ft}^2}\right)(0.349\,\text{ft}^2)}$$

$$v_8 = 7.3\,\frac{\text{ft}}{\text{sec}}$$

KINEMATIC QUANTITIES OF FLOW FIELDS

FLOW LINES OF A VELOCITY FIELD

The considerations carried out in for fluid elements for the com-putation of path lines and sweepings paths have to be separated strictly from considerations for the determination of the flow lines of a flow field. Although for stationary flows flow lines, path lines and sweeping paths are identical, this does not justify to neglect the fundamental differences or even worse to assume that they do not exist. Only by a clear separation of the con-siderations it becomes generally comprehensible why under the steady-state condition of a flow field the above-stated identities of flow lines, path lines and sweeping paths exist.

When one considers the non-stationary flow field $U_j\,(x_i, t)$, flow lines can be defined for this field at any moment in time t so that space curves run parallel to the velocity vector which are defined as flow lines at each point. In the case that one considers initially a two-dimensional flow field, for which all velocity vectors lie parallel to the plane $x_1 - x_2$, then the above-indicated definition of the flow line leads to the following defining equation:

$$\frac{d(x_2)_\psi}{d(x_1)_\psi} = \frac{U_2}{U_1} \quad \text{for the time } t = \text{const}$$

The relation means that the gradient of the flow line is equal to the tangent of the angle formed by the velocity vector with the axis x_1 at the moment in time t. The index ψ taken up in the equation indicates that the stated coordinates $x_1 - x_2$ describe the flow line ψ.

When one considers the defining equation for the flow line of a two dimensional flow field, it becomes comprehensible that in general for each moment in time t the quotient of U_2 and U_1 is a function of x_1 and x_2.

The thus resulting differential equation has to be solved in order to conserve the equation for the flow line. This will be explained on the basis of an example that starts from the following two-dimensional velocity field:

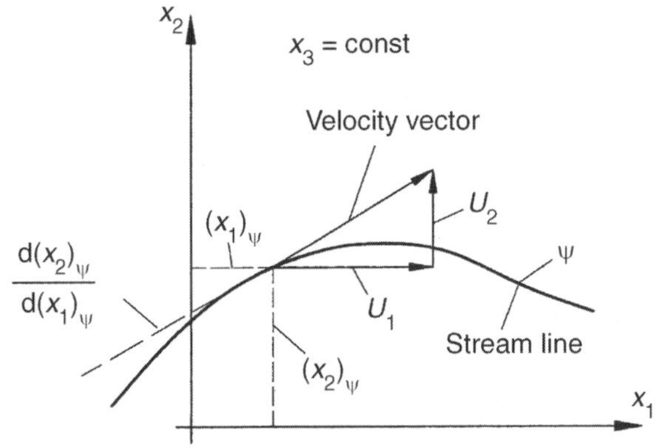

Fig. Sketch for clarifying the defining equation for the flow line of a flow field

$$U_1 = x_1\,(1 + 2t);\ U_2 = x_2 \text{ and } U_3 = 0$$

Introduced into the defining equation for the flow line one obtains:

$$\frac{d(x_2)_\psi}{d(x_1)_\psi} = \frac{U_2}{U_1} = \frac{x_2}{x_1(1+2t)}$$

or

$$\frac{d(x_2)_\psi}{x_2} = \frac{d(x_1)_\psi}{x_1}\frac{1}{2+2t}$$

By integration results:

$$\ln(x_2)_\psi = C + \frac{1}{1+2t}\ln(x_2)_\psi$$

Considering the flow line, passing at time $t = 0$ at the point (1, 1, 0) C = 0 results and thus:

$$(x_2)_\psi = [(x_1)_\psi]^{\frac{1}{1+2t}}$$

In the case that there is a three-dimensional velocity field, the above considerations, that were indicated for the projections of the flow lines into the planes $x_1 - x_2$., can be accepted in an analogous way. For the projections into the planes $x_1 - x_3$ – and $x_2 - x_3$. relations analogous to the defining equation result:

$$\frac{d(x_3)_\psi}{d(x_1)_\psi} = \frac{U_3}{U_1}$$

$$\frac{d(x_3)_\psi}{d(x_2)_\psi} = \frac{U_3}{U_2}$$

Thus the defining equations of the flow lines of a velocity field can be stated as follows:

$$\frac{d(x_1)_\psi}{U_1} = \frac{d(x_2)_\psi}{U_2}, \quad \frac{d(x_1)_\psi}{U_1} = \frac{d(x_3)_\psi}{U_3}, \quad \frac{d(x_2)_\psi}{U_2} = \frac{d(x_3)_\psi}{U_3}$$

or rewritten as:

$$\frac{d(x_1)_\psi}{U_1} = \frac{d(x_2)_\psi}{U_2} = \frac{d(x_3)_\psi}{U_3}$$

These differential relations for the flow line of a velocity field hold at each moment in time t. Their solution leads to a relation $(x_3)_\psi = \psi(x_1, x_2)$, which describes a space curve, the three-dimensional flow line.

Probably the most simple way to solve the law of differential equations is to seek a parameter solution $(x_1)_\psi = x_1(s)$, here s is a parameter, whose value at a certain reference point of the flow line is equal to zero and which takes increasing values along the flow line and in flow direction. When all values $-\infty < s < \infty$ are passed, through a presentation of the entire flow line is obtained.

By introducing s one obtains:

$$\frac{d(x_j)_\psi}{ds} = U_j(x_i, t) \text{ for } t = \text{const and } j = s$$

a relation which represents for each coordinate $(x_j)_\psi$ a differential equation $j = 1, 2, 3$ describing the flow lines in space for $t = \text{const}$. If the flow line passing through the space point $[x_0]_j$ at time t is sought, $s = 0$, results from integrating the three differential equations, when $x_j(t) = x_{j,0}$ From this results the entire flow-line field as:

$$(x_j)_\psi = \psi_j(x_{0,j}, t, s)$$

In order to demonstrate the way of proceeding in determining three-dimensional flow-line fields, the following velocity field is to be considered again:

$$U_1 = x_1(1 + t),\ U_2 = -x_2 \text{ and } U_3 = -x_3 t$$

Thus a law/principle/theorem of differential equations for the flow lines of this velocity field results:

$$\frac{d(x_1)_\psi}{ds} = x_1(1 + t) \quad \frac{d(x_2)_\psi}{ds} = -x_2 \quad \frac{d(x_3)_\psi}{ds} = -x_3 t$$

The integration of these equations yields:

$(x_1)_\psi = C_1 \exp[(1 + t)s]$,

$(x_2)_\psi = C_2 \exp[-s]$,

$(x_3)_\psi = C_3 \exp[-ts]$.

Searching for the flow line that passes through the point(1, 1, 1) one can chose this point as a point of reference and set $s = 0$ for $(x_i)_\psi = 1$ From this the following results for the integration constants:

$C_1 = C_2 = C_3 = 1$

One obtains thus:

$(x_1)_\psi = \exp[(1 + t)s],$

$(x_2)_\psi = \exp[-s],$

$(x_3)_\psi = \exp[-ts].$

For the time $t = 0$ a flow line path results

$(x_1)_\psi = \exp[s],\ (x_2)_\psi = \exp[-s],\ (x_3)_\psi = 1$

So one obtains a flow line passing in the plane $x_3 = 1$ and described there by the relation

$$(x_2)_\psi = \frac{1}{(x_1)_\psi}$$

The entire flow line field is obtained when one introduces for $s = 0$ arbitrary position coordinates $(x_{0,i})$ so that for the position coordinates holds:

$(x_1)_\psi = (x_1)_{\psi,0} \exp[(1 + t)s],$

$(x_2)_\psi = (x_2)_{\psi,0} \exp[-s],$

$(x_3)_\psi = (x_3)_{\psi,0} \exp[-ts].$

This set of equations indicates at each time t the flow lines passing through the point $\{x_j\}_{\psi,0}$ There the parameter s has the value $s = 0.$ and for all other points of the flow line a value equal to zero. A clear assignment is thus occurring which guarantees that flow lines never intersect, as otherwise two velocities would exist simultaneously at this point of intersection.

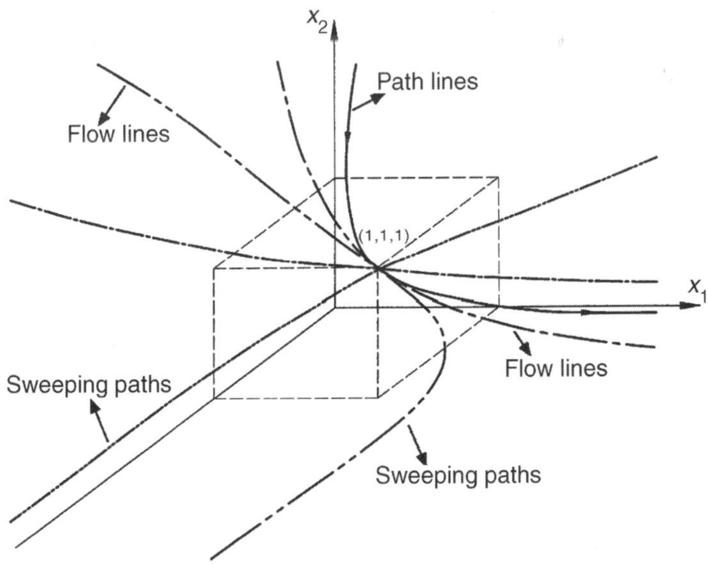

Fig. Comparison of flow lines, path lines and sweeping paths

thus occurring which guarantees that flow lines never intersect, as otherwise two velocities would exist simultaneously at this point of intersection. This is excluded because of the existence of clear velocity fields (except for stagnation points and singularities).

In the preceding section it was emphasized that in general flow lines, path lines and sweeping lines are not identical and the computed examples have made this clear have confirmed this. For stationary flows all three lines are identical, *i.e.*

- For stationary flow fields marked fluid elements move along flow lines, *i.e.* flow lines are equal to the path lines
- For stationary flow fields flow lines can be made visible by locally injected tracer particles, *i.e.* flow lines are equal to the sweeping paths

For non-stationary flows generally flow lines, path lines and sweeping paths are different space curves.

STREAM FUNCTION AND FLOW LINES OF TWO-DIMENSIONAL FLOW FIELDS

For two-dimensional incompressible velocity fields $\{U_j\} = \{U_1, U_2, 0\}$ the stream function can be introduced as a field quantity. It is defined as follows:

$$U_1 = \frac{\partial_\psi}{\partial x_2} \text{ und } U_2 = \frac{\partial_\psi}{\partial x_1}$$

and is computed from the velocity field over the following line integrals:

$$\psi - \psi_0 = \int_{x_2,0}^{x_2} U_1 dx_2 \text{ for } x_1 = \text{const}$$

$$\psi - \psi_0 = \int_{x_1,0}^{x_1} U_1 dx_1 \text{ for } x_2 = \text{const}$$

The above equations show that the difference existing between two stream-

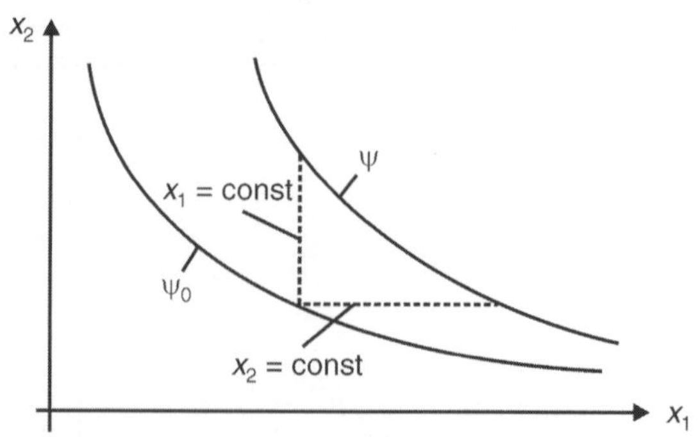

Fig. Explanation of the defining equations of stream function

function values is a measure for the volume flow that runs between two lines each with a constant stream-function value each, where a depth of the linear measure 1 is considered vertically to the plane $x_1 - x_2$. Accordingly the integrations along the line drawn and stated in the equations x_1 = const and x_2 = const yield identical volume-flow values, when the upper integration limits x_1 and x_2 are chosen such that the difference $\psi - \psi_0$ is equally large in both integrals. For a fluid with ρ = const, *i.e.* for an ideal fluid, the identity of the integrals represents a mass-conservation relation. When one carries out the integrations stated in the equations along a line ψ = const *i.e.* for $d\psi = 0$, one obtains:

$$U_1 d(x_2)_\psi = U_2 d(x_1)\psi$$

or transcribed:

$$\frac{d(x_1)_\psi}{U_1} = \frac{d(x_2)_\psi}{U_2}$$

From the above indicated relation it can be said that the stream function by a comparison with the defination equation for the flow line stated in equation that the stream function defined for two-dimensional flow fields yields flow lines ψ for ψ = const.

While for the kinematic considerations is arbitrary mathematics set-ups for the velocity field could be used, the intro duction of the stream function requires a reduction of the considerations to those velocity fields which have to fulfill physical conditions. Physically possible flow fields have to fulfill the mass-conservation law which can be formulated for ideal fluids (ρ = const) as follows:

$$\frac{\partial U_i}{\partial x_i} = \frac{\partial U_1}{\partial x_1} + \frac{\partial U_2}{\partial x_2} + \frac{\partial U_3}{\partial x_3} = 0$$

That a stream function fulfills the mass-conservation law for two-dimensional flows automatically, can easily be checked by inserting the definition in equation where the law/principle of Schwarz has to be applied.

When a flow field does not fulfill the mass-conservation law, the integrations to be carried out according to the equations result in solutions contradicting one another. This can be shown for the employed two-dimensional flow field which does not fulfill the mass-conservation law (*i.e.* the continuity equation):

$$U_1 = x_1;\ U_2 = x_2(1 + 2t) \text{ and } U_3 = 0$$

When one carries out the integration stated in equation in spite of this, from the defining equations for the stream function, *i.e.* from the relations results:

$$\frac{\partial_\psi}{\partial x_2} = x_1 \text{ and } \frac{\partial_\psi}{\partial x_1} = -x_2(1+2t)$$

By integration of these equations one obtains for the stream function:

$$\psi = x_1 x_2 + F(x_1, t) \quad \psi = -x_1 x_2(1 + 2t) + G(x_2, t)$$

or expressed otherwise:

$$x_1 x_2 + F(x_1, t) \neq -x_1 x_2(1 + 2t) + G(x_2, t)$$

$$\text{or } 2(t + 1)x_1 x_2 \neq G(x_2, t) - F(x_1, t)$$

A comparison of the results of both the integrations, which yield the equations shows the contradiction resulting for the stream function.

The time dependence showing on the right side of the relation is missing on the left side. This results from the fact that the velocity field indicated according to although mathematically clearly defined, cannot exist physically; the velocity field does not fulfill the requirements determined by the mass-conservation law for ρ = const When one considers on the other hand the velocity field:

$U_1 = \exp[x_1(1 + t)]$; $U_2 = -x_2(1 + t)\exp[\text{x1}\ (1 + \text{t})]$; $U_3 = 0$ for which the relation is fulfilled, where it holds:

$$\frac{\partial U_1}{\partial x_1} = (1 + t)\exp[x_1(1 + t)] \text{ and}$$

$$\frac{\partial U_2}{\partial x_2} = -(1 + t)\exp[x_1(1 + t)]$$

one obtains from the defining equations for the stream function :

$$\frac{\partial \psi}{\partial x_2} = \exp[x_1(1 + t)] \text{ and}$$

$$\frac{\partial \psi}{\partial x_1} = x_2(1 + t)\exp[x_1(1 + t)]$$

the following solution for the stream function ψ:

$$\psi = x_2 \exp[x_1(1 + t)] + C$$

If one knows the value of the stream function for $x_{1,0}$ and $x_{2,0}$, it holds

$$\psi_0 = x_2, 0 \exp[x_1, 0\ (1 + t)] + C$$

$$\psi - \psi_0 = x_2 \exp[x_1(1 + t)] - x_{2,0} \exp[x_{1,0}\ (1 + t)]$$

The obtained result can also be computed for $x_1 = x_{1,0}$ = const . from the relation:

$$\psi - \psi_0 = (x_2 - x_{2,0})\exp[x_{1,0}(1 + t)] \text{ for } x_1 = x_{1,0}$$

This is indicated by the distribution $\psi(x_2)$ at the location $x_{1,0}$ an.

When one wants to determine the flow line path in the plane $x_1 - x_2$ – one has to consider y as a parameter in equation and derive the relation $x_2 - x_1$ – for this parameter from equation. Here ψ_0, x_1,0 and x_2, 0 are constants that can be chosen freely.

Divergence of a Flow Field

In this section mathematics operators shall be explained that are known from vector analysis and that can be applied to flow fields. Their derivations will also be repeated and considered as concerns their physical meaning. Divergence of the stream field $\vec{U}$:

$$\frac{\partial U_i}{\partial x_i} = \frac{\partial U_1}{\partial x_1} + \frac{\partial U_2}{\partial x_2} + \frac{\partial U_3}{\partial x_3}$$

The above defined divergence of a flow field is to be considered as a scalar field which, in the presence of a steady velocity field, is defined at each point in space and can be computed from the velocity field that can be assumed to be known.

When one wants to perceive the physical meaning of the operator U_i applied to the velocity field $\frac{\partial}{\partial x_i} x_i$ the consideration of a fluid element, as stated in Fig.4.9, is recommended.

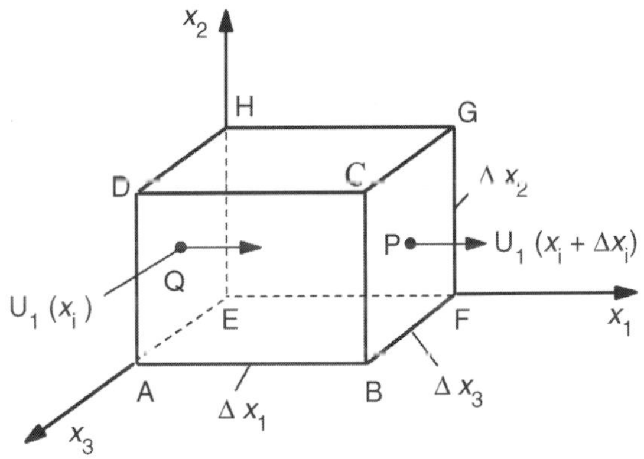

Fig. Fluid element for explaining the physical meaning of $\frac{\partial U_i}{\partial x_i}$

The edge lengths Δx_i were assumed to be very small, so that a velocity vector can be assigned to each surface such that this vector indicates with which velocity the considered surface of a fluid element moves. Accordingly, the surface AEHD moves in the direction x_1 with the velocity component of the velocity field present in point Q *i.e.* with $U_1\ (x_i) = U_1\ (x_1, x_2, x_3)$. In comparison, the surface BFGC moves with the velocity component $U_1(x_i + Dx_i, x_2, x_3)$.present in point *P*. This velocity component can be expressed by a Taylor series expansion as follows:

$$U_1(x_1 + Dx_1, x_2, x_3) = U_1(x_1) + \left(\frac{\partial U_1}{\partial x_1}\right) \Delta x_1 + \frac{1}{2}\left(\frac{\partial^2 U_1}{\partial x_1^2}\right) \Delta x_1^2 + \ldots$$

The difference velocity between the surfaces AEHD and BFGC can thus be computed as:

$$\Delta U_1(x_i, \Delta x_i) = \left(\frac{\partial U_1}{\partial x_1}\right)\Delta x_1 + \frac{1}{2}\left(\frac{\partial^2 U_1}{\partial x_1^2}\right)\Delta x_1^2 + \ldots$$

As a consequence of this velocity difference a volume increase or a volume decrease results, depending on which signs the derivations of the velocity fields, which can be stated as follows in a first approximation by multiplication with the surface $\Delta x_2\, \Delta x_3$, neglecting the terms of second and higher order in Δx_1:

$$\frac{\mathrm{d}}{\mathrm{d}t}(\delta V_1)_{\Re} = \Delta U_1(x_i)(\Delta x_2 \Delta x_3) = \frac{\partial U_1}{\partial x_1}\Delta x_1(\Delta x_2 \Delta x_3)$$

On the basis of simultaneously existing gradients of the velocity field in the directions x_2 – and x_3 in addition volume changes occur in the time unit, which again can be stated in a first approximation as follows:

$$\frac{\mathrm{d}}{\mathrm{d}t}(\delta V_2)_{\Re} = \frac{\partial U_2}{\partial x_2}\Delta x_2(\Delta x_1 \Delta x_3) \text{ and}$$

$$\frac{\mathrm{d}}{\mathrm{d}t}(\delta V_3)_{\Re} = \frac{\partial U_3}{\partial x_3}\Delta x_3(\Delta x_1 \Delta x_2)$$

so that the entire volume change that can be expected in a flow field for a fluid element in the time unit can be indicated as follows:

$$\frac{\mathrm{d}}{\mathrm{d}t}(\delta V_3)_{\Re} = \sum_{\alpha=1}^{3}\frac{\mathrm{d}}{\mathrm{d}t}(\delta V_\alpha)_{\Re} = \frac{\partial U_i}{\partial x_i}(\delta V_{\Re})$$

or can be written as:

$$\frac{\partial U_i}{\partial x_i} = \frac{1}{\delta V_{\Re}}\frac{\mathrm{d}}{\mathrm{d}t}(\delta V_{\Re})$$

This relation makes emphasizes the physical significance of the divergence of a velocity field [3]. In accordance the divergence of a velocity field states how big the volume change of a fluid element is that occurs per time and volume unit at a certain position in a flow field. At such locations of the flow field where the divergence of a velocity vector is equal to zero, there is no temporal volume change locally for a fluid element moving in the velocity field. When the divergence in sub-domains of the velocity field is computed negatively, a fluid element experiences volume decreases in these domains.

When one carries out considerations concerning the physical significance of the divergence of a velocity field at a stationary volume element of a fluid, inflows and outflows occur through the surfaces of the considered volume because of the existing velocity field, to an extent that the volume flowing in per time unit can be stated:

$$\dot{V}_{\text{inflow}} = U_i \Delta x_j \Delta_{xk} \; i = j, k$$

For the volume flowing out it can be computed:

$$\dot{V}_{\text{Out flow}} = \left[U_i + \frac{\partial U_i}{\partial x_i} \right] \Delta x_j \Delta x_k \;\; i = j, k$$

The relation makes clear that the sumation for $i = 1$ to 3 is stated in a sufficient6ly comprehensible way by the double index in $\partial U_i / \partial x_i$

The difference of inflows and outflows, considering DV = Dx_1, Dx_2 , Dx_3 can be computed as:

$$\Delta \dot{V} = \dot{V}_{\text{In flow}} - \dot{V}_{\text{Out flow}} = -\frac{\partial U_i}{\partial x_i} \Delta V$$

This relation makes clear that the presence of a positive divergence of the velocity field in volume is equal to a source, as more "'fluid volume"' is flowing out than flowing in. When, however, the divergence of a velocity field is negative, a sink occurs, as then the inflow in "'volume"' has to be larger than the outflow.

TRANSLATION, DEFORMATION AND ROTATION OF FLUID ELEMENTS

Analogous to considerations in solid-state mechanics, the deformations of fluid elements that occur due to existing velocity gradients are of interest in some flow-mechanics considerations. When one includes the translatory motion and the rotation of a fluid element in the fluid deformations, the entire local state of motion and deformation can be stated by four "'geometrically easily separable"' sub-states. The pure translatory motion sketched in Fig. 4.10 leads to a change of position of the fluid element marked to an extent that it holds:

$$d(x_j) = (U_j)_{\Re} \, dt = U_j dt$$

This relation expresses that the locally existing velocity field is responsible for the translatory motion of a fluid element, *i.e.* fluid elements move at each moment in time with the locally existing velocity vector.

In order to state or compute the rotation of a fluid element one has to describe both the angles $\Delta\Theta_1$ and $d\Delta\Theta_2$:

$$\Delta\Theta_1 = \left[\tan \frac{\frac{\partial U_2}{\partial x_1} (\Delta x_1)_{\Re} \Delta t}{(\Delta x_1)_{\Re}} \right] = \frac{\partial U_2}{\partial x_1} \Delta t$$

As a rotational speed of the fluid element the positive change of angle of the diagonal of the element occurring per time unit is defined:

Translation, deformation and rotation of fluid elements 107

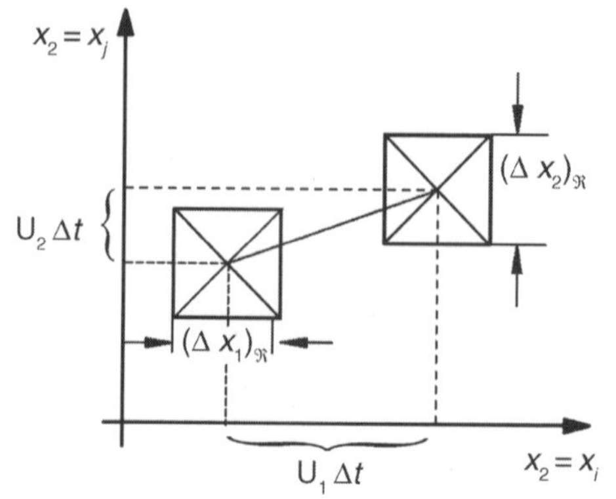

Fig. Pure translatory motion; considerations of the projection into the planes $x_1 - x_2$–

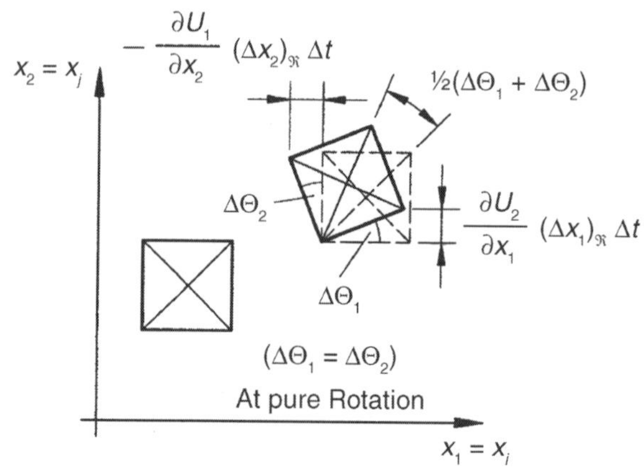

Fig. Translation and rotation of a fluid element in a flow field

$$\dot{\Theta}_{21}^{R} = \left(\frac{d\Theta^R}{dt}\right) = \frac{1}{2}\left(\frac{d\Theta_1}{dt} + \frac{d\Theta_2}{dt}\right) = \frac{1}{2}\left(\frac{\partial U_2}{\partial x_1} - \frac{\partial U_1}{\partial x_2}\right)$$

Thus generally the components of the rotational speed vector $\{\omega_k\}$ for the locally occurring rotation of a fluid element per time unit can be stated as follows:

$$2\dot{\Theta}_{ij}^{R} = 2\omega_k = \frac{\partial U_i}{\partial x_j} = rot(\vec{U})$$

The quantity ω_k states the double rotational speed of the axis of the fluid element occurring in positive direction. The second diagonal of the considered fluid element rotates with the same angular speed. $\{\omega_k\}$ is an important kinematic quantity of the velocity field. It is defined as swirling/rotational power

(Wirbelstrke) and is computed as half the rotation of the velocity field. Thus it is a field quantity of its own $\omega_k\,(x_i, t)$, for which holds:

$\omega_k(x_i, t) = 0$ when a flow field is free of rotation.

When $\omega_k(x_i, t) \neq 0$, flows subjected to rotations are present, whose swirling properties are best accessible when expressing the conservation laws for mass, impulse and energy in ω_k.

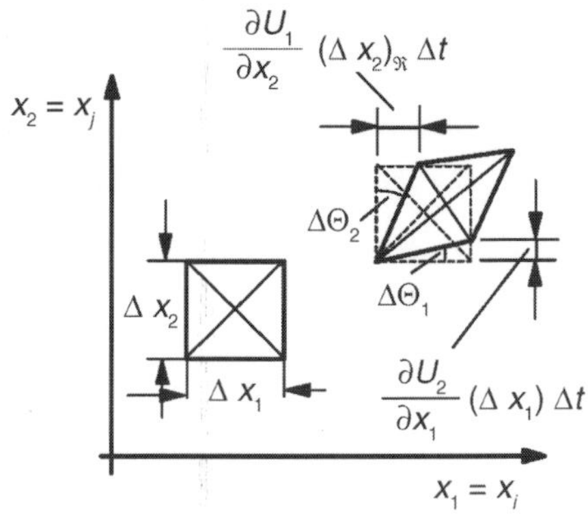

Fig. Translation and angle deformation of a fluid element in a flow field due to velocity gradients

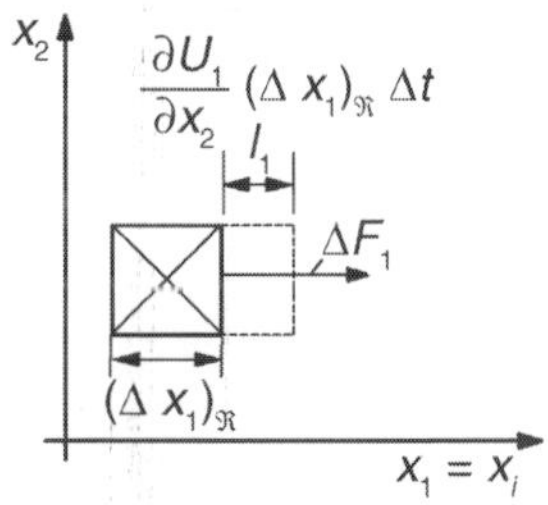

Fig. Elongation of a volume element due to velocity gradients in the flow field

When one considers next the angle deformation of a fluid element , one can see that for the angle deformation holds:

$$\dot{\Theta}_{21}^{D} = \frac{1}{2}\left(\frac{d\Theta_1}{dt} - \frac{d\Theta_2}{dt}\right) \quad \text{Deformation angular speed}$$

Thus for the angular deformation in the plane $x_1 - x_2$ holds generally

$$\dot{\Theta}_{21}^{D} = \frac{1}{2}\left(\frac{\partial U_2}{\partial x_1} + \frac{\partial U_1}{\partial x_2}\right)$$

Or generally:

$$\dot{\Theta}_{ij}^{D} = \frac{1}{2}\left(\frac{\partial U_j}{\partial x_i} + \frac{\partial U_i}{\partial x_j}\right) \quad i \neq j$$

Analogous to considerations in solid-state mechanics the symmetry of the deformation tensor holds:

$$\dot{\Theta}_{ij}^{D} = \dot{\Theta}_{ij}^{D}$$

Finally one has to consider the dilatation of a fluid element which experiences strain rates due to the velocity gradient existing in a flow field. The linear deformation/change in length occurring due to an existing velocity gradient in direction x_1 can be stated as follows:

$$\frac{\mathrm{d}l}{\mathrm{d}t} = \lim_{\Delta t \to 0}\left(\frac{\partial U_1}{\partial x_1}\right)\frac{(\delta x_1)_{\Re}\,\Delta t}{\Delta t} = \left(\frac{\partial U_1}{\partial x_1}\right)(\delta x_1)_{\Re}$$

From this is computed the linear deformation occurring per length and time unit:

$$\frac{1}{(\Delta x_1)}\frac{\mathrm{d}(l_1)}{\mathrm{d}t} = \frac{\partial U_1}{\partial x_1}$$

When multiplying the linear deformation with the transverse area the apper-taining volume change results:

$$\frac{\mathrm{d}(\delta V_1)_{\Re}}{\mathrm{d}t} = \left(\frac{\partial U_1}{\partial x_1}\right)(\delta V_1)_{\Re}$$

Summed up over all three axis directions one obtains for the entire volume change per time unit:

$$\frac{1}{(\delta V)_{\Re}}\frac{\mathrm{d}(\delta V)_{\Re}}{\mathrm{d}t} = \frac{\partial U_i}{\partial x_i}$$

i.e. the divergence of the velocity field indicates how the volume of a fluid element at a point in space changes with time.

It is customary in literature to combine elongations of fluid elements and their angular deformations to a deformation tensor in such a way that it holds:

$$\varepsilon_{ij} = \frac{1}{2}\left(\frac{\partial U_j}{\partial x_i} + \frac{\partial U_i}{\partial x_j}\right) \quad \text{for } i \neq j$$

so that for the deformation tensor holds:

$$\{\varepsilon_{ij}\} = \begin{Bmatrix} \varepsilon_{11} & \varepsilon_{12} & \varepsilon_{13} \\ \varepsilon_{21} & \varepsilon_{22} & \varepsilon_{23} \\ \varepsilon_{31} & \varepsilon_{32} & \varepsilon_{33} \end{Bmatrix} \text{und } \varepsilon_{ij} = \varepsilon_{ij}$$

From the above considerations the following relation results:

$$\frac{\partial U_j}{\partial x_i} = \frac{1}{2}\left(\frac{\partial U_j}{\partial x_i} + \frac{\partial U_i}{\partial x_j}\right) + \frac{1}{2}\left(\frac{\partial U_j}{\partial x_i} + \frac{\partial U_i}{\partial x_j}\right)$$

i.e. it holds:

$$\frac{\partial U_j}{\partial x_i} = \varepsilon_{ij} + \frac{\mathrm{d}\Theta_{ij}^R}{\partial x_i} = \varepsilon_{ij} + \frac{1}{2}\varepsilon_{ijk}\frac{\partial U_j}{\partial x_i}$$

The gradients existing in velocity fields are linked to deformations and rotations of fluid elements, the gradients allowing to state corresponding rates of deformation (normalized deformation per unit time) and rotational angle velocities. This has to be considered when employing the analogy between solid-state mechanics and fluid mechanics, in order to transfer considerations on deformations of elastic bodies, carried out in solid-state mechanics to rates of deformation of fluid elements occurring in fluid mechanics, due to existing gradients of the flow field.

5

Wave Motions in Fluid Flows

WAVE MOTIONS

The wave motions represent introducing considerations that experience extensions in relevant manuals with emphasis on wave motions. Nonetheless, manuals with general considerations on wave motions in fluids are missing, *i.e.* the treatment of wave motions is always limited to the treatment of very special wave motions. The following wave motions in fluids are dealt with:

- Gerstner waves
- Solitar waves
- Rossby-waves
- Stokes-waves
- Cnoidal-waves
- Axisymmetric waves

So if one wants to find the introducing literature on the wave motion observed in nature, it is necessary to perceive the physical causes o the considered wave motion. Thus one observes for example that a long body which is moved transversely to its linear expansion near the free surface forms waves mainly in its wake. In front of the body one observes viewed from the amplitude smaller surface waves, when the dimensions of the bodies in flow direction are smaller as compared to$(\sigma/\rho g)^{1/2}$ Otherwise the gravity waves occurring behind the body dominate and the capillary waves that can be observed in front of the body are negligible. So when one has recognized the nature of the observed wave motions.

FLUID FLOWS

Fluid flows were considered whose analytical treatment was possible by employing simplified forms of the generally valid basic equations of fluid mechanics. The solution methods required for this are known, *i.e.* they are at disposal and it is known that they can be employed. The application of methods was shown which permit the solutions of the basic equations of fluid mechanics in order to obtain one-dimensional and two-dimensional flows.

In particular, potential flows were dealt with whose given properties were chosen such that methods of the theory of functions can be employed to treat analytically two-dimensional and irrotational flow problems.

The special properties of potential flows made it possible to take over a fully developed domain of mathematics into flow mechanics and to employ it for computing potential and streamline fields. From these fields velocity fields of the treated potential flows could be derived and the employment of the mechanical energy equation and its integral form, respectively finally lead to pressure distributions in the considered flow fields.

The latter again lead to the computation of forces and moments for control volume which are of particular interest for the solution of engineering problems. Simplifications of the flow properties by introducing two-dimensionality and irrotationality thus have permitted a closed treatment of flow problems with known mathematical methods.

The treatment of wave motions of fluids is planned. As with all mechanical wave motions, motions are concerned that occur around a medium rest position, *i.e.* the fluid particles involved in the wave motion experience no change of position in the time average.

Thus in the case of wave motions in fluids only the vibration state characterizing a wave continues and not the fluid itself. This holds independently from whether with wave motions in fluids longitudinal or transversal waves form shows the oscillation motion of fluid particles for both wave modes.

One can infer that the considered wave motions are periodical, with regard to space as well as time. Oscillations on the other hand are periodical either concerning time or space.

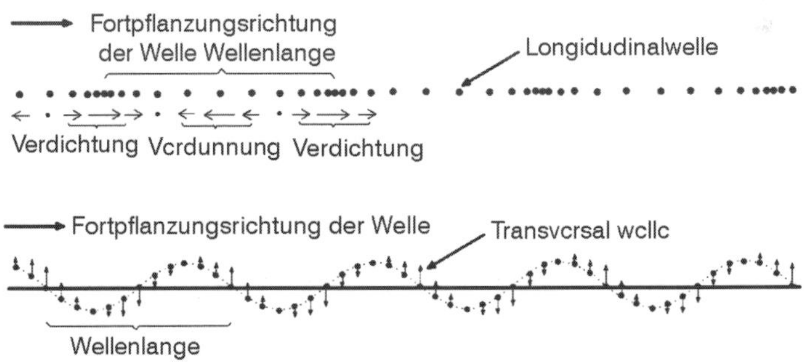

Fig. Instantaneous Image of a Progressing Longitudinal and Transversal Wave

Mechanical longitudinal waves, which are characterized by condensations and dilutions, *i.e.* by specific changes in volume or density, can exist in all matters having "volume elasticity", *i.e.* react with elastic counter forces in case of occurring volume changes. Such counter forces form in gases, as volume changes are coupled to pressure changes, so that for an ideal gas at T = const holds

$$P \cdot dv = -v \cdot dP,$$

and thus longitudinal waves can occur in isothermal gases, which are not possible in thermodynamically ideal fluids because of ñ = $1/v$ = const. makes also clear that the formation of transversal waves is dependent on the presence of "shear forces", *i.e.* lateral forces, in order to permit the wave motion of "particles" diagonally to the direction of propagation. With this mechanical transversal waves only occur in solid matters which can build up elastic transversal forces in the case of corresponding stress. In purely viscose fluids no transversal waves are possible. At first sight his statement seems to be a contradiction to observations of water waves whose development and propagation can be observed easily when one throws an object in a water container.

A transversal wave develops which, however, proves to be a wave motion restricting itself to a small area on the water surface. In the interior of the fluid the wave motion cannot be observed. Moreover can be seen that the observed wave on the surface does not form owing to "shear forces" but that the presence of gravity or the occurrence of surface tensions are responsible for the wave motion.

In fluids the most different wave motions are possible, whose initiation and conservation are connected to an energy input into the fluid. For the generation of a wave and its conservation a certain work e ort is necessary which then propagates in the space as energy of the wave.

With this it is essential that two different energy modes are possible, and also occur, between which an exchange of energy can take place in periodical sequence.

This makes it clear that an essential characteristic of a wave motion in a fluid is that energy is transported without a mass transport taking place. Depending on the form of the wave flows, *i.e.* the form of the source of the wave motion, one distinguishes wave modes, namely plane waves, spherical waves and cylindrical waves. For the velocity field of such waves it can be stated:

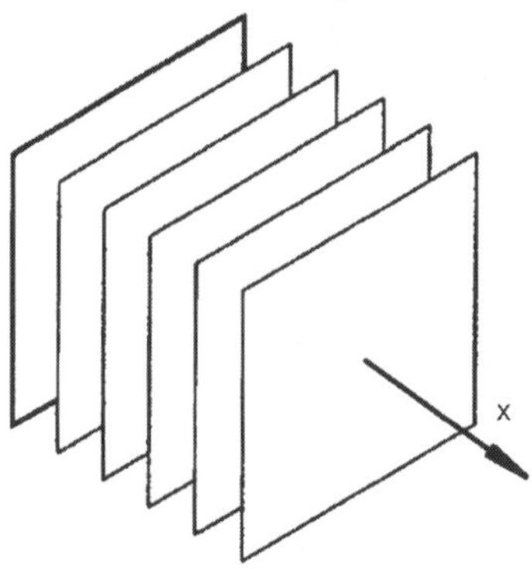

Fig. Diagram of a Two-dimensional Wave and Statements Concerning the Nomenclature

Plane waves:

$$u'(x, t) = u_A \cdot \sin\left[2\pi\left(\frac{t}{T} \mp \frac{x}{\lambda}\right)\right]$$

In the case of a plane wave the mean energy density is constant, as a considered surface of a wave does not change along the propagation direction x. In the above relation T is the oscillation period of the wave motion and l is the wave length. The periodicity of the plane wave in the propagation direction x and the time t can be seen from the sinusoidal term.

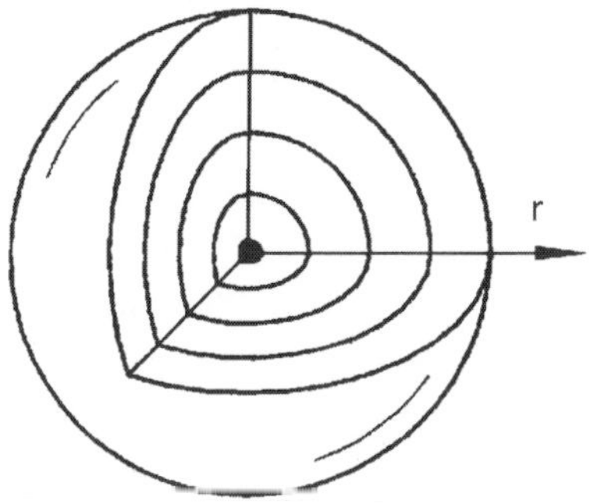

Fig. Diagram of a Spherical wave and Statements Concerning the Nomenclature

Spherical wave:

$$u'(x, t) = \frac{u_A}{r} \cdot \sin\left[2\pi\left(\frac{t}{T} \mp \frac{r}{\lambda}\right)\right]$$

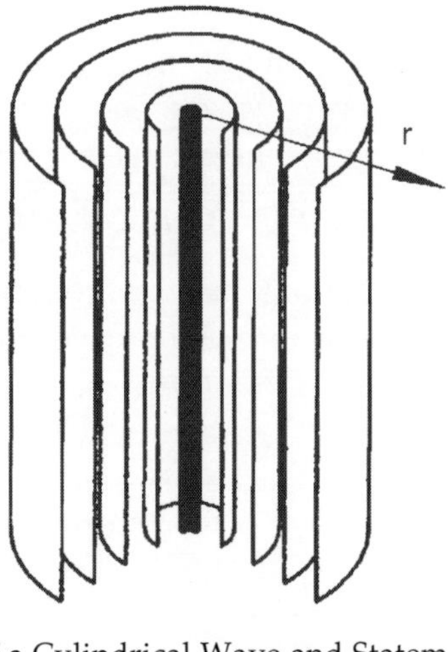

Fig. Diagram of a Cylindrical Wave and Statements Concerning the Nomenclature

As concerns spherical waves in fluids, the energy density decreases with the square of the distance of point $r = 0$ ab, as the surface increases with the square of the distance. In point $r = 0$ the exciter of the spherical wave is located; the entire energy of the wave is concentrated at this place. Here the above-stated equation of the wave only holds for $r \neq 0$. The negative sign in front of the r/λ- Term indicates a diverging wave that runs/moves away from the exciter centre and the positive sign indicates a diverging wave running/moving towards the exciter centre.

Cylindrical wave:

$$u'(x, t) = \frac{u_A}{\sqrt{r}} \cdot \sin\left[2\pi\left(\frac{t}{T} \mp \frac{r}{\lambda}\right)\right]$$

Cylindrical waves propagate radially from the exciter line located in the center , *i.e.* $r = 0$, such that the wave surface increases in a linear way with the distance r Thus the energy density decreases in a linear way with the distance r Therefore then the amplitude of the wave is in inverse ratio proportional to the root of the distance r from the exciter line. Again the negative sign in front of the r/λ-Term indicates a wave moving from the exciter line in positive r-direction, whereas the positive sign describes a wave moving towards the exciter line.

Many general properties of wave motions known from physics can be transferred to wave propagations in fluids. Nevertheless, in a book meant as an introduction into fluid mechanics special considerations are required, in particular is it necessary to arouse the deeper comprehension of the causes of the considered wave motions and to show how to deal with them on the basis of the Navier-Stokes equations.

The derivations of properties of theses wave motions shall show in which way to proceed in fluid mechanics to derive the properties from the basic equations of fluid mechanics. The aim of the descriptions/explications thus is not a broad consideration of different wave motions in fluids, but an introduction into the mathematical treatment of longitudinal and transverse waves in fluids.

PLANE STANDING WAVES

When considering wave motions, where the fluid particles move only in parallel to the $x_1 - x_2$–plane, *i.e.* where the pressure P and the velocity U_j are independent of x_3 , so that the fluid motions in all areas parallel to the $x_1 - x_2$–plane take place in the same way, a plane wave motion with the following potential results:

$$\phi(x, y, t) = \phi(x, y) \cos(\varphi t + \in)$$

For the case of a standing wave to be dealt with in this section it stated:

$$\phi(x, y) = \frac{P(y)}{\rho} \sin[k(x - \xi)]$$

The potential ϕ fulfills the Laplace equation.

$$\frac{\partial^2 \phi}{\partial x^2}+\frac{\partial^2 \phi}{\partial y^2} = 0$$

Mit $\rho\frac{\partial^2 \phi}{\partial x^2} = -P(y)k^2 \sin[k(x-\xi)]$ and $\rho\frac{\partial^2 \phi}{\partial y^2} = P\frac{d^2 P}{dy^2}\sin[k(x-\xi)]$

one obtains the following differential equation:

$$\frac{d^2 P}{dy^2}-k^2 P = 0$$

the solution of which is as follows:

$$P(y) = C_1 \exp(ky) + C_2 \exp(-ky)$$
$$\text{fur } -\infty \leq y \leq 0 - \infty \leq +x \leq +\infty$$

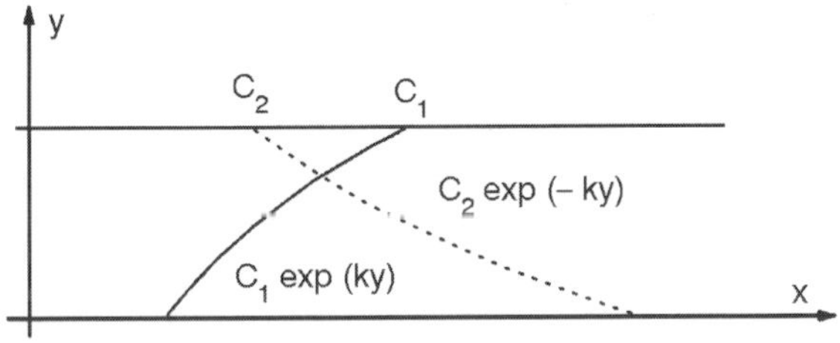

Fig. Illustration of the Decrease for $y \to -\infty$

The integration constant C_2, when carrying out more precise considerations, results as $C_2 = 0$, as otherwise for large depths $y =\to -\infty$ the P (y)-term would become very big, so that the solution:

$$\phi(x, y) = \frac{C_1}{\rho}\exp(ky)\sin[k(x-\xi)]$$

can be obtained, or

$$\phi(x, y, t) = \frac{C_1}{\rho}\exp(ky)\sin[k(x-\xi)]\cos(\varphi t+\in)$$

When starting from the assumption that the occurring fluid motion is slow, the equation

$$\frac{\partial \phi}{\partial t}+\frac{P}{\rho}+\frac{1}{2}U_j^2+g\eta = 0 \text{ for } y = \varsigma$$

can be written as follows:

$$\frac{\partial \phi}{\partial t}+\frac{P}{\rho}+g\eta = 0 \text{ for } y = \varsigma$$

A derivation to the time yields the relation indicated below, as the pressure along the free surface does not change:

$$\frac{\partial^2\phi}{\partial t^2}+g\frac{\partial\eta}{\partial t} = \frac{\partial^2\phi}{\partial t^2}+gu_2$$

With $u2 = \frac{\partial\phi}{\partial y}$ it results:

When employing to one obtains:

$$\frac{\partial^2\phi}{\partial t^2} = -\frac{C_1}{\rho}\exp(ky)\sin[k(x-\xi)]\varphi^2\cos(\varphi+\in)$$

and

$$\frac{\partial\phi}{\partial y} = +\frac{C_1}{\rho}\exp(ky)\sin[k(x-\xi)]\cos(\varphi+\in)$$

$$\varphi^2 = kg$$

Thus for the remaining considerations the following fluid motion has to be examined which for the sake of simplicity is considered for $\xi = 0$ and $\in = 0$.

$$\phi(x, y, t) = \frac{C_1}{\rho}\exp(ky)\sin(kx)\cos(\varepsilon t)$$

For the free surface it is computed from equation:

$$ç = -\frac{1}{g}\frac{\partial\phi}{\partial t} = -\frac{1}{g}\frac{\partial}{\partial t}\phi(x,0,t)$$

or

$$ç = \frac{C_1\varphi}{g}\sin(kx)\sin(\varphi t)$$

With $A = \frac{C_1\varphi}{\rho g}\sin(\varphi t)$ it holds:

$$ç = \text{A}\sin(kx)$$

For $x = \frac{m\cdot\pi}{k}$ result for m $= 0, \pm1, \pm2, \pm\cdots$ nodal points of a standing wave. In the middle between these nodes are the "antinodal points" of the wave motion. The wavelength of the sinusoidal fluid motion can be computed as:

$$ë = \frac{2\pi}{k}$$

The amplitude of the wave motion is $\frac{C_1\varphi}{g}\cdot\sin(\varphi t) = A$ quency of the wave motion it holds:

$$f = \frac{\varphi}{2\pi} = \frac{1}{T}$$

Taking into consideration and one obtains:

$$T = \frac{1}{f} = \sqrt{\frac{2\pi k}{g}} \quad \text{bzw.} \quad \lambda = \frac{\gamma\tau^2}{2\pi}$$

or

$$\ddot{e} = \frac{g}{2\pi f^2}$$

The above relations show that the wavelength of standing fluid waves decreases with increasing frequency of the motion.

PLANE PROGRESSING WAVES

For the considerations stated below it is assumed that the considered fluid takes up the space as follows:

$$-\infty \leq y \leq 0 \quad \text{and} -\infty \leq x \leq +\infty$$

and at the point $y = 0$ occupies a free surface which for the considerations carried out possesses a finite surface. The equations required for the treatment of progressing waves can be stated as follows:

$$\frac{\partial^2 g}{\partial x^2} + \frac{\partial^2 \phi}{\partial y^2} = 0$$

With $y = \varsigma(x_1, t)$ for the free surface it holds:

$$\frac{D}{Dt}(y - \eta) = 0 \rightsquigarrow u_2 = \left(\frac{\partial}{\partial t} + u_1 \frac{\partial}{\partial x_1}\right)\eta$$

Neglecting the term of second order it results:

$$\frac{\partial \phi}{\partial y} = \frac{\partial \eta}{\partial t}$$

For the pressure at the free surface it can be stated:

$$P = -\sigma\left[\frac{1}{R_1} + \frac{1}{R^2}\right]$$

where R_1 and R_2 represent the main radiuses of curvature of the disturbed surface and σ indicates the surface tension. Linearized it can be written:

$$P = -\sigma \frac{\partial^2 \eta}{\partial x^2}$$

where the pressure above the free surface is formulated with $P = 0$ Otherwise P is to be replaced by $(P = P_0)$.

For plane progressing waves the following potential can be stated:

$$\phi(x, y, t) = C \exp(ky) \cos [k (x - ct)]$$

with $\phi = 0$ for $y = -\infty$ The formulation for. fulfills the continuity equation in the form (11.80).

The Bernouilli equation can be stated as follows:

$$\frac{P}{\rho} = -\frac{\partial\phi}{\partial t} - gy .$$

Or transcribed:

$$\frac{\partial\phi}{\partial t} = -\frac{P}{\rho} - g\eta = \frac{\sigma}{\rho}\frac{\partial^2\eta}{\partial x^2} - g\eta .$$

From this follows:

$$\frac{\partial^2\phi}{\partial t^2} = \frac{\sigma}{\rho}\frac{\partial^2}{\partial x^2}\left(\frac{\partial\eta}{\partial t}\right) - g\frac{\partial\eta}{\partial t} .$$

and in consideration of

$$\frac{\partial^2\phi}{\partial t^2} = \left[\frac{\sigma}{\rho}\frac{\partial^2}{\partial x^2} - g\right]\frac{\partial\phi}{\partial g}$$

For the left side of it can be written with:

$$\frac{\partial^2\phi}{\partial t^2} = -Ck^2c^2 \exp(ky) \cos [k(x - ct)] = -k^2c^2\phi$$

so that it holds:

$$k^2c^2\phi = \left[g - \frac{\sigma}{\rho}\frac{\partial^2}{\partial x^2}\right]\frac{\partial\phi}{\partial y}$$

With $\frac{\partial\phi}{\partial y} = k\phi$ and $\frac{\partial^2\phi}{\partial x^2} = -k^2\phi$ results from for the velocity of the progressing wave:

$$c^2 = \frac{g}{k} + \frac{k\sigma}{\rho}$$

With $k = \frac{2\pi}{\lambda}$ it can be seen that for long waves the influence of the gravity dominates:

$$c = \sqrt{\frac{g}{k}} \quad \text{shear waves}$$

For waves with small wavelengths the capillary e ects dominate:

$$c = \sqrt{\frac{k\sigma}{\rho}} \quad \text{capillary waves}$$

Concerning wavelengths, often the path curves of the fluid particles are also of interest, which occur at the water surface or in certain depths below the water surface. In this respect the following considerations can be carried out, where x_0 and y_0 are introduced as the coordinates which

$$u_x = \frac{\partial \phi}{\partial x} = \frac{dx}{dt} = Ck\exp(ky)\sin[k(x-ct)]$$

$$u_y = \frac{\partial \phi}{\partial y} = \frac{dy}{dt} = Ck\exp(ky)\cos[k(x-ct)]$$

From this result:

$$x = x_0 + C \cdot k \exp(ky) \cos [k(x - ct)] \left(\frac{+1}{Ck}\right)$$

$$y = y_0 + ck \exp(ky) \sin [k(x - ct)] \left(\frac{-1}{ck}\right)$$

Or transcribed:

$$(x - x_0)^2 + (y - y_0)^2 = \left(\frac{C}{c}\right)^2 \exp(2ky_0)$$

The path curves of the fluid particles are computed as circles whose radius gets smaller with increasing water depth. For the water surface the radius of the circular path is equal to the amplitude of the surface wave, while in a depth of a wavelength it has already decreased to a 1/535th of the wave amplitude at the water surface. This makes it clear that the considered wave motion remains limited to an area in immediate proximity of the water surface.

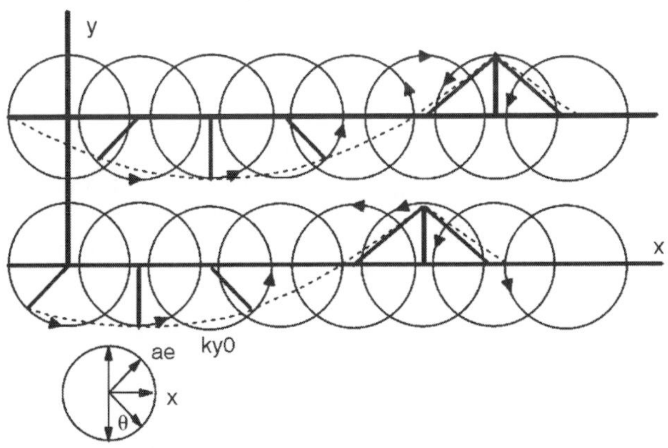

Fig. Circular Paths of the Fluid Particle Motion

Shows the circular paths described by fluid particles. These will run in an anticlockwise direction, so that it holds:

$$(x - x_0) = a \exp(ky_0) \sin \Theta$$
$$(y - y_0) = -a \exp(ky_0) \cos \Theta$$

so that for Θ it holds:

$$\Theta = kx_0 + kct \cos(yt + \in)$$

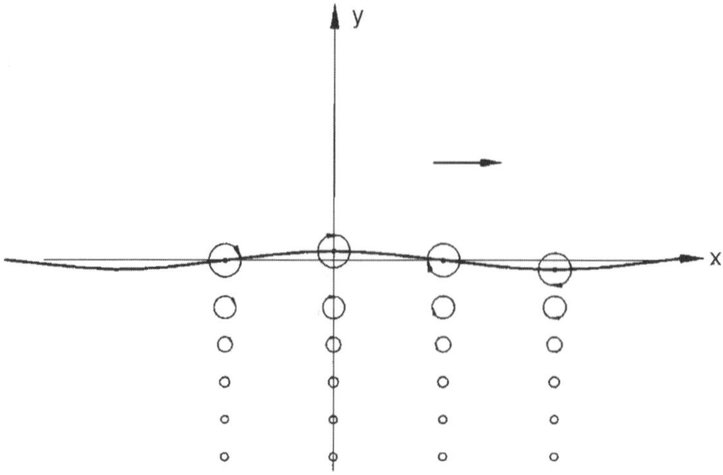

Fig. Path Lines in a Plane Gravity Wave

As concerns the change of the motions of the fluid particles with water depth, the strong decrease of the radius with fluid depth was not taken into consideration.

In order to be able to investigate gravity waves with free surfaces in fluids with a finite depth h a mean surface position at point $y = 0$ is assumed. At position $y = -h$ a wall is considered as being given, so that a mean fluid film thickness with the height h occurs. To fulfill now the continuity equation:

$$\frac{\partial^2 \phi}{\partial x^2} + \frac{\partial^2 \phi}{\partial y^2} = 0$$

by a wave with the wave number k., the following potential formulation is carried out:

$$\phi = C \cosh k(y + h) \cos k(x - ct)$$

which does not only fulfill the continuity equation, but permits also to fulfill the boundary condition at the bottom of the fluid layer:

$$\frac{\partial \phi}{\partial y} = 0 \text{ for } y = -h$$

It then results a condition for the free surface that can be stated as below indicated:

$$kc^2 \cosh kh = \left(g + \frac{k^2 T}{\rho} \right) \sinh kh,$$

Or resolved to the wave velocity one obtains:

$$c^2 = \left(g + \frac{k^2 T}{\rho} \right) \frac{\tanh kh}{k}$$

For long-waved waves, *i.e.* for small values of the wave number k one obtains for the wave velocity:

$$c^2 = gh.$$

The waves moving with this velocity are essentially gravitation waves, as the surface curvature is so small that the influences of the surface tension at the wave motion are not perceived.

For very short waves, *i.e.* for large values of the wave time k one obtains on the other hand:

$$c^2 = \frac{kT}{\rho}$$

This is the propagation velocity of the capillary waves which this equation shows for the velocity of the capillary waves, so that their propagation velocity is not influenced by the height of the fluid layer.

SURFACE WAVES

GENERAL SOLUTION SET-UP

On the free surfaces of fluids wave appearances can occur, *i.e.* propagation of transversal waves owing to introduced disturbances. These can be two- or three-dimensional, however, the analytical treatment of surface waves presented here concentrates on two-dimensional surfaces. By linearization of the basic equations written in potential form one obtains the partial differential equations solved normally for surface waves. These indicate that the field of propagation of surface waves belongs to the potential theory.

Their treatment takes place separately nevertheless, as a special problem is concerned, *i.e.* a special class of flow appearances whose treatment correspondingly requires a special methodology. The latter is shown below in an introducing way.

The relations stated in the following can again be derived from the basic equations, which can be stated as follows for a fluid-mechanically ideal fluid, *i.e.* a fluid free from viscosity:

$$\frac{\partial U}{\partial t} + U_i \cdot \frac{\partial U_j}{\partial x_i} = -\frac{1}{\rho}\frac{\partial P}{\partial x_j} + g_j$$

When integrating this equation over a period of time τ, one obtains

$$\bar{U}_j + \int_0^\tau U_i \frac{\partial U_j}{\partial x_i} dt = -\frac{1}{\rho}\frac{\partial}{\partial x_j}\int_0^\tau P\, dt + \int_0^\tau g_j\, dt$$

This equation can now be interpreted with $\pi = \int_0^\tau P\, dt$ as the pressure impulse during the time interval τ, for small time intervals τ as follows for $\rho =$ const:

$$\bar{U}_j = \frac{\partial}{\partial x_j}\frac{P}{\rho} \text{ mit } \int_0^\tau U_i \frac{\partial U_j}{\partial x_i} dt \approx 0 \text{ and } \int_0^\tau g_j \cdot dt \approx 0$$

Thus the fluid motion generated as a result of pressure impulses on free surfaces is described by a velocity potential, by $\bar{U}_j = U_j$:

$$\bar{U}_j = -\frac{\partial \phi}{\partial x_j} \quad \text{mit} \quad \phi = \frac{P}{\rho}$$

The motion thus is irrotational. Strictly speaking all this holds only at the free surface and the determination of ϕ in the entire flow area requires further considerations still. The continuity equation can be written as follows for ϕ.

$$\frac{\partial^2 \phi}{\partial x_i \cdot \partial x_i} = 0 = \frac{\partial^2 \phi}{\partial x_1^2} + \frac{\partial^2 \phi}{\partial x_2^2} + \frac{\partial^2 \phi}{\partial x_3^2}$$

The momentum equation (11.42) can be written as stated below:

$$\frac{DU_j}{Dt} = -\frac{1}{\rho} \cdot \frac{\partial P}{\partial x_j} + g_j / \cdot U_j$$

or can be transcribed after multiplication by U_j as follows:

$$\frac{D}{Dt}\left(\frac{1}{2}U_j^2\right) - \frac{1}{\rho}\frac{DP}{Dt} \quad \frac{1}{\rho}\frac{\partial P}{\partial t} \quad \frac{DG}{Dt}$$

With $g_j = -\rho \frac{DG}{Dt}$ for $\frac{\partial g}{\partial t} = 0$,

$$\frac{D}{Dt}\left(\frac{1}{2}U_j^2\right) = -\frac{1}{\rho}\frac{DP}{Dt} - \frac{1}{\rho}\frac{\partial P}{\partial t} - \frac{DG}{Dt}$$

or transcribed:

$$\frac{\partial \phi}{\partial t} + \frac{P}{\rho} + \frac{1}{2}U_j^2 + G = F(t)$$

The function $F(t)$ introduced by the integration can be included into the potential ϕ, so that it holds:

$$\frac{\partial \phi}{\partial t} + \frac{P}{\rho} + \frac{1}{2}U_j^2 + G = 0$$

Represents a two-dimensional surface wave whose deflection, measured from the position of rest $x_2 = 0$ can be stated a follows:

$$x_2 = y = h(x_1, t) = h(x, t)$$

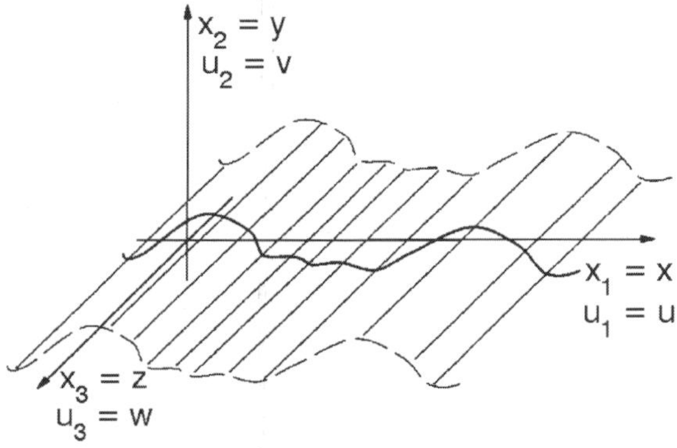

Fig. Two-dimensional Surface Wave

The kinematic boundary condition of the flow problem to be solved can thus be stated as follows:

$$y = \eta(x, t) = 0$$

This means that a fluid particle which belonged to the fluid surface at a point in time t will always belong to the free surface. From equation results with u_i as fluid velocity of the considered wave motion

$$\frac{D}{Dt}(y-\eta) = 0 = \frac{\partial}{\partial t}(y-\eta) + u_i \frac{\partial}{\partial x_i}(y-\eta) = 0$$

or the deflections carried out:

$$-\frac{\partial \eta}{\partial t} - u_1 \frac{\partial \eta}{\partial x_1} + u_2 - u_3 \frac{\partial \eta}{\partial x_3} = 0$$

When introducing now the potential function ϕ with

$$u_1 = \frac{\partial \phi}{\partial x_1}, \quad u_2 = \frac{\partial \phi}{\partial x_2} \quad \text{and} \quad u_3 \frac{\partial \phi}{\partial x_3}$$

the following relation results for the free surface with $x_1 = x, x_2 = y$ and $x_3 = z$:

$$\frac{\partial \phi}{\partial y} = \frac{\partial \eta}{\partial x} \cdot \frac{\partial \eta}{\partial x} + \frac{\partial \phi}{\partial z} \cdot \frac{\partial \eta}{\partial z}$$

In the entire area of the flow the potential function fulfills the continuity equation which thus can be stated in two-dimensional form as follows:

$$\frac{\partial^2 \phi}{\partial x^2} + \frac{\partial^2 \phi}{\partial y^2} = 0$$

On the prerequisite of absence of viscosity the Bernoulli equation can be employed in the form indicated by equation (11.51).

$$\frac{\partial\phi}{\partial t}+\frac{P}{\rho}+\frac{1}{2}U_j^2+G\ =0$$

This is equivalent to the assumption that typically the pressure along the free surface is constant and corresponds to the atmospheric pressure over the surface. When now including the solid bottom in a certain position $y = -h$, one obtains as a boundary condition at this point:

$$\frac{\partial\phi}{\partial y}=0 \quad \text{for} \quad y=-h$$

Thus one obtains the following set of equations, which are to be fulfilled in order to treat the propagation of waves on free surfaces analytically.

$$\frac{\partial^2\phi}{\partial x^2}+\frac{\partial^2\phi}{\partial y^2}\ =0$$

$$\frac{\partial\eta}{\partial t}+\frac{\partial\phi}{\partial x}\frac{\partial\eta}{\partial x}+\frac{\partial\phi}{\partial z}+\frac{\partial\eta}{\partial z}\ =\ \frac{\partial\phi}{\partial y}\ \text{for}\ y=\eta$$

$$\frac{\partial\phi}{\partial t}+\frac{P}{\rho}+\frac{1}{2}U_j^2+g\eta\ =0\ \text{for}\ y=\eta$$

$$\frac{\partial\phi}{\partial y}\ =0\ \text{for}\ y=-h$$

Here the last equations are to be understood as boundary conditions. Thus it becomes clear that the problem when solving wave problems for fluids with free surfaces is characterized by the imposed kinematic and dynamical boundary conditions. It proves to be a peculiarity here that the main problem when solving problems concerning the wave motion in fluids with free surfaces is the introduction of the boundary conditions and not the solution of the differential equations describing the fluid motion.

Considerable simplifications of the equation system result further for the assumption of surface waves of small amplitudes. Assuming that the amplitude of the wave is smaller than all other linear dimensions of the problem, *i.e.* smaller than the depth of the water h, and the wavelength ë, it results that η is small and $\frac{\partial\eta}{\partial x}$ also can be assumed to be small. The latter is the gradient of the course of the water surface. Moreover it hold that can be assumed to be small, as surface waves have no high frequencies and the assumption of small amplitudes is also valid here. Thus it holds for two-dimensional waves:

$$\frac{\partial\eta}{\partial t}=\frac{\partial\phi}{\partial\eta} \qquad \text{for}\ y=ç$$

This equation still contains the problem that the boundary condition under lying it for surface waves has to be imposed at the point y = ç However, when one expands $\frac{\partial \phi}{\partial \eta}$ in a Taylor series

$$\frac{\partial \phi}{\partial y}(x,\eta,t) = \frac{\partial \phi}{\partial y}(x,0,t)+\eta\frac{\partial^2 \phi}{\partial \eta^2}(x,0,t)+\cdots$$

it can be seen that the second term on the right side can be neglected because of the assusmed small ç-values. In an analogous way it holds

$$\frac{\partial \phi}{\partial y}(x,\eta,t)+\frac{P(x,t)}{\rho}+g\cdot\eta(x,t) = F(t)$$

and for small velocities the following relation:

$$\frac{\partial \phi}{\partial t}(x,0,t)+\frac{P(x,t)}{\rho}+g\cdot\eta(x,t) = 0$$

where the function $F(t)$ was included into the potential $\phi(x, y, z)$. The differentiation of 11.64 to t yields:

$$\frac{\partial^2 \phi}{\partial t^2}+\frac{1}{\rho}\frac{\partial P}{\partial t}+g\frac{\partial \eta}{\partial t} = \frac{\partial^2 \phi}{\partial t^2}+(x,0,z)+\frac{1}{\rho}\frac{\partial P(x,t)}{\partial t}+g\cdot\frac{\partial \phi}{\partial y}(x,0,z) = 0$$

so that one obtains the following simplified set of equations for the treatment of surfaces of small amplitudes:

$$\frac{\partial^2 \phi}{\partial x^2}+\frac{\partial^2 \phi}{\partial y^2} = 0$$

$$\frac{\partial \phi}{\partial y}(x,0,t) = \frac{\partial \eta}{\partial t}(x,t) \quad (\text{for } y = ç)$$

$$\frac{\partial^2 \phi}{\partial t^2}(x,0,t)+\frac{1}{\rho}\frac{\partial P(x,t)}{\partial t}+g\frac{\partial \phi}{\partial y}(x,0,t) = 0 \ (\text{for } y = ç)$$

$$\frac{\partial \phi}{\partial y}(x,-h,t) = 0 \ (\text{for } y = -h)$$

With the above listed equations gravitational waves and capillary waves can be treated, which usually represent small amplitudes.

ROTATIONAL FLOWS

A rotational fluid flow can contain streamlines that loop back on themselves. Hence, fluid particles following such streamlines will travel along closed paths. Bounded (and hence nonuniform) viscous fluids exhibit rotational flow, typically within their boundary layers. Since all real fluids are viscous to some amount, all real fluids exhibit a level of rotational flow somewhere in their domain. Regions of rotational flow correspond to the regions of viscous losses in a fluid. Inviscid fluid flows can also be rotational, but these are special nonphysical cases. For an inviscid fluid flow to be rotational, it must be set up that way by initial conditions. The amount of rotation (called the *velocity circulation*) in an inviscid fluid flow is conserved, provided that the fluid is also barotropic and subject only to conservative body forces. This conservation is known as *Kelvin's Theorem* of constant circulation.

ROTATIONAL SPEED

Rotational speed (sometimes called speed of revolution) indicates, for example, how fast a motor is running. Rotational speed is equivalent to angular speed, but with different units. Rotational speed tells how many complete rotations (*i.e. revolutions* or *cycles*) there are per time unit. It is therefore a cyclic frequency, measured in *hertz* (*revolutions per second*) in the SI System. The units revolutions per minute (rpm or 1/min) are more common in everyday life. Angular speed, however, tells the change in angle per time unit, which is measured in radians per second in the SI system. Since there are 2Π radians per cycle, or 360 degrees per cycle, we can convert angular speed to rotational speed by:

$$w_{cyc} = w_{rad} / 2\pi$$

and,

$$W_{cyc} = W_{ang}/360$$

where,

- W_{cyc} is rotational speed (cycles per second)
- W_{rad} is angular speed (radians per second)
- W_{cyc} is angular speed (degrees per second).

For example, a stepper motor might turn exactly one complete revolution each second. Its angular speed is 360 degrees per second (360°/s), or 2Π radians per second (2Π rad/s), while the rotational speed is 60 rpm.

Rotational speed is not to be confused with tangential speed, despite some relation between the two concepts. Imagine a rotating merry-go-round. No matter how close or far you stand from the axis of rotation, your rotational speed will remain constant. However, your tangential speed does not remain constant. If you stand two meters from the axis of rotation, your tangential speed will be double the amount if you were standing only one meter from the axis of rotation.

Stream Function

The stream function is defined for two-dimensional flows of various kinds. The stream function can be used to plot stream lines, which are perpendicular to equipotential lines. In most cases, the stream function is the imaginary part of the complex potential, while the potential function is the real part.

Considering the particular case of fluid dynamics, the difference between the stream function values at any two points gives the volumetric flow rate (or flux) through a line connecting the two points. Note that since streamlines are tangent to the flow, the value of the stream function must be the same along a streamline. If there were a flux across a line, it would necessarily not be tangent to the flow, hence would not be a streamline.

The usefulness of the stream function lies in the fact that the velocity components in the *x*- and *y*- directions at a given point are given by the partial derivatives of the stream function at that point. A stream function may be defined for any flow of dimensions greater than two, however the two dimensional case is generally the easiest to visualize and derive.

Taken together with the velocity potential, the stream function may be used to derive a complex potential for a fluid flow.

Two Dimensional Stream Function

The stream function ψ for a two dimensional flow is defined such that the flow velocity can be expressed as:

$$u = \nabla \times \psi$$

Where $\psi = (0, 0, \psi)$ if the velocity vector u = (u, v, 0).

In Cartesian coordinate system this is equivalent to,

$$u = \frac{\partial \psi}{\partial y}, \qquad v = -\frac{\partial \psi}{\partial x}$$

Where *u* and *v* are the velocities in the $\hat{x}$ and $\hat{y}$ directions, respectively.

This formulation of the stream function satisfies the two dimensional continuity equation:

$$\frac{\partial u}{\partial x} + \frac{\partial v}{\partial y} = 0$$

Derivation of the Two Dimensional Stream Function

Consider two points A and B in two dimensional plane flow. If the distance between these two points is very small: δn, and a stream of flow passes between these points with an average velocity, q perpendicular to the line AB, the volume flow rate per unit thickness, δΨ is given by:

$$\delta\psi = q\delta n$$

As δn → 0, rearranging this expression, we get:

$$q = \frac{\delta\psi}{\partial n}$$

Now consider two dimensional plane flow with reference to a coordinate system. Suppose an observer looks along an arbitrary axis in the direction of increase and sees flow crossing the axis from *right to left*. A sign convention is adopted such that the velocity of the flow is *positive*. However, this sign convention is not universal and thus ought to be taken with caution. A couple of examples ought to clarify this point:

Flow in Cartesian Coordinates

By observing the flow into an elemental square in an x-y Cartesian coordinate system, we have:

$$\delta\psi = -u\delta y$$
$$\delta\psi = v\delta x$$

where u is the velocity parallel to and in the direction of the x-axis, and v is the velocity parallel to and in the direction of the y-axis. In this case, we have a negative velocity of -u since the flow crosses the y-axis from left to right. we have:

$$u = -\frac{\delta\psi}{\partial y}$$
$$v = \frac{\delta\psi}{\partial x}$$

Flow in Polar Coordinates

By observing the flow into an elemental square in an r-θ Polar coordinate system, we have:

$$\delta\psi = -u_r\left(r\delta\theta\right)$$
$$\delta\psi = -v_\theta\delta r$$

where v_r is the velocity parallel to and in the direction of the r-axis, and v_θ is the velocity parallel to and in the direction of the θ-axis. In this case, we have a negative velocity of $-v_r$ since the flow crosses the θ-axis from left to right. Thus as $\delta n \to 0$ and by rearranging, we have:

$$v_r = -\frac{1}{r}\frac{\delta\psi}{\partial\theta}$$
$$v_\theta = \frac{\delta\psi}{\partial r}$$

Continuity: The Derivation

Consider two dimensional plane flow within a Cartesian coordinate system. Continuity states that if we consider flow into an elemental square, the flow into that small element must equal the flow out of that element.

The total flow into the element is given by:

$$\delta\psi_{in} = u\delta y + u\delta x$$

The total flow out of the element is given by:

$$\delta\psi_{out} = \left(u + \frac{\partial u}{\partial x}\delta x\right) + \delta y\left(v + \frac{\partial v}{\partial y}\delta y\right)\delta x$$

Thus we have:

$$\delta\psi_{in} = \delta\psi_{out}$$

$$u\delta y + u\delta x = \left(u + \frac{\partial u}{\partial x}\delta x\right)\delta y + \left(v + \frac{\partial v}{\partial y}\delta y\right)\delta x$$

and simplifying to:

$$\frac{\partial u}{\partial x} + \frac{\partial v}{\partial y} = 0$$

(NB: Whilst we have ignored the earlier sign convention for Ψ, the mathematics ought work out fine but this proof is perhaps slightly tidier and easier to follow).

Substituting the expressions of the stream function into this equation, we have:

$$\frac{\partial^2 \psi}{\partial x \partial y} + \frac{\partial^2 \psi}{\partial y \partial x} = 0$$

Vorticity

In Cartesian coordinates, the stream function can be found from vorticity using Poisson's equation:

$$\vec{\nabla}^2 \psi = -w$$

where,

$$\vec{w} = (0,0,w) \text{and} \vec{w} = \vec{\nabla} \times \vec{v}.$$

ROTATIONAL AND IRROTATIONAL FLOW

From now on, we will only deal with steady, incompressible, inviscid flow. We will neglect all body forces, even gravity.

Previously we have derived an expression for vorticity vector:

$$\vec{\xi} = \vec{D} \times \vec{V}$$

We mentioned that vorticity is a measure of the angular velocity of the fluid elements.

A flow in which there is a lot of vorticity is called a rotational flow. In rotational flows, the fluid elements will rotate as they move from upstream to downstream. This is like a snow ball tumbling down a steep hill side.

A flow in which there is very little vorticity is called an irrotational flow. The fluid particles may move up and down, laterally etc. However, like boxes placed on an escalator, they will maintain their upright position (whatever it is) as they move from point to point.

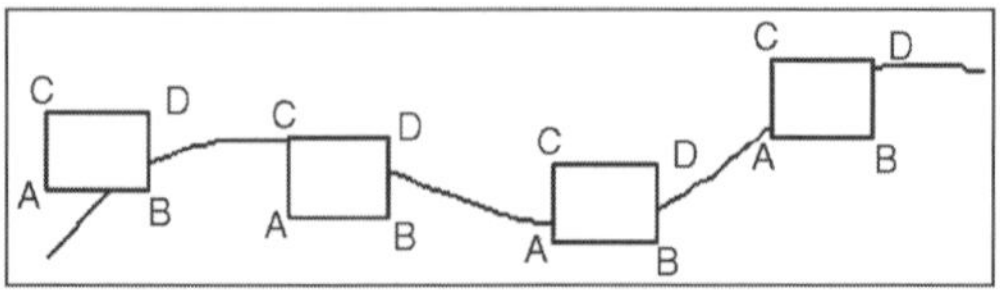

What causes fluid particles to rotate? A number of physical phenomena can.

Phenomena

- Viscous stresses can pull the fluid particles in different direction. Consider a fluid particle right next to a solid surface. The bottom side of the particle may want to stick to the surface due to viscosity. The top side of the particle may be dragged by the particles above in the general direction of the flow. The result is a rotation of the particle.

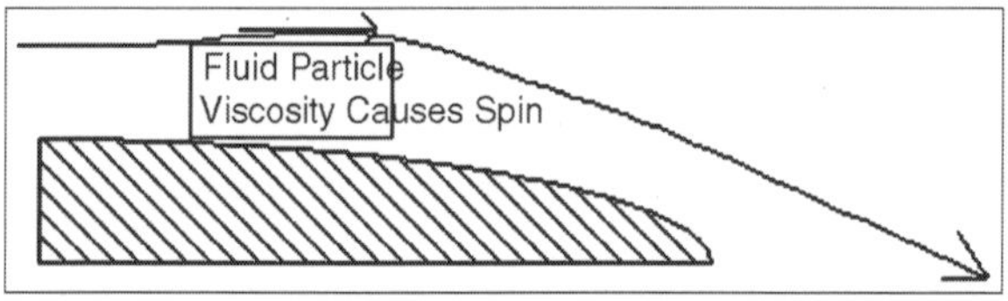

Thus, boundary layers and viscous regions are rotational, and filled with vorticity.

- Velocity differences at the edge of a jet can cause the top part of a fluid particle to move faster than the lower part, or vice versa.

Result*:* Rotation

- Shock waves can slow down the flow. The strength of shock waves depends on the angle they make with the flow. There is usually a curved "bow shock" ahead of a blunt body. Even though the fluid particles ahead of the shock may be uniform, different fluid particles may impact the shock at different angles, and slowed down differently behind the shock wave.
  Result: Rotation.
- Non-uniform body forces (example, forces exerted by a spinning propeller on air particles) can cause different fluid particles to be accelerated differently, causing a non-uniform velocity field.

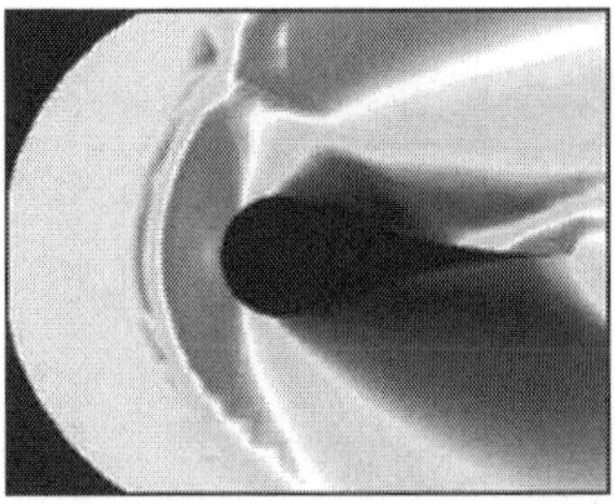

Result: ***Rotation***

Although most flows encountered in nature are rotational, if we stay away from viscous regions (boundary.layers), curved shock waves or non-uniform body forces, in all other regions the flow may be assumed to be irrotational.

Assuming the flow to be irrotational greatly simplifies the mathematics, and the eventual solution of the flow on the computer. We will be invoking the irrotational flow assumption soon.

***Circulation*:**

Consider a closed curve C in space. It is made of tiny straight line segments 'ds' each which may be thought of as a short vector. Then circulation over this contour C is defined as.

$$r=\oint_{c}\vec{V}\bullet d\vec{s}$$

There is a theorem called Stokes Theorem which states that the circulation is related to the vorticity vectors crossing the surface A enclosed by the closed curve C as follows:

$$r=\oint_{c}\vec{V}\bullet d\vec{s}=\iint_{s}\vec{\xi}\bullet d\vec{A}$$

The area A may be assumed to be made of several infinitesimal elements dA. The vector $d\vec{A}$ has the magnitude of the area dA, and is directed normal to dA.

Stokes theorem says that if there is circulation, then there is vorticity, and vice versa. There will be no circulation over a contour C if the area captured by that contour does not contain any vorticity.

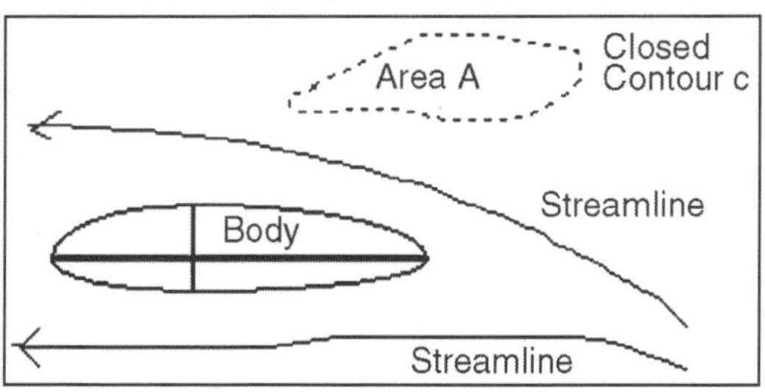

What happens if the contour C encloses the body? Then, since the body has a boundary layer, and the boundary layer contains vorticity, the circulation over any contour that encloses the body will usually be nonzero. The exception is a symmetric body (*e.g.* symmetric airfoil) which contains clockwise vorticity on one side, say the upper side boundary layer and counterclockwise vorticity on the lower side.

Potential Flow

A irrotational flow (that is, a flow in which vorticity is zero) is also called a potential flow. This is because we can define a function called the velocity potential F.

If u, v and w are defined as derivatives of Φ as shown, then vorticity in that flow is zero.

This is because,

$$\vec{\xi}=\vec{\nabla}\times\vec{V}=\begin{vmatrix}\vec{i} & \vec{j} & \vec{k}\\ \dfrac{\partial}{\partial x} & \dfrac{\partial}{\partial y} & \dfrac{\partial}{\partial z}\\ \dfrac{\partial\phi}{\partial x} & \dfrac{\partial\phi}{\partial y} & \dfrac{\partial\phi}{\partial z}\end{vmatrix}$$

$$\vec{i}\left(\frac{\partial^2\phi}{\partial y\partial z}-\frac{\partial^2\phi}{\partial z dy}\right)+\vec{j}\left(\frac{\partial^2\phi}{\partial z\partial x}-\frac{\partial^2\phi}{\partial x dz}\right)+\vec{k}\left(\frac{\partial^2\phi}{\partial x\partial y}-\frac{\partial^2\phi}{\partial y dx}\right)=0$$

If we use equation in continuity equation, we get,

$$\vec{\nabla}\bullet\vec{V}=0$$

$$\text{Or,}\left(\vec{i}\frac{\partial}{\partial x}+\vec{j}\frac{\partial}{\partial x}+\vec{k}\frac{\partial}{\partial z}\right)\bullet\left(\vec{i}\frac{\partial\phi}{\partial x}+\vec{j}\frac{\partial\phi}{\partial x}+\vec{k}\frac{\partial\phi}{\partial z}\right)=\frac{\partial^2\phi}{\partial x^2}+\frac{\partial^2\phi}{\partial z^2}=0$$

or,

$$\tilde{N}^2 f = 0$$

The operator ∇^2 is called the Laplacian operator.

It is simply three second derivatives added together as shown below:

$$\nabla^2=\frac{\partial^2}{\partial x^2}+\frac{\partial^2}{\partial y^2}+\frac{\partial^2}{\partial z^2}$$

Equation above is called Laplace's equation for the velocity potential Φ. Why did we introduce a new variable called Φ? It is because we would rather solve a single linear PDE for Φ (given by equation 7) than solve 4 nonlinear PDEs - conservation of mass, u– v–, w-momentum etc. If we can somehow solve equation, we can find u, v and w from equation. Finally, we can find p from the Bernoulli equation.

We will spend much of our remaining time in this course solving Bernoulli equation. This simple equation may be used to model flow over airfoils, wings, bodies etc. It is truely useful. Of course, it works only for inviscid, incompressible, irrotational flows. It can not be used in regions of large vorticity, *e.g.* boundary layers.

STREAM FUNCTION

A less useful new variable is called stream function y. It exists only in 2-D flows, or axi-symmetric flows (*e.g.* flow around a body of revolution such as a sphere or a bullet).

The stream function is defined as:

$$u = \frac{\partial \psi}{\partial y}$$

$$v = \frac{\partial \psi}{\partial x}$$

We define u and v this way to automatically satisfy the continuity equation for 2-D incompressible flows.

$$\frac{\partial u}{\partial x} + \frac{\partial u}{\partial y} = 0$$

For 2-D flows we only need to worry about the z- component of vorticity given by,

$$\frac{\partial v}{\partial x} - \frac{\partial u}{\partial y}.$$

You may recall that we derived this component by considering the angular velocity of a fluid element in a previous hand-out titled "Pathlines, Streamlines, Vorticity and Bernoulli's Equation". In irrotational flow, vorticity is zero. Setting,

$$\frac{\partial v}{\partial x} - \frac{\partial u}{\partial y}$$

to zero, and using the definitions given in equation we get:

$$\tilde{N}^2 y = 0$$

The above equation is Laplace's equation for stream function. It is linear, and easy enough to solve. After we somehow solve for y we can get u and v from equation. We can finally compute p from Bernoulli's equation.

Equation of a Streamline

Recall that a streamline is defined as:

$$\frac{dx}{u} = \frac{dy}{v}$$

or,

$$udy - vdx = 0$$

Use the definitions of u and v from equation. Then equation becomes:

$$\frac{\partial \psi}{\partial y}dy+\frac{\partial y}{\partial x}\partial x=0$$

The above equation implies that y is constant along a streamline.

DETERMINATION OF THE MOTION OF A PARTICLE

Motion of a charged particle in the simultaneous presence of both electric and magnetic fields has variety of manifestations ranging from straight line motion to the cycloid and other complex motion. Both electric and magnetic fields impart acceleration to the charged particle. But, there is a qualification for magnetic field as acceleration due to magnetic field relates only to the change of direction of motion. Magnetic force being always normal to the velocity of the particle tends to move the particle about a circular trajectory.

On the other hand, electric force is along electric field and is capable to bring about change in both direction and magnitude depending upon the initial direction of velocity of the charged particle with respect to electric field. If velocity and electric vectors are at an angle then the particle follows a parabolic path. One of the important orientations of electric and magnetic fields is referred as "crossed fields".

We use the term "crossed fields" to mean simultaneous presence of electric and magnetic fields at right angle. The behaviour of charged particles such as electrons under crossed fields has important significance in the study of electromagnetic measurement and application (determination of specific charge of electron, cyclotron etc.).

Before we proceed, we should understand that elementary charged particles have mass of the order of 10^{-28} kg or less. Therefore, even small electric or magnetic force is capable to generate very high acceleration of the order of $10^{12}\ m/s^2$ or more. Under proper set up, these particles achieve velocity comparable to speed of light. In order to keep our discussion in the simple classical context, however, we shall confine our discussion limited to the cases which are less complicated and which neglect relativistic effects.

Some of the important applications or phenomena associated with simultaneous presence of two fields include:

- Motion of a charged particle in electric and magnetic fields
- Measurement of specific charge of an electron (J.J.Thomson experiment)
- Acceleration of charged particles (cyclotron)

We have already studied motion of charged particle in individual fields. Here, we shall combine the effects of two fields. Few of the interesting cases are discussed here.

CHARGED PARTICLE IS MOVING ALONG PARALLEL ELECTRIC AND MAGNETIC FIELD

The velocity, electric and magnetic vectors are in in the same direction. Let they are aligned along x-axis. Since magnetic field and velocity vectors are parallel, there is no magnetic force.

$$FM = v_0 qB\sin 0° = 0$$

where v_0 is initial speed of the particle. The charged particle is, however, acted upon by electric field. It is accelerated or decelerated depending on the polarity of charge and direction of electric field. Considering positive charge, the electric force on the charge is given as:

$$FE = qE.$$

The acceleration of particle carrying charge in x-direction is:

$$a_y = \frac{F_E}{m} = \frac{qE}{m}.$$

The displacement along x-axis after time "t" is given by:

$$x = v_0 t + \frac{1}{2} a_y t^2$$

$$x = v_0 t + \frac{qEt^2}{2m}.$$

CHARGE IS MOVING PERPENDICULAR TO PARALLEL ELECTRIC AND MAGNETIC FIELDS

However, electric field in y-direction imparts acceleration in that direction. The particle, therefore, acquires velocity in y-direction and resulting motion is a helical motion. But since particle is accelerated in y–direction, the linear distance between consecutive circular elements of helix increases. In other words, the resulting motion is a helical motion with increasing pitch.

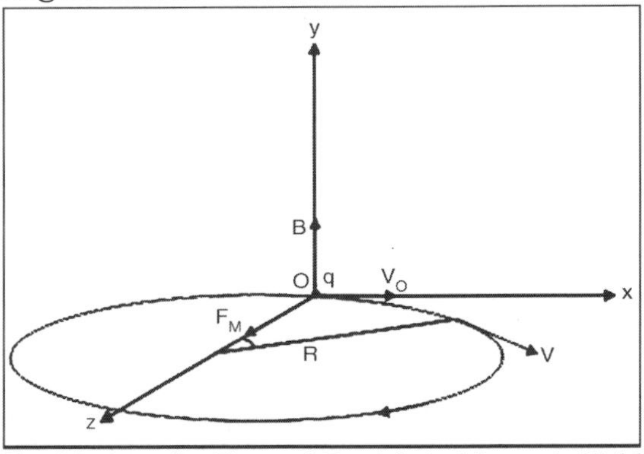

Fig. Motion of a Charged Particle in Magnetic Field

Let electric and magnetic fields align along y-direction and velocity vector is aligned along positive x-direction. Let the charge be positive and initial velocity be $v0$.In this case, velocity and magnetic field vectors are perpendicular to each other. Applying Right hand vector cross product rule, we determine that magnetic force is acting in positive z-direction. If electric field is not present, then the particle revolves along a circle in xz plane as shown in the figure.

Motion of a Charged Particle in Electric and Magnetic Fields

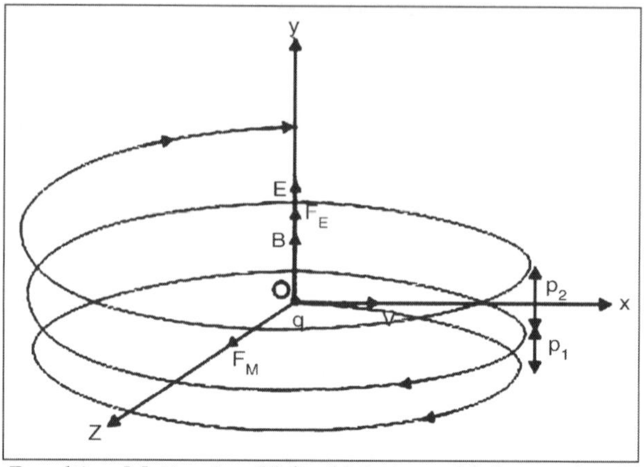

Fig. Resulting Motion is a Helical Motion with Increasing Pitch.

The radius of each of the circular element and other periodic attributes like time period, frequency and angular frequency are same as for the case of circular motion of charged particle in perpendicular to magnetic field.

$$R = -\frac{v}{\alpha B};\ T = \frac{2\pi}{\alpha B};$$
$$\nu = \alpha B / 2\pi;\ \ \omega = \alpha B.$$

Velocity of the Charged Particle

The velocity of the particle in xz plane (as also derived in the module Motion of a charged particle in magnetic field) is:

$$\mathrm{v} = v_x i + v_z j = v_0 \cos \omega t \mid v_0 \sin \omega t k$$

$$\Rightarrow \mathrm{v} = v_0 \cos(\alpha B t) i + v_0 \sin(\alpha B t) k$$

where α is specific charge. We know that magnetic force does not change the magnitude of velocity. It follows then that magnitude of velocity is xy plane is a constant given as:

$$v_x^2 + v_z^2 = v_{xy}^2.$$

But, there is electric field in y-direction. This imparts linear acceleration to the charged particle. As such, the particle which was initially having no component in y direction gains velocity with time as electric field imparts acceleration to the particle in y direction. The velocity components in xz plane, however, remain same.

The acceleration in y-direction due to electric field is:

$$\Rightarrow a_y = \frac{F_E}{m} = \frac{qE}{m} = \alpha E.$$

Since initial velocity in y-direction is zero, the velocity after time t is:

$$\Rightarrow v_y = a_y t = \alpha E t.$$

The velocity of the particle at a time t, therefore, is given in terms of component velocities as:

$$\mathrm{v} = v_x i + v_y j + v_j k$$

$$\Rightarrow \mathrm{v} = v_0 \cos(\alpha B t) i + \alpha E t j + v_0 \sin(\alpha B t) k$$

DISPLACEMENT OF THE CHARGED PARTICLE

Component of displacement of the charged particle in xz plane is given (see module Motion of a charged particle in magnetic field) as:

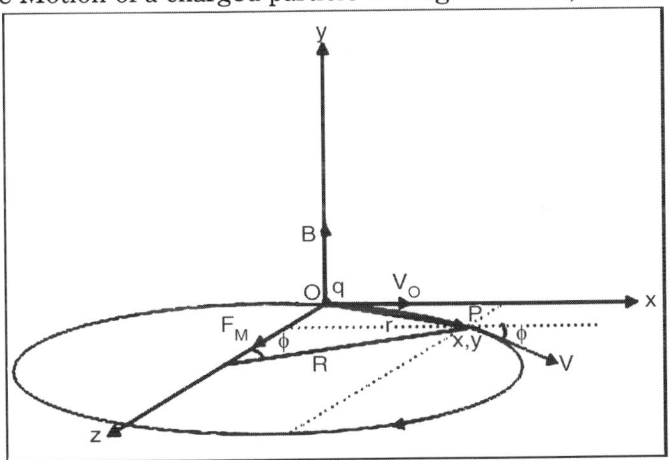

Fig. Displacement of the Charged Particle in xz Plane

$$x = R \sin(\alpha B t) = \frac{v_o}{\alpha B} \sin(\alpha B t)$$

$$z = R\left[1 - \cos(\alpha B t)\right] = \frac{v_0}{\alpha B}\left[1 - \cos(\alpha B t)\right]$$

The motion in y-direction is due to electric force. Let the displacement in this direction be y after time t. Then:

$$y = \frac{1}{2} a_y t^2$$

$$=\frac{1}{2}Et^2.$$

The position vector of the particle after time t is:

$$r = x_i + y_j + z_k$$

$$r = \frac{v_0}{\alpha B}\sin(\alpha Bt)i + \frac{1}{2}\alpha Et^2 j + \frac{v_0}{\alpha B}\left[1-\cos(\alpha Bt)\right]k$$

CHARGE IS PLACED AT REST IN CROSSED ELECTRIC AND MAGNETIC FIELDS

Let electric and magnetic fields are aligned along z and x directions and charge is placed at the origin of coordinate system. Initially, there is no magnetic force as charge is at rest. However, there is electric force, which accelerates the charge in z-direction. As the particle acquires velocity in z-direction, the magnetic force comes into play and tries to rotate the particle in xz plane about a center on x-axis.

Cycloid Motion

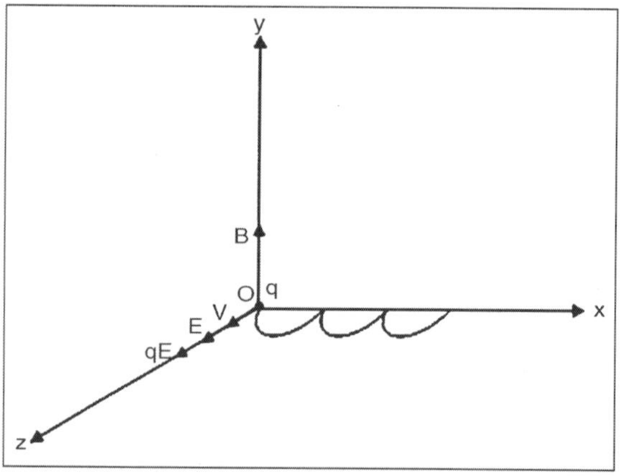

Fig. Cycloid Motion

However, z-component of velocity keeps increasing with time due to electric force in that direction. The magnetic force though draws the charged particle away from z-axis along a curved path. This action of magnetic force is countered by electric force in z-direction. The velocity of charged particle ultimately reduces to zero at x-axis.

This cycle repeats itself forming cycloid motion. The cycloid path is generated by a point on the circumference of a rolling wheel. Here, we shall skip the mathematical derivation and limit ourselves to a descriptive analysis only.

DETERMINATION OF SPECIFIC CHARGE OF ELECTRON (J.J.THOMSON'S EXPERIMENT)

The specific charge of an electron is ratio of charge and mass of electron. The specific charge (α) of electron is measured employing crossed fields on a beam of electrons. The beam of electrons emerging from cathode plate passes through a very narrow slit in anode plate. The electrons are accelerated between cathode and anode due to applied electrical potential V. The kinetic energy of the electron emerging from the slit is given by:

$$\frac{1}{2}mv^2 = eV$$

$$mv^2 = 2eV.$$

where v is the velocity of electron moving into the region of force fields.

Two parallel plates connected to an electric source produce a uniform electric field E from positive plate to negative plate. The electrical force works in the direction opposite to the direction of field E as charge on electron is negative. In the figure, electric field is directed in downward direction. Hence, electric force acts in upward direction.

On the other hand, the magnetic field is produced by a solenoid in a circular region covering the plate as shown in the figure. Its direction is chosen such that it applies a force in the opposite direction to that applied by the electrical field. For a magnetic field into the plane of drawing as shown by uniformly distributed cross signs, the magnetic field applies a upward magnetic force on a positive charge. However, as the charge on the electron is negative, the magnetic force acts in downward direction.

J.J.Thomson's Experimental Set Up

The beam of electrons hit the center of fluorescent screen, producing light as electrons collide with it when electric and magnetic fields are switched off. The point on the fluorescent screen is noted. Then, the electric field is switched on which moves the electron beam in upward direction following a parabolic path. Finally, magnetic field is turned and its magnitude is adjusted such that electric and magnetic forces acting in opposite directions balance each other and the electron is brought to hit original spot as noted earlier for the fields in switched off condition. In this situation:

$$eE = evB$$

$$v = \frac{E}{B}.$$

Note that maximum magnetic force applies as velocity and magnetic field vectors are perpendicular to each other. Substituting expression of v in the kinetic energy equation obtained earlier, we have:

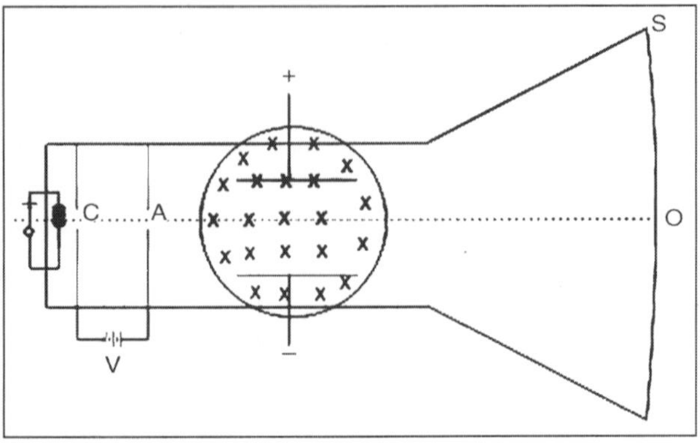

Fig. Measurement of Specific Charge of Electron

$$\frac{mE^2}{B^2} = 2eV$$

$$\alpha = \frac{e}{m} = \frac{E^2}{2VB^2}.$$

All the quantities on the right hand side of the equation are measurable, allowing us to measure the specific charge of electron. As a matter of fact, the determination of specific charge of particles composing cathode ray by J.J.Thomson is considered to be the discovery of electron. It can also be easily inferred that he could determine the nature of charge of an electron by studying direction of deviation (upward or downward) when only either of the fields operate.

In the derivation above, we measure potential difference applied to accelerate particle between cathode and anode. We should, however, realise that we can determine specific charge measuring some other quantities as well. We can measure the deflection of electron beam when either of two fields operates and use the data to determine specific charge of an electron.

Problem: The D.C. voltage applied to accelerate particle between cathode and anode and the d.c. voltage applied to the plates to produce electric field perpendicular to electrons beam are equal in the Thomson's experimental set up. If each of the two D.C. voltages as applied are doubled, then by what factor should the magnetic field be changed to keep the electron beam un-deflected.

Solution: Let V_1, E_1 and B_1 be the potential difference, electric field and magnetic field for un-deflected condition. Then, the specific charge is given by:

$$\alpha = \frac{e}{m} = \frac{E_1^2}{2V_1B_1^2}.$$

Here, the electric field can be expressed in terms of potential difference provided we know the separation between plates. Let the separation be d.

$$E_1 = \frac{V_1}{d}.$$

Putting in the equation above, we have:

$$\alpha = \frac{V_1^2}{2d^2 V_1 B_1^2} = \frac{V_1}{2d^2 B_1^2}.$$

Let B_2 be the new magnetic field when two potential differences as applied are doubled. Here,

$$V_2 = 2V_1$$

Putting new values in the expression for specific charge (note that specific charge of electron is a constant),

$$\alpha = \frac{2V_1}{2d^2 B_2^2}.$$

Combining two equations,

$$\frac{2V_1}{2d^2 B_2^2} = \frac{V_1}{2d^2 B_1^2}$$

$$2V_1 d^2 B_2^2 = 4V_1 d^2 B_1^2$$

$$\frac{B_2^2}{B_1^2} = 2$$

$$\frac{B_2}{B_1} = \sqrt{2}.$$

MEASUREMENT OF DEFLECTION BY MAGNETIC FIELD

Once the magnetic and electric forces are balanced, electric field is switched off and electron beam is allowed to be deviated due to magnetic field. The magnetic force acts always perpendicular to the direction of motion. The particle, therefore, moves along a circular path inside the region of magnetic field. When electron moves out of the magnetic field, it moves along the straight line and hits the fluorescent screen. If R be the radius of curvature, then:

$$\frac{mv^2}{R} = evB.$$

Substituting $v=E/B$ as obtained earlier,

$$\alpha = \frac{e}{m} = \frac{E}{RB^2}.$$

We measure R using geometry. We see that the angles enclosed between pairs of two perpendicular lines are equal. Hence,

J.J.Thomson's Experimental Set Up

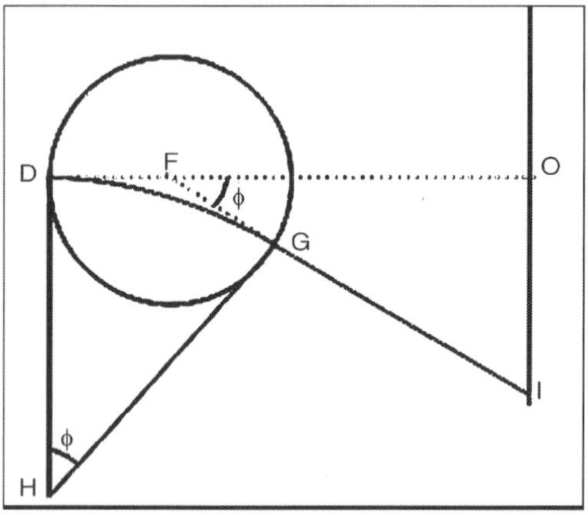

Fig. Deviation due to Only Magnetic Field

$$\varphi = \frac{DG}{R} = \frac{OI}{FO}$$

$$R = \frac{FOXDG}{OI}.$$

We approximate DG to be equal to the width of magnetic region.

MEASUREMENT OF DEFLECTION BY ELECTRIC FIELD

In this case, once the magnetic and electric forces are balanced, electric field is switched off and electron beam is allowed to be deviated due to electric field. The electron beam moving into the region of electric field experiences an upward force. The force in upward (y-direction) imparts acceleration in y-direction. The particle, however, moves with same velocity in x-direction. As a result, path of motion is parabolic. Let the length of plate be L and y be the deflection inside the plate. Then, time to travel through the plate is:

J.J.Thomson's experimental set up

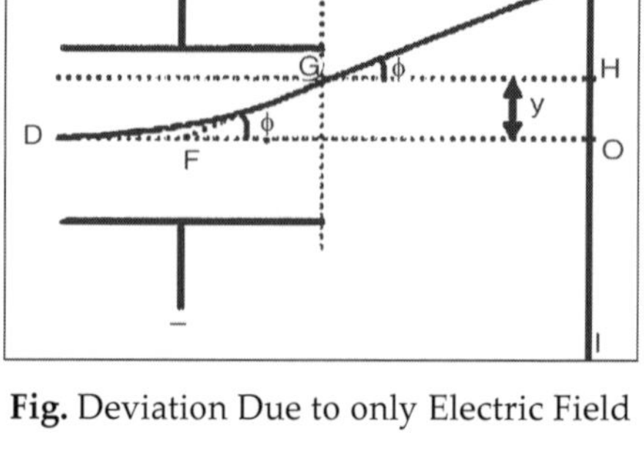

Fig. Deviation Due to only Electric Field

$$t = \frac{L}{v}.$$

and acceleration of the particle in y-direction is:

$$a_y = \frac{F_E}{m} = \frac{eE}{m}$$

The vertical displacement is:

$$y = \frac{1}{2} a_y t^2.$$

Substituting for time and acceleration, we have:

$$y = \frac{1}{2} X \frac{eE}{m} X \frac{L^2}{v^2} = \frac{eEL^2}{2mv^2}.$$

Substituting v=E/B as obtained earlier,

$$y = \frac{eEL^2}{2m} X \left(\frac{B}{E}\right)^2 = \frac{\alpha EL^2}{2} X \left(\frac{B}{E}\right)^2$$

$$\alpha = \frac{e}{m} = \frac{2yE}{B^2 L^2}.$$

It is clear that measuring GH and HI, we can determine angle φ and then y as required.

6

Applications of Thermodynamics

THERMODYNAMICS

The term "compressibility" is also used in thermodynamics to describe the deviance in the thermodynamic properties of a real gas from those expected from an ideal gas.

The compressibility factor is defined as,

$$Z = \frac{p\tilde{V}}{RT}$$

where p is the pressure of the gas, T is its temperature, and $\tilde{V}$ is its molar volume. In the case of an ideal gas, the compressibility factor Z is equal to unity, and the familiar ideal gas law is recovered:

$$p = \frac{RT}{\tilde{V}}$$

Z can, in general, be either greater or less than unity for a real gas. The deviation from ideal gas behaviour tends to become particularly significant (or, equivalently, the compressibility factor strays far from unity) near the critical point, or in the case of high pressure or low temperature. In these cases, a generalized Compressibility chart or an alternative equation of state better suited to the problem must be utilized to produce accurate results.

Earth Sciences

Table. Vertical, Drained Compressibilities

Material	β (m²/N)
Plastic clay	$2\times10^{-6} - 2.6\times10^{-7}$
Stiff clay	$2.6\times10^{-7} - 1.3\times10^{-7}$
Medium-hard clay	$1.3\times10^{-7} - 6.9\times10^{-8}$
Loose sand	$1\times10^{-7} - 5.2\times10^{-8}$
Dense sand	$2\times10^{-8} - 1.3\times10^{-8}$
Dense, sandy gravel	$1\times10^{-8} - 5.2\times10^{-9}$
Rock, fissured	$6.9\times10^{-10} - 3.3\times10^{-10}$
Rock, sound	$<3.3\times10^{-10}$
Water at 25°C (undrained)	4.6×10^{-10}

Compressibility is used in the Earth sciences to quantify the ability of a soil or rock to reduce in volume with applied pressure.

This concept is important for specific storage, when estimating groundwater reserves in confined aquifers. Geologic materials are made up of two portions: solids and voids (or same as porosity).

The void space can be full of liquid or gas. Geologic materials reduces in volume only when the void spaces are reduced, which expel the liquid or gas from the voids.

This can happen over a period of time, resulting in settlement. It is an important concept in geotechnical engineering in the design of certain structural foundations. For example, the construction of high-rise structures over underlying layers of highly compressible bay mud poses a considerable design constraint, and often leads to use of driven piles or other innovative techniques.

BULK MODULUS

The ratio of the compressive or tensile force applied to a substance per unit surface area to the change in volume of the substance per unit volume.

Also known as bulk modulus; compression modulus; hydrostatic modulus; modulus of compression; modulus of volume elasticity.. The bulk modulus (K) of a substance essentially measures the substance's resistance to uniform compression. It is defined as the pressure increase needed to effect a given relative decrease in volume. As an example, suppose an iron cannon ball with bulk modulus 160 GPa (gigapascal) is to be reduced in volume by 0.5 per cent. This requires a pressure increase of $0.005 \times 160\text{GPa}$ = 0.8 GPa. If the cannon ball is subjected to a pressure increase of only 100 MPa, it will decrease in volume by a factor of 100MPa/160GPa = 0.000625, or 0.0625 per cent.

The bulk modulus K can be formally defined by the equation:

$$K = -V\frac{\partial p}{\partial V}$$

where p is pressure, V is volume, and $\partial p/\partial V$ denotes the partial derivative of pressure with respect to volume. The inverse of the bulk modulus gives a substance's compressibility. Other moduli describe the material's response (strain) to other kinds of stress: the shear modulus describes the response to shear, and Young's modulus describes the response to linear strain.

For a fluid, only the bulk modulus is meaningful. For an anisotropic solid such as wood or paper, these three moduli do not contain enough information to describe its behaviour, and one must use the full generalized Hooke's law. Strictly speaking, the bulk modulus is a thermodynamic quantity, and it is necessary to specify how the temperature varies in order to specify a bulk modulus: constant-temperature (K_T), constant-enthalpy (adiabatic K_S), and other variations are possible.

In practice, such distinctions are usually only relevant for gases.

Table. Bulk Modulus Values for Some Example Substances

Water	2.2×10^9 Pa (value increases at higher pressures)
Air	1.42×10^5 Pa (adiabatic bulk modulus)
Air	1.01×10^5 Pa (constant temperature bulk modulus)
Steel	1.6×10^{11} Pa
Solid helium	5×10^7 Pa (approximate)

For a gas, the adiabatic bulk modulus K_S is approximately given by,

$$Ks = kp$$

where,

k is the adiabatic index, sometimes called Y.

p is the pressure.

In a fluid, the bulk modulus K and the density ρ determine the speed of sound c (pressure waves), according to the formula,

$$C = \sqrt{\frac{K}{\rho}}.$$

Solids can also sustain transverse waves, for these one additional elastic modulus, for example the shear modulus, is needed to determine wave speeds.

NEWTONIAN FLUID

A Newtonian fluid (named for Isaac Newton) is a fluid that flows like water—its stress versus rate of strain curve is linear and passes through the origin. The constant of proportionality is known as the viscosity. A simple equation to describe Newtonian fluid behaviour is,

$$\tau = \mu \frac{du}{dx}.$$

where,

T is the shear stress exerted by the fluid ("drag") [Pa]

μ is the fluid viscosity - a constant of proportionality [Pa·s]

$\frac{du}{dx}$ is the velocity gradient perpendicular to the direction of shear [s^{-1}]

In common terms, this means the fluid continues to flow, regardless of the forces acting on it.

For example, water is Newtonian, because it continues to exemplify fluid properties no matter how fast it is stirred or mixed.

Contrast this with a non-Newtonian fluid, in which stirring can leave a "hole" behind (that gradually fills up over time - this behaviour is seen in materials such as pudding, oobleck, or, to a less rigourous extent, sand), or cause the fluid to become thinner, the drop in viscosity causing it to flow more (this is seen in non-drip paints, which brush on easily but become more viscous when on walls).

For a Newtonian fluid, the viscosity, by definition, depends only on temperature and pressure (and also the chemical composition of the fluid if the fluid is not a pure substance), not on the forces acting upon it.

If the fluid is incompressible and viscosity is constant across the fluid, the equation governing the shear stress, in the Cartesian coordinate system, is,

$$\tau_{ij} = \mu\left(\frac{\partial u_i}{\partial x_j} + \frac{\partial u_j}{\partial x_i}\right)$$

with comoving stress tensor $\mathbb{P}$ (also written as σ)

$$\mathbb{P}_{ij} = -p\delta_{ij} = \mu\left(\frac{\partial u_i}{\partial x_j} + \frac{\partial u_j}{\partial x_i}\right)$$

where, by the convention of tensor notation,

T_{ij} is the shear stress on the ith face of a fluid element in the jth direction
u_i is the velocity in the ith direction
x_j is the jth direction coordinate

If a fluid does not obey this relation, it is termed a non-Newtonian fluid, of which there are several types.

Comparison of Newtonian and Non-Newtonian Fluids

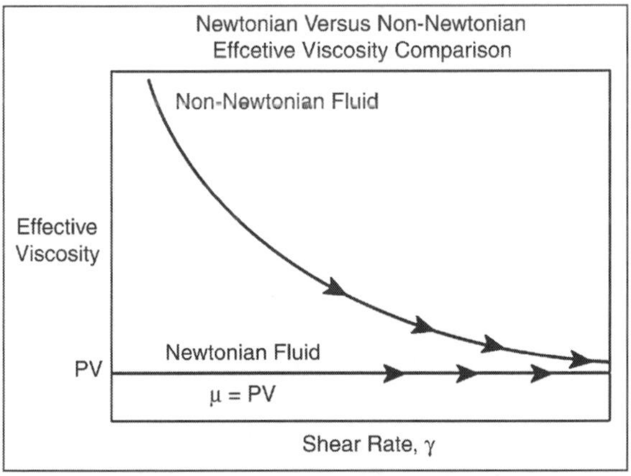

Non-Newtonian fluid. The viscosity of non-Newtonian fluids varies with shear rate, as shown by the red curve in this plot of effective viscosity versus shear rate.

NON-NEWTONIAN FLUID

A non-Newtonian fluid is a fluid in which the viscosity changes with the applied strain rate. As a result, non-Newtonian fluids may not have a well-defined viscosity. Although the concept of viscosity is commonly used to characterize a material, it can be inadequate to describe the mechanical behaviour of a substance, particularly non-Newtonian fluids.

They are best studied through several other rheological properties which relate the relations between the stress and strain tensors under many different flow conditions, such as oscillatory shear, or extensional flow which are measured using different devices or rheometers.

The rheological properties are better studied using tensor-valued constitutive equations, which are common in the field of continuum mechanics.

COMMON EXAMPLES

An inexpensive, non-toxic sample of a non-Newtonian fluid sometimes known as oobleck can be made very easily by adding corn starch (cornflour) to a cup of water. Add the starch in small portions and stir it in slowly. When the suspension nears the critical concentration - becoming like single cream (light cream) in consistency - the so called "shear thickening" property of this non-Newtonian fluid becomes apparent.

The application of force - for example by stabbing the surface with a finger, or rapidly inverting the container holding it - leads to the fluid behaving like a solid rather than a liquid. More gentle treatment, such as slowly inserting a spoon, will leave it in its liquid state. Trying to jerk the spoon back out again, however, will trigger the return of the temporary solid state. A person moving quickly and/or applying sufficient force with their feet can literally walk across such a liquid, as shown here and here Shear thickening fluids of this sort are being researched for bullet resistant body armor, useful for their ability to absorb the energy of a high velocity projectile impact but remain soft and flexible while worn.

A familiar example of the opposite, a shear-thinning fluid, is paint: one wants the paint to flow readily off the brush when it is being applied to the surface being painted, but not to drip excessively.

Classification Types

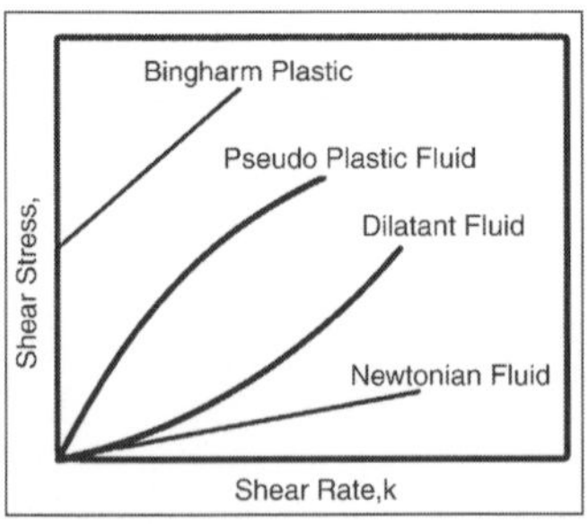

PARAMAGNETISM MICROSCOPIC SYSTEM

The simplest microscopic system which we can analyse using the Boltzmann distribution is one which has only two possible states (there would clearly be little point in analysing a system with only one possible state). Most elements, and some compounds, are paramagnetic: *i.e.*, their constituent atoms, or molecules, possess a permanent magnetic moment due to the presence of one or more unpaired electrons. Consider a substance whose constituent particles contain only one unpaired electron.

Such particles have spin 1/2, and consequently possess an intrinsic magnetic moment m. According to quantum mechanics, the magnetic moment of a spin 1/2 particle can point either parallel or antiparallel to an external magnetic field B. Let us determine the mean magnetic moment $\bar{\mu}_B$ (in the direction of B) of the constituent particles of the substance when its absolute temperature is T. We assume, for the sake of simplicity, that each atom (or molecule) only interacts weakly with its neighbouring atoms. This enables us to focus attention on a single atom, and treat the remaining atoms as a heat bath at temperature T.

Our atom can be in one of two possible states: the (+) state in which its spin points up (*i.e.*, parallel to B), and the (–) state in which its spin points down (*i.e.*, antiparallel to B). In the (+)state, the atomic magnetic moment is parallel to the magnetic field, so that $\mu_\beta = \mu$. The magnetic energy of the atom is $\in_+ = -\mu B$. In the (–) state, the atomic magnetic moment is antiparallel to the magnetic field, so that $\mu_B = -\mu$. The magnetic energy of the atom is $\in_- = \mu_B$.

According to the Boltzmann distribution, the probability of finding the atom in the (+) state is:

$$P_+ = C\exp(-\beta\in_+) = C\exp(\beta\,\mu\,\beta),$$

where C is a constant, and b = $(kT)^{-1}$. Likewise, the probability of finding the atom in the (–) state is,

$$P_- = C\exp(-\beta\in_-) = C\exp(-\beta\,\mu\,B).$$

Clearly, the most probable state is the state with the lowest energy [*i.e.*, the (+) state]. Thus, the mean magnetic moment points in the direction of the magnetic field (*i.e.*, the atom is more likely to point parallel to the field than antiparallel).

It is clear that the critical parameter in a paramagnetic system is,

$$y = \beta\mu B = \frac{\mu B}{kT}.$$

This parameter measures the ratio of the typical magnetic energy of the atom to its typical thermal energy.

If the thermal energy greatly exceeds the magnetic energy then $y << 1$,, and the probability that the atomic moment points parallel to the magnetic field is about the same as the probability that it points antiparallel. In this situation, we expect the mean atomic moment to be small, so that $\bar{\mu}_0 \simeq 0$.

On the other hand, if the magnetic energy greatly exceeds the thermal energy then $y >> 1$, and the atomic moment is far more likely to point parallel to the magnetic field than antiparallel. In this situation, we expect $\bar{\mu}_n \simeq \mu$. Let us calculate the mean atomic moment $\bar{\mu}_n$. The usual definition of a mean value gives,

$$\bar{\mu}_B = \frac{P_+\mu + P_-(-\mu)}{P_+ + P_-} = \mu \frac{\exp(\beta\mu\beta) - \exp(-\beta\mu\beta)}{\exp(\beta\mu\beta) + \exp(-\beta\mu\beta)}.$$

This can also be written,

$$\bar{\mu}_B = \mu \tan h \frac{\mu B}{kT},$$

where the hyperbolic tangent is defined,

$$\tanh y \frac{\exp(y) - \exp(-y)}{\exp(y) + \exp(-y)}.$$

For small arguments $y << 1$,

$$\tanh y \simeq y - \frac{y^3}{3} + \cdots,$$

whereas for large arguments $y >> 1$,

$$\tanh \simeq 1.$$

It follows that at comparatively high temperatures, $kT >> \text{m}B$,

$$\bar{\mu}_B \simeq \frac{\mu^2 B}{kT},$$

whereas at comparatively low temperatures, $kT << \text{m}B$,

$$\bar{\mu}_B \simeq \mu.$$

Suppose that the substance contains N_0 atoms (or molecules) per unit volume. The magnetization is defined as the mean magnetic moment per unit volume, and is given by,

$$\bar{M}_0 = N_0\, \bar{\mu}_B.$$

At high temperatures, $kT >> \text{m}B$, the mean magnetic moment, and, hence, the magnetization, is proportional to the applied magnetic field, so we can write,

$$\bar{M}_0 \simeq \chi B,$$

where χ is a constant of proportionality known as the magnetic susceptibility. It is clear that the magnetic susceptibility of a spin 1/2 paramagnetic substance takes the form,

$$\chi = \frac{N_0 \mu^2}{kT}.$$

The fact that $\chi \propto T^{-1}$ is known as Curie's law, because it was discovered experimentally by Pierre Curie at the end of the nineteenth century. At low temperatures, $kT << mB$,

$$\bar{M}_0 \to N_0 \mu,$$

so the magnetization becomes independent of the applied field. This corresponds to the maximum possible magnetization, where all atomic moments are lined up parallel to the field. The breakdown of the $\bar{M}_0 \propto B$ law at low temperatures (or high magnetic fields) is known as saturation. The above analysis is only valid for paramagnetic substances made up of spin one-half (J = 1/2) atoms or molecules. However, the analysis can easily be generalized to take account of substances whose constituent particles possess higher spin (*i.e.*, J > 1/2).

Compares the experimental and theoretical magnetization versus field-strength curves for three different substances made up of spin 3/2, spin 5/2, and spin 7/2 particles, showing the excellent agreement between the two sets of curves. Note that, in all cases, the magnetization is proportional to the magnetic field-strength at small field-strengths, but saturates at some constant value as the field-strength increases.

The previous analysis completely neglects any interaction between the spins of neighbouring atoms or molecules. It turns out that this is a fairly good approximation for paramagnetic substances. However, for ferromagnetic substances, in which the spins of neighbouring atoms interact very strongly, this approximation breaks down completely. Thus, the above analysis does not apply to ferromagnetic substances.

Mean Values

Consider a system in contact with a heat reservoir. The systems in the representative ensemble are distributed over their accessible states in accordance with the Boltzmann distribution. Thus, the probability of occurrence of some state t with energy E_r is given by,

$$P_r = \frac{\exp(-\beta E_r)}{\sum_r \exp(-\beta E_r)}.$$

The mean energy is written,

$$\bar{E} = \frac{\sum_r \exp(-\beta E_r) E_r}{\sum_r \exp(-\beta E_r)},$$

where the sum is taken over all states of the system, irrespective of their energy. Note that,

$$\sum_r \exp(-\beta E_r) E_r = -\sum_r \frac{\partial}{\partial \beta} \exp(-\beta E_r) = -\frac{\partial Z}{\partial \beta},$$

where,

$$Z-\sum_r \exp(-\beta E_r).$$

It follows that,

$$\bar{E}=-\frac{1}{Z}\frac{\partial Z}{\partial\beta}=-\frac{\partial \ln Z}{\partial\beta}.$$

The quantity Z, which is defined as the sum of the Boltzmann factor over all states, irrespective of their energy, is called the partition function. We have just demonstrated that it is fairly easy to work out the mean energy of a system using its partition function.

In fact, as we shall discover, it is easy to calculate virtually any piece of statistical information using the partition function.

Let us evaluate the variance of the energy. We know that,

$$\overline{(\Delta E)^2}=\overline{E^2}-\bar{E}^2.$$

Now, according to the Boltzmann distribution,

$$\overline{E^2}=\frac{\sum_r \exp(-\beta E_r)\, E_r{}^2}{\sum_r \exp(-\beta E_r)}.$$

However,

$$\sum_r \exp(-\beta E_r)E_r{}^2=-\frac{\partial}{\partial\beta}\left[\sum_r \exp(-E_r)E_r\right]=\left(-\frac{\partial}{\partial\beta}\right)^2\left[\sum_r \exp(-\beta E_r)\right].$$

Hence,

$$\overline{E^2}=\frac{1}{Z}\frac{\partial^2 Z}{\partial\beta^2}.$$

We can also write,

$$\overline{E^2}=\frac{\partial}{\partial\beta}\left(\frac{1}{Z}\frac{\partial Z}{\partial\beta}\right)+\frac{1}{Z^2}\left(\frac{\partial Z}{\partial\beta}\right)^2=-\frac{\partial\bar{E}}{\partial\beta}+\bar{E}^2,$$

where use has been made of above equations. It follows from previous equation that,

$$\overline{(\Delta E)^2}=-\frac{\partial\bar{E}}{\partial\beta}=\frac{\partial^2 \ln Z}{\partial\beta^2}.$$

Thus, the variance of the energy can be worked out from the partition function almost as easily as the mean energy. Since, by definition, a variance can never be negative, it follows that $\partial\bar{E}/\partial\beta \le 0$, or, equivalently, $\partial\bar{E}/\partial T \ge 0$. Hence, the mean energy of a system governed by the Boltzmann distribution always increases with temperature.

Suppose that the system is characterized by a single external parameter x (such as its volume).

The generalization to the case where there are several external parameters is obvious. Consider a quasi-static change of the external parameter from x to $x + dx$. In this process, the energy of the system in state r changes by,

$$\delta E_r = -\frac{\partial E_r}{\partial x} dx.$$

The macroscopic work $đW$ done by the system due to this parameter change is,

$$đW = \frac{\sum_r \exp(-E_r)(-\partial E_r / \partial x \, dx)}{\sum_r \exp(-\beta E_r)}.$$

In other words, the work done is minus the average change in internal energy of the system, where the average is calculated using the Boltzmann distribution. We can write,

$$\sum_r \exp(-\beta E_r) \frac{\partial E_r}{\partial x} = -\frac{1}{\beta}\frac{\partial}{\partial x}\left[\sum_r \exp(-\beta E_r)\right] = -\frac{1}{\beta}\frac{\partial Z}{\partial x},$$

which gives,

$$đW = \frac{1}{\beta Z}\frac{\partial Z}{\partial x} dx = \frac{1}{\beta}\frac{\partial \ln Z}{\partial x} dx.$$

We also have the following general expression for the work done by the system,

$$đW = \overline{X} dx,$$

where,

$$\overline{X} = -\overline{\frac{\partial E_r}{\partial x}}$$

is the mean generalized force conjugate to x. It follows that,

$$\overline{X} = -\frac{1}{\beta}\frac{\partial \ln Z}{\partial x}.$$

Suppose that the external parameter is the volume, so $x = V$. It follows that,

$$đW = \overline{p} dV = \frac{1}{\beta}\frac{\partial \ln Z}{\partial V} dV.$$

and,

$$\overline{p} = \frac{1}{\beta}\frac{\partial \ln Z}{\partial V}.$$

Since the partition function is a function of b and V(the energies E_r depend on V), it is clear that the above equation relates the mean pressure $\bar{p}$ to T (via $\beta = 1/kT$) and V. In other words, the above expression is the equation of state. Hence, we can work out the pressure, and even the equation of state, using the partition function.

Partition Functions

It is clear that all important macroscopic quantities associated with a system can be expressed in terms of its partition function Z. Let us investigate how the partition function is related to thermodynamical quantities. Recall that Z is a function of both b and x (where x is the single external parameter). Hence, $Z = Z(\beta, x)$, and we can write,

$$d\ln Z = \frac{\partial \ln Z}{\partial x} dx + \frac{\partial \ln Z}{\partial x} d\beta.$$

Consider a quasi-static change by which x and bchange so slowly that the system stays close to equilibrium, and, thus, remains distributed according to the Boltzmann distribution. If follows from Eqs. that,

$$d\ln Z = \beta \,đW - \bar{E}\, d\beta.$$

The last term can be rewritten,

$$d\ln Z = \beta đW - d(\bar{E}\,\beta) + \beta\, d\bar{E},$$

giving,

$$d(\ln Z + \beta\bar{E}) = \beta(đW + d\bar{E}) \equiv \beta đQ.$$

The above equation shows that although the heat absorbed by the system $đQ$ is not an exact differential, it becomes one when multiplied by the temperature parameter β. This is essentially the second law of thermodynamics.

In fact, we know that,

$$dS = \frac{đQ}{T}.$$

Hence,

$$S \equiv k(\ln Z + \beta\bar{E}).$$

This expression enables us to calculate the entropy of a system from its partition function.

Suppose that we are dealing with a system $A^{(0)}$ consisting of two systems A and $A¢$which only interact weakly with one another. Let each state of A be denoted by an index tand have a corresponding energy E_r. Likewise, let each state of $A¢$ be denoted by an index s and have a corresponding energy $E's$.

A state of the combined system $A^{(0)}$ is then denoted by two indices t and s. Since A and $A¢$ only interact weakly their energies are additive, and the energy of state ts is,

$$E_{\tau s}^{(0)} = E_r + E_s'.$$

By definition, the partition function of $A^{(0)}$ takes the form,

$$Z^{(0)} = \sum_{\tau,s} \exp[-\beta\, E_{\tau s}^{(0)}]$$

$$= \sum_{\tau,s} \exp(-\beta\,[E_r + E_s'])$$

$$= \left[\sum_{\tau} \exp(-\beta\, E_r)\right]\left[\sum_{s} \exp(-\beta\, E_s')\right].$$

Hence,

$$Z^{(0)} = ZZ',$$

giving,

$$\ln Z^{(0)} = \ln Z + \ln Z',$$

where Z and $Z¢$ are the partition functions of A and $A¢$, respectively. It follows from equation. that the mean energies of $A^{(0)}$, A, and $A¢$ are related by,

$$\bar{E}^{(0)} = \bar{E} + \bar{E}'.$$

It also follows from equation that the respective entropies of these systems are related via:

$$S^{(0)} = S + S'.$$

Hence, the partition function tells us that the extensive thermodynamic functions of two weakly interacting systems are simply additive.

It is clear that we can perform statistical thermodynamical calculations using the partition function Zinstead of the more direct approach in which we use the density of states Ω. The former approach is advantageous because the partition function is an unrestricted sum of Boltzmann factors over all accessible states, irrespective of their energy, whereas the density of states is a restricted sum over all states whose energies lie in some narrow range.

In general, it is far easier to perform an unrestricted sum than a restricted sum. Thus, it is generally easier to derive statistical thermodynamical results using Zrather than Ω, although Ω has a far more direct physical significance than Z.

BOLTZMANN DISTRIBUTIONS

We have gained some understanding of the macroscopic properties of the air around us. For instance, we know something about its internal energy and specific heat capacity. How can we obtain some information about the statistical properties of the molecules which make up air? Consider a specific molecule: it constantly collides with its immediate neighbour molecules, and occasionally bounces off the walls of the room.

These interactions "inform" it about the macroscopic state of the air, such as its temperature, pressure, and volume. The statistical distribution of the molecule over its own particular microstates must be consistent with this macrostate. In other words, if we have a large group of such molecules with similar statistical distributions, then they must be equivalent to air with the appropriate macroscopic properties. So, it ought to be possible to calculate the probability distribution of the molecule over its microstates from a knowledge of these macroscopic properties.

We can think of the interaction of a molecule with the air in a classroom as analogous to the interaction of a small system A in thermal contact with a heat reservoir A'.

The air acts like a heat reservoir because its energy fluctuations due to any interactions with the molecule are far too small to affect any of its macroscopic parameters. Let us determine the probability P_r of finding system A in one particular microstate $r$of energy E_r when it is thermal equilibrium with the heat reservoir A'.

As usual, we assume fairly weak interaction between A and $A¢$, so that the energies of these two systems are additive. The energy of A is not known at this stage. In fact, only the total energy of the combined system $A^{(0)} = A + A¢$ is known. Suppose that the total energy lies in the range $E^{(0)}$ to $E^{(0)} + dE$. The overall energy is constant in time, since $A^{(0)}$ is assumed to be an isolated system, so,

$$E_r + E' = E^{(0)},$$

where $E¢$ denotes the energy of the reservoir $A¢$. Let $\Omega'(E¢)$ be the number of microstates accessible to the reservoir when its energy lies in the range E' to $E' + \delta E$ Clearly, if system A has an energy E_r then the reservoir A' must have an energy close to $E' = E^{(0)} - E_r$. Hence, since A is in one definite state (*i.e.*, state r), and the total number of states accessible to $A¢$ is $\Omega'(E^{(0)} - E_r)$ it follows that the total number of states accessible to the combined system is simply $\Omega'(E^{(0)} - E_r)$. The principle of equal a priori probabilities tells us the the probability of occurrence of a particular situation is proportional to the number of accessible microstates. Thus,

$$P_r = C' \Omega' (E^{(0)} - E_r),$$

where C' is a constant of proportionality which is independent of r. This constant can be determined by the normalization condition

$$\sum_r P_r = 1,$$

where the sum is over all possible states of system A, irrespective of their energy.

Let us now make use of the fact that system A is far smaller than system $A¢$. It follows that $E_r << E^{(0)}$, so the slowly varying logarithm of P_r can be Taylor expanded about $E¢ = E^{(0)}$. Thus,

$$\ln \text{Pr} = \ln C' + \ln \Omega'(E^{(0)}) - \left[\frac{\partial \ln \Omega'}{\partial E'}\right]_0 E_r + \cdots,$$

Note that we must expand ln P_r, rather than P_r itself, because the latter function varies so rapidly with energy that the radius of convergence of its Taylor series is far too small for the series to be of any practical use. The higher order terms in previous equation can be safely neglected, because $E_r <<$ $E^{(0)}$. Now the derivative

$$\left[\frac{\partial \ln \Omega'}{\partial E'}\right]_0 \equiv \beta$$

is evaluated at the fixed energy $E¢ = E^{(0)}$, and is, thus, a constant independent of the energy E_r of A. In fact, we know that this derivative is just the temperature parameter $\beta = (kT)^{-1}$ characterizing the heat reservoir $A¢$. Hence, equation becomes,

$$\ln P_r = \ln C' + \ln \Omega'(E^{(0)}) - \beta E_r,$$

giving,

$$P_r = C \exp(-\beta E_r),$$

where C is a constant independent of τ. The parameter C is determined by the normalization condition, which gives,

$$C^{-1} = \sum_r \exp\,(-\beta E_r),$$

so that the distribution becomes,

$$P_r = \frac{\exp(-\beta E_r)}{\sum_r \exp(-\beta E_r)}.$$

This is known as the Boltzmann probability distribution, and is undoubtably the most famous result in statistical physics.

The Boltzmann distribution often causes confusion. People who are used to the principle of equal a priori probabilities, which says that all microstates are equally probable, are understandably surprised when they come across the Boltzmann distribution which says that high energy microstates are markedly less probable then low energy states. However, there is no need for any confusion. The principle of equal a priori probabilities applies to the whole system, whereas the Boltzmann distribution only applies to a small part of the system.

The two results are perfectly consistent. If the small system is in a microstate with a comparatively high energy E_r then the rest of the system (*i.e.*, the reservoir) has a slightly lower energy $E¢$ than usual (since the overall energy is fixed). The number of accessible microstates of the reservoir is a very strongly increasing function of its energy.

It follows that when the small system has a high energy then significantly less states than usual are accessible to the reservoir, and so the number of microstates accessible to the overall system is reduced, and, hence, the configuration is comparatively unlikely.

The strong increase in the number of accessible microstates of the reservoir with increasing $E¢$ gives rise to the strong (*i.e.*, exponential) decrease in the likelihood of a state t of the small system with increasing E_r. The exponential factor $\exp(-\beta E_r)$ is called the Boltzmann factor.

The Boltzmann distribution gives the probability of finding the small system A in one particular state t of energy E_r. The probability $P(E)$ that A has an energy in the small range between E and $E + \delta E$ is just the sum of all the probabilities of the states which lie in this range. However, since each of these states has approximately the same Boltzmann factor this sum can be written,

$$P(E) = C\Omega(E)\exp(-\beta E),$$

where $\Omega(E)$is the number of microstates of A whose energies lie in the appropriate range. Suppose that system $A$is itself a large system, but still very much smaller than system $A¢$. For a large system, we expect $\Omega(E)$ to be a very rapidly increasing function of energy, so the probability $P(E)$ is the product of a rapidly increasing function of E and another rapidly decreasing function (*i.e.*, the Boltzmann factor).

This gives a sharp maximum of $P(E)$ at some particular value of the energy. The larger system A, the sharper this maximum becomes. Eventually, the maximum becomes so sharp that the energy of system A is almost bound to lie at the most probable energy. As usual, the most probable energy is evaluated by looking for the maximum of ln P, so,

$$\frac{\partial \ln P}{\partial E} = \frac{\partial \ln \Omega}{\partial E} - \beta = 0,$$

giving,

$$\frac{\partial \ln \Omega}{\partial E} = \beta.$$

Of course, this corresponds to the situation in which the temperature of A is the same as that of the reservoir. This is a result which we have seen before. Note, however, that the Boltzmann distribution is applicable no matter how small system A is, so it is a far more general result than any we have previously obtained.

CONSIDERATIONS OF THERMODYNAMICS

The thermodynamic state equations of fluids are often used in supplement for the solution of flow problems. However in the present text only "simple fluids", *i.e.* for homogenous liquids and gases for which the thermodynamic state can be expressed by a relation between pressure, temperature and density are considered. The statements are possible for substantial as well as for field quantities, *i.e.* it holds,

$$P_{\Re} = f\,(T_{\Re}, \mathrm{r}_{\Re}) \text{ or } P = f\,(T, \mathrm{r})$$

Thermodynamic state equations are known to be:

$$\frac{P_{\Re}}{\rho_{\Re}} = RT_{\Re} \text{ (thermodynamically ideal gases)}$$

$\rho_{\Re}$ = const (thermodynamically ideal liquids)

If one defines with $\alpha_{\Re} = P_{\Re}, T_{\Re}, \rho_{\Re}, e_{\Re}$ and with $\alpha = P, T, \rho, e...$, the following relation holds, when the fluid element $\Re$ is located at the time t at position x_i:

$$\alpha_{\Re}(t) = \alpha(x_i, t) \rightsquigarrow \frac{d\alpha_{\Re}}{dt} = \frac{\partial \alpha}{\partial t} + U_i \frac{\partial \alpha}{\partial x_i} = \frac{D\alpha}{Dt}$$

The second part of equation states how temporal changes of substantial, thermodynamic quantities can be computed from the substantial derivative of corresponding field quantities.

In addition to the above introduced thermodynamic state properties $P_{\Re}$, $T_{\Re}$, $\rho_{\Re}$, $e_{\Re}$..., other state properties can be defined whose introduction is of advantage in certain thermodynamic considerations. Some of them are:

- Specific volume: $u_{\Re} = \dfrac{1}{\rho_{\Re}}$
- Enthalpy: $h_{\Re} = e_{\Re} + P_{\Re} u_{\Re}$
- Free energy: $f_{\Re} = e_{\Re} - T_{\Re} s_{\Re}$ (Helmholtz potential)
- Free enthalpy: $g_{\Re} = h_{\Re} - T_{\Re} s_{\Re}$ (Gibb potential)

Accordingly, it is possible to apply certain mathematical operators in order to define "new" thermodynamic quantities. However, their introduction makes sense only when advantages result from the introduction into the thermodynamic considerations that are to be carried out with the new quantities.

In one of the above definitions for thermodynamic potentials, the entropy was used whose definition is given by a differential relation:

$$T_{\Re} ds_{\Re} = de_{\Re} + P_{\Re} du_{\Re} \quad \text{(Gibb relation)}$$

Integrating one obtains:

$$s_{\Re} = s(\Re)_0 + \int_{(e_{\Re})_0}^{e_{\Re}} \frac{1}{T_{\Re}} de_{\Re} + \int_{(u_{\Re})_0}^{u_{\Re}} \frac{P_{\Re}}{T_{\Re}} du_{\Re}$$

The above relations can be understood as identical definition of equations for the entropy s of a fluid element. When employing the relation one obtains with,

$$\frac{ds_{\Re}}{dt}=\frac{Ds}{Dt}\quad \frac{de_{\Re}}{dt}=\frac{De}{dt}\quad \text{and}\quad \frac{du_{\Re}}{dt}=u_{\Re}\frac{\partial U_i}{\partial x_i}3$$

the following relation:

$$T_{\Re}\frac{ds_{\Re}}{dt}=\frac{de_{\Re}}{dt}+P_{\Re}\frac{du_{\Re}}{dt}\quad \rightsquigarrow =T\frac{Ds}{Dt}=\frac{De}{dt}+P\frac{1}{\rho}\frac{\partial U_i}{\partial x_i}$$

or:

$$T\left(\frac{\partial s}{\partial t}+U_i\frac{\partial s}{\partial x_i}\right)=\left(\frac{\partial e}{\partial t}+U_i\frac{\partial e}{\partial x_i}\right)+\frac{P}{\rho}\left(\frac{\partial U_i}{\partial x_i}\right)$$

When one applies the mass conservation equation of the law of differential equations, it can be rearranged further: $\left(\frac{\partial U_i}{\partial x_i}\right)=-\frac{1}{\rho}\frac{D}{Dt}$ of the mass conservation equation inserted in yields:

$$T\frac{Ds}{Dt}=\frac{De}{Dt}-\frac{P}{\rho^2}\frac{D\rho}{Dt}$$

From this relation further relations can be derived that are of importance for fluid mechanics considerations, *e.g.* for $s_{\Re}$ = const.:

$$\left(\frac{De}{Dt}\right)_s=\frac{P}{\rho^2}\frac{D\rho}{Dt}\rightsquigarrow\left(\frac{de_{\Re}}{dt}\right)_{s\Re}=\frac{P_{\Re}}{\rho_{\Re}^2}=\left(\frac{de_{\Re}}{d\rho_{\Re}}\right)_{s\Re}$$

For $\rho_{\Re}$ = const. or $u_{\Re}$ = const it holds:

$$T\left(\frac{Ds}{Dt}\right)_\rho=\left(\frac{De}{Dt}\right)_\rho\rightsquigarrow T_{\Re}=\left(\frac{de_{\Re}}{ds_{\Re}}\right)_{\rho\Re}$$

It holds further for $e_{\Re}$ = const.:

$$T\left(\frac{Ds}{Dt}\right)_e=-\frac{P}{\rho^2}\left(\frac{D\rho}{Dt}\right)_e\rightsquigarrow P_{\Re}=T_{\Re}\left(\frac{\partial s_{\Re}}{\partial \upsilon_{\Re}}\right)_{e\Re}=-T_{\Re}\rho_{\Re}^2\left(\frac{\partial s_{\Re}}{\partial \rho_{\Re}}\right)_{e\Re}$$

Further significant relations known from thermodynamics are needed in the following:

- Specific heat capacity of a fluid at constant volume

$$c_\upsilon=\left(\frac{\partial e_{\Re}}{\partial T_{\Re}}\right)_{\upsilon\Re}=T_{\Re}\left(\frac{\partial s_{\Re}}{\partial T_{\Re}}\right)_{\upsilon\Re}$$

- Specific heat capacity of a fluid at constant pressure

$$c_p=\left(\frac{\partial h_{\Re}}{\partial T_{\Re}}\right)_{P\Re}=T_{\Re}\left(\frac{\partial s_{\Re}}{\partial T_{\Re}}\right)_{P\Re}$$

where is $h_{\Re}=e_{\Re}+P_{\Re}\upsilon_{\Re}$

- Isothermal compressibility coeffcient

$$\alpha = -\frac{1}{\upsilon_{\Re}}\left(\frac{\partial \upsilon_{\Re}}{\partial P_{\Re}}\right)_{T\Re} = \frac{1}{\rho\Re}\left(\frac{\partial \rho_{\Re}}{\partial P_{\Re}}\right)_{T\Re}$$

- Thermal expansion coeffcient

$$\beta = \frac{1}{\upsilon_{\Re}}\left(\frac{\partial \upsilon_{\Re}}{\partial T_{\Re}}\right)_{P\Re} = -\frac{1}{\rho_{\Re}}\left(\frac{\partial \rho_{\Re}}{\partial T_{\Re}}\right)_{P\Re}$$

When one takes into consideration that the following relation holds

$$d_{r\hat{A}} = \left(\frac{\partial \rho_{\Re}}{\partial T_{\Re}}\right)_{P\Re} dT_{\Re} + \left(\frac{\partial \rho_{\Re}}{\partial P_{\Re}}\right)_{T\Re} dP_{\Re}$$

the following relation can be formulated for all fluids: $\frac{1}{\rho_{\Re}} d\rho_{\Re} = \alpha dP_{\Re} = \beta dT_{\Re}$

Or rearranged in terms of field variables:

$$\frac{1}{\rho}\frac{D\rho}{Dt} = \alpha\frac{DP}{Dt} - \beta\frac{DT}{Dt}$$

This relation allows the statement that all fluids of constant density, *i.e.* fluids having the property $\rho_{\Re}$ = const. or $(D\rho/Dt) = 0$, can be designated as incompressible. They react neither to pressure variations ($\alpha = 0$) nor to temperature variations ($\beta = 0$) for changes in volume or density.

For any fluids, the difference of the heat capacities results:

$$(c_p - c_\upsilon) = \frac{T_{\Re}}{\rho_{\Re}}\frac{\beta^2}{\alpha} = -\frac{T_{\Re}}{\rho_{\Re}} \cdot \beta\left(\frac{\partial P_{\Re}}{\partial T_{\Re}}\right)_{\rho\Re} = T_{\Re}\left(\frac{\partial P_{\Re}}{\partial T_{\Re}}\right)_{\rho\Re}\left(\frac{\partial \upsilon_{\Re}}{\partial T_{\Re}}\right)_{P\Re}$$

The above general relations can now be employed to derive the special relations that hold for the two thermodynamic ideal fluids that receive special attention in this manual, namely the ideal gas and the ideal fluid. For an ideal gas holds,

$$\frac{P_{\Re}}{\rho_{\Re}} = RT_{\Re} \text{ consequently } \frac{P}{\rho} = RT$$

and in addition $\left(\frac{\partial e_{\Re}}{\partial \upsilon_{\Re}}\right)_{T\Re} = \left(\frac{\partial e_{\Re}}{\partial P_{\Re}}\right)_{T\Re} = 0$ and c_υ = const, *i.e.* the internal energy of an ideal gas is a pure function of the temperature. For the isothermal compressibility coeffcient a and the thermal expansion coeffcient β yields,

$$\alpha = \frac{1}{\rho_{\Re}}\left(\frac{\partial e_{\Re}}{\partial P_{\Re}}\right)_{T\Re} = \frac{1}{\rho_{\Re}}\frac{1}{RT_{\Re}} = \frac{1}{P_{\Re}}$$

$$\beta = -\frac{1}{\rho_{\Re}}\left(\frac{\partial \rho_{\Re}}{\partial T_{\Re}}\right)_{P\Re} = -\frac{1}{\rho_{\Re}}\left(\frac{P_{\Re}}{RT_{\Re}^2}\right) = \frac{1}{T_{\Re}}$$

and thus for the difference of the specific heat capacities:

$$c_p - c_u = \frac{T_{\Re}}{\rho_{\Re}} \frac{1}{T_{\Re}^2} P_{\Re} = \frac{P_{\Re}}{\rho_{\Re} T_{\Re}} = R$$

It can further be formulated for the change in density:

$$\frac{d\rho_{\Re}}{\rho_{\Re}} = \frac{dP_{\Re}}{P_{\Re}} - \frac{dT_{\Re}}{T_{\Re}}$$

As a further fluid of significance, we introduce the thermodynamic ideal liquid that distinguishes itself by $\alpha = 0$ and $\beta = 0$, *i.e.*

$$\frac{d\rho_{\Re}}{\rho_{\Re}} = 0 \quad \text{(fluid of constant Density)}$$

For the difference of the heat capacities it can be computed:

$$c_p - c_\upsilon = -\frac{T_{\Re}}{\rho_{\Re}} \beta \left(\frac{\partial P_{\Re}}{\partial T_{\Re}} \right)_{\rho_{\Re}} \quad \text{bei } \beta = 0 \rightsquigarrow c_p = c_\upsilon .$$

When one employs the Gibb relation, $d\rho_{\Re} = d\upsilon_{\Re} = 0$ yields:

$$\left(\frac{\partial s_{\Re}}{\partial e_{\Re}} \right)_{\rho_{\Re}} = \frac{1}{T_{\Re}}$$

Because $s_{\Re} \neq s_{\Re}(\rho_{\Re})$, the pressure in an ideal liquid is not taken into account as a thermodynamic quantity. It exists as a mechanical quantity, however, for an ideal fluid it is not part of a thermodynamic state equation.

A further physical property of a fluid, which is of significance when dealing with some of the flow problems presented in this book, is the velocity propagation of small pressure perturbations, the so-called sound velocity:

$$c^2 = \left(\frac{\partial P_{\Re}}{\partial \rho_{\Re}} \right)_{s_{\Re}}$$

This quantity is defined as pressure modification due to the changes in density, the entropy being maintained constant, *i.e.* the propagation of small acoustic perturbations takes place isentropically.

When one takes into account the following relation for the cited sequence of partial derivations,

$$1 = \left(\frac{\partial T_{\Re}}{\partial e_{\Re}} \right)_{\rho_{\Re}} \left(\frac{\partial e_{\Re}}{\partial s_{\Re}} \right)_{\rho_{\Re}} \left(\frac{\partial s_{\Re}}{\partial T_{\Re}} \right)_{\rho_{\Re}}$$

and if one considers,

$$T_{\Re} = \left(\frac{\partial e_{\Re}}{\partial s_{\Re}} \right)_{\rho_{\Re}} c_\upsilon \left(\frac{\partial e_{\Re}}{\partial T_{\Re}} \right)_{\rho_{\Re}}$$

it holds:

$$c_{\upsilon} = T_{\Re}\left(\frac{\partial s_{\Re}}{\partial T_{\Re}}\right)_{\rho_{\Re}}$$

When taking into account the Maxwell relations,

$$\left(\frac{\partial T_{\Re}}{\partial \rho_{\Re}}\right)_{s_{\Re}} = \frac{1}{\rho_{\Re}^2}\left(\frac{\partial P_{\Re}}{\partial s_{\Re}}\right)_{\rho_{\Re}}$$

$$\left(\frac{\partial \rho_{\Re}}{\partial s_{\Re}}\right)_{T\Re} = -\rho_{\Re}^2\left(\frac{\partial T_{\Re}}{\partial P_{\Re}}\right)_{\rho_{\Re}}$$

it can be formulated,

$$\left(\frac{\partial s_{\Re}}{\partial T_{\Re}}\right)_{\rho_{\Re}} \cdot \left(\frac{\partial T_{\Re}}{\partial \rho_{\Re}}\right)_{\rho_{\Re}} \left(\frac{\partial \rho_{\Re}}{\partial s_{\Re}}\right)_{T_{\Re}} = -1$$

and it can also be written for the quantity c_{υ}

$$c_{\upsilon} = -T_{\Re}\left(\frac{\partial \rho_{\Re}}{\partial T_{\Re}}\right)_{s_{\Re}} \left(\frac{\partial s_{\Re}}{\partial \rho_{\Re}}\right)_{T_{\Re}}$$

Similarly it can be derived:

$$c_{p} = -T_{\Re}\left(\frac{\partial P_{\Re}}{\partial T_{\Re}}\right)_{s_{\Re}} \left(\frac{\partial s_{\Re}}{\partial P_{\Re}}\right)_{T_{\Re}}$$

For the relation of the heat capacities can be formulated,

$$k = \frac{c_p}{c_{\upsilon}} \frac{\left(\frac{\partial P_{\Re}}{\partial T_{\Re}}\right)_{s_{\Re}} \left(\frac{\partial s_{\Re}}{\partial P_{\Re}}\right)_{T_{\Re}}}{\left(\frac{\partial \rho_{\Re}}{\partial T_{\Re}}\right)_{s_{\Re}} \left(\frac{\partial s_{\Re}}{\partial \rho_{\Re}}\right)_{T_{\Re}}} = \left(\frac{\partial P_{\Re}}{\partial \rho_{\Re}}\right)_{s_{\Re}} \left(\frac{\partial \rho_{\Re}}{\partial P_{\Re}}\right)_{T_{\Re}}$$

Under consideration of the definition equations for the sound velocity and for the isothermal compressibility coeffcient one obtains:

$$c^2 = \kappa\left(\frac{\partial P_{\Re}}{\partial \rho_{\Re}}\right)_{T_{\Re}} = \frac{\kappa}{\rho_{\Re}\alpha}$$

For the ideal gas with $\alpha = \frac{1}{P_{\Re}}$, considering the ideal gas equation, yields,

$$c = \sqrt{\kappa R_{\Re} T_{\Re}}$$

For an ideal liquid with $\alpha \to 0$ holds:

$$c \to \infty$$

i.e. for a fluid with constant density an infinite large sound velocity results.

HARMONIC OSCILLATORS

Our proof of the equipartition theorem depends crucially on the classical approximation. To see how quantum effects modify this result, let us examine a particularly simple system which we know how to analyse using both classical and quantum physics: *i.e.*, a simple harmonic oscillator. Consider a one-dimensional harmonic oscillator in equilibrium with a heat reservoir at temperature T. The energy of the oscillator is given by,

$$E = \frac{p^2}{2m} + \frac{1}{2}kx^2,$$

where the first term on the right-hand side is the kinetic energy, involving the momentum p and mass m, and the second term is the potential energy, involving the displacement x and the force constant k. Each of these terms is quadratic in the respective variable. So, in the classical approximation the equipartition theorem yields:

$$\frac{\overline{p^2}}{2m} = \frac{1}{2}kT,$$

$$\frac{1}{2}k\overline{x^2} = \frac{1}{2}kT.$$

That is, the mean kinetic energy of the oscillator is equal to the mean potential energy which equals (1/2)kT. It follows that the mean total energy is,

$$\overline{E} = \frac{1}{2}kT + \frac{1}{2}kT = kT.$$

According to quantum mechanics, the energy levels of a harmonic oscillator are equally spaced and satisfy,

$$E_n = (n + 1/2)\,\hbar\omega,$$

where n is a non-negative integer, and,

$$\omega = \sqrt{\frac{k}{m}}.$$

The partition function for such an oscillator is given by,

$$Z = \sum_{n=0}^{\infty} \exp(-\beta E_n) = \exp[-(1/2)\beta\hbar\omega] \sum_{n=0}^{\infty} \exp(-n\beta\hbar\omega).$$

Now,

$$\sum_{n=0}^{\infty} \exp(-n\beta\hbar\omega) = 1 + \exp(-\beta\hbar\omega) + \exp(-2\,\beta\hbar\omega) + \cdots$$

is simply the sum of an infinite geometric series, and can be evaluated immediately,

$$\sum_{n=0}^{\infty} \exp(-n\beta\hbar\omega) = \frac{1}{1-\exp(-\beta\hbar\omega)}.$$

Thus, the partition function takes the form,

$$Z = \frac{\exp[-(1/2)\beta\hbar\omega]}{1-\exp(-\beta\hbar\omega)},$$

and,

$$\ln Z = -\frac{1}{2}\beta\hbar\omega - \ln[1-\exp(-\beta\hbar\omega)].$$

The mean energy of the oscillator is given by equation,

$$\bar{E} = -\frac{\partial}{\partial\beta}\ln Z = -\left[-\frac{1}{2}\hbar\omega - \frac{\exp(-\beta\hbar\omega)\,\hbar\omega}{1-\exp(-\beta\hbar\omega)}\right],$$

or,

$$\bar{E} = \hbar\omega\left[-\frac{1}{2} + \frac{1}{\exp(\beta\hbar\omega)-1}\right].$$

Consider the limit,

$$\beta\hbar\omega = \frac{\hbar\omega}{kT} \ll 1,$$

in which the thermal energy kT is large compared to the separation $\hbar\omega$ between the energy levels. In this limit,

$$\exp(\beta\hbar\omega) \simeq 1 + \beta\hbar\omega,$$

so,

$$\bar{E} \simeq \hbar\omega\left[\frac{1}{2} + \frac{1}{\beta\hbar\omega}\right] \simeq \hbar\omega\left[\frac{1}{\beta\hbar\omega}\right],$$

giving,

$$\bar{E} \simeq \frac{1}{\beta} = kT.$$

Thus, the classical result holds whenever the thermal energy greatly exceeds the typical spacing between quantum energy levels.

Consider the limit,

$$\beta\hbar\omega\frac{\hbar\omega}{kT} >> 1,$$

in which the thermal energy is small compared to the separation between the energy levels. In this limit,

$$\exp(\beta\hbar\omega) >> 1,$$

Thus, if the thermal energy is much less than the spacing between quantum states then the mean energy approaches that of the ground-state (the so-called zero point energy). Clearly, the equipartition theorem is only valid in the former limit, where $kT >> \hbar\omega$, and the oscillator possess sufficient thermal energy to explore many of its possible quantum states.

THERMODYNAMIC ENTROPY

The first opportunity for confusion arises when we introduce the term *entropy* into the mix. Clausius invented the term in 1865. He had noticed that a certain ratio was constant in reversible, or ideal, heat cycles. The ratio was heat exchanged to absolute temperature. Clausius decided that the conserved ratio must correspond to a real, physical quantity, and he named it "entropy". Surely not every conserved ratio corresponds to a real, physical quantity. Historical accident has introduced this term to science. On another planet there could be physics without the concept of entropy. It completely lacks intuitive clarity. Even the great physicist James Clerk Maxwell had it backward for a while . Nevertheless, the term has stuck.

The *American Heritage Dictionary* gives as the first definition of entropy, "For a closed system, the quantitative measure of the amount of thermal energy not available to do work." So it's a negative kind of quantity, the opposite of available energy.

Today, it is customary to use the term entropy to state the second law: Entropy in a closed system can never decrease. As long as entropy is defined as unavailable energy, the paraphrasing just given of the second law is equivalent to the earlier ones above. In a closed system, available energy can never increase, so (because energy is conserved) its complement, entropy, can never decrease.

A familiar demonstration of the second law is the flow of heat from hot things to cold, and never vice-versa. When a hot stone is dropped into a bucket of cool water, the stone cools and the water warms until each is the same temperature as the other. During this process, the entropy of the system increases. If you know the initial temperatures of the stone and the water, and the final temperature of the water, you can quantify the entropy increase in calories or joules per degree.

You may have noticed the words "closed system" a couple of times above. Consider simply a black bucket of water initially at the same temperature as the air around it. If the bucket is placed in bright sunlight, it will absorb heat from the sun, as black things do. Now the water becomes warmer than the air around it, and the available energy has increased.

Has entropy*decreased*? Has energy that was previously unavailable become available, in a closed system? No, this example is only an apparent violation of the second law. Because sunlight was admitted, the local system was not closed; the energy of sunlight was supplied from outside the local system. If we consider the larger system, including the sun, available energy has decreased and entropy has increased as required.

LOGICAL ENTROPY

Entropy is also used to mean disorganization or disorder. J. Willard Gibbs, the nineteenth century American theoretical physicist, called it "mixedupness." The *American Heritage Dictionary* gives as the second definition of entropy, "a measure of disorder or randomness in a closed system." Again, it's a negative concept, this time the opposite of organization or order. The term came to have this second meaning thanks to the great Austrian physicist Ludwig Boltzmann.

In Boltzmann's day, one complaint about the second law of thermodynamics was that it seemed to impose upon nature a preferred direction in time. Under the second law, things can only go one way. This apparently conflicts with the laws of physics at the molecular level, where there is no preferred direction in time — an elastic collision between molecules would look the same going forward or backward. In the 1880s and 1890s, Boltzmann used molecules of gas as a model, along with the laws of probability, to show that there was no real conflict. The model showed that, no matter how it was introduced, heat would soon become evenly diffused throughout the gas, as the second law required.

The model could also be used to show that two different kinds of gasses would become thoroughly mixed. The reasoning he used for mixing is very similar to that for the diffusion of heat, but there is an important difference. In the diffusion of heat, the entropy increase can be measured with the ratio of physical units, joules per degree. In the mixing of two kinds of gasses already at the same temperature, if no energy is dissipated, the ratio of joules per degree — thermodynamic entropy — is irrelevant. The non-dissipative mixing process is related to the diffusion of heat only by analogy . Nevertheless, Boltzmann used a factor, k, now called Boltzmann's constant, to attach physical units to the latter situation. Now the word entropy has come to be applied to the simple mixing process, too. (Of course, Boltzmann's constant has a legitimate use — it relates the average kinetic energy of a molecule to its temperature.)

Entropy in this latter sense has come to be used in the growing fields of information science, computer science, communications theory, etc. The story is often told that in the late 1940s, John von Neumann, a pioneer of the computer age, advised communication-theorist Claude E. Shannon to start using the term "entropy" when discussing information because "no one knows what entropy really is, so in a debate you will always have the advantage" (6). Richard Feynman knew there is a difference between the two meanings of entropy. He discussed thermodynamic entropy is called "Entropy" using physical units, joules per degree, and over a dozen equations.

So we now have to talk about what we mean by disorder and what we mean by order. Suppose we divide the space into little volume elements. If we have black and white molecules, how many ways could we distribute them among the volume elements so that white is on one side and black is on the other? On the other hand, how many ways could we distribute them with no restriction on which goes where? Clearly, there are many more ways to arrange them in the latter case. We measure "disorder" by the number of ways that the insides can be arranged, so that from the outside it looks the same. *The logarithm of that number of ways is the entropy.* The number of ways in the separated case is less, so the entropy is less, or the "disorder" is less.

This is Boltzmann's model again. Notice that Feynman does not use Boltzmann's constant. He assigns no physical units to this kind of entropy, just a number (a logarithm.) Notice another thing. The "number of ways" can only be established by first artificially dividing up the space into little volume elements. This is not a small point. In every real physical situation, counting the number of possible arrangements requires an arbitrary parceling.

There is, however, nothing to tell us how fine the [parceling] should be. Entropies calculated in this way depend on the size-scale decided upon, in direct contradiction with thermodynamics in which entropy changes are fully objective. This sort of entropy is clearly different. Physical units do not pertain to it, and (except in the case of digital information) an arbitrary convention must be imposed before it can be quantified. To distinguish this kind of entropy from thermodynamic entropy, let's call it *logical entropy*.

In spite of the important distinction between the two meanings of entropy, the rule as stated above for thermodynamic entropy seems to apply non-etheless to the logical kind: entropy in a closed system can never decrease. And really, there would be nothing mysterious about this law either. It's similar to saying *things never organize themselves*. (The original meaning of organize is "to furnish with organs.") Only this rule has little to do with thermodynamics.

It is true that crystals and other regular configurations can be formed by unguided processes. And we are accustomed to saying that these configurations are "organized." But crystals have not been spontaneously "furnished with organs."

The correct term for such regular configurations is "ordered." The recipe for a crystal is already present in the solution it grows from the crystal lattice is prescribed by the structure of the molecules that compose it. The formation of crystals is the straightforward result of chemical and physical laws that do not evolve and that are, compared to genetic programmes, very simple.

The rule that things never organize themselves is also upheld in our everyday experience. Without someone to fix it, a broken glass never mends. Without maintenance, a house deteriorates. Without management, a business fails. Without new software, a computer never acquires new capabilities.

Charles Darwin understood this universal principle. It's common sense. That's why he once made a note to himself pertaining to evolution, "Never use the words higher or lower". Even today, if you assert that a human is more highly evolved than a flatworm or an *amoeba*, there are darwinists who'll want to fight about it. They take the position, apparently, that evolution has not necessarily shown a trend towards more highly organized forms of life, just different forms.

LIFE IS ORGANIZATION

Seen in retrospect, evolution as a whole doubtless had a general direction, from simple to complex, from dependence on to relative independence of the environment, to greater and greater autonomy of individuals, greater and greater development of sense organs and nervous systems conveying and processing information about the state of the organism's surroundings, and finally greater and greater consciousness. You can call this direction progress or by some other name.

Life is organization. From prokaryotic cells, eukaryotic cells, tissues, and organs, to plants and animals, families, communities, ecosystems, and living planets, life is organization, at every scale. The evolution of life is the increase of biological organization, if it is anything. Clearly, if life originates and makes evolutionary progress without organizing input from outside, then something has organized itself. Logical entropy in a closed system has decreased. This is the violation that people are getting at, when they say that life violates the second law of thermodynamics. This violation, the *decrease* of logical entropy in a closed system, must happen continually in the darwinian account of evolutionary progress.

Most darwinists just ignore this staggering problem. When confronted with it, they seek refuge in the confusion between the two kinds of entropy. Entropy [logical] has not decreased, they say, because the system is not closed. Energy such as sunlight is constantly supplied to the system. If you consider the larger system that includes the sun, entropy has increased, as required.

HEAT ENGINE

Heat engine is defined as a device that converts heat energy into mechanical energy or more exactly a system which operates continuously and only heat and work may pass across its boundaries. The operation of a heat engine can best be represented by a thermodynamic cycle. Some examples are: Otto, Diesel, Brayton, Stirling and Rankine cycles.

Forward Heat Engine

- LTER = Low Temperature Energy Reservoir
- HTER = High Temperature Energy Reservoir

A forward heat engine has a positive work output such as Rankine or Brayton cycle. Applying the first law of thermodynamics to the cycle gives:

$$Q1 - Q2 - W = 0$$

The second law of thermodynamics states that the thermal efficiency of the cycle, η, has an upper limit (the thermal efficiency of the Carnot cycle), *i.e.*

It can be shown that:

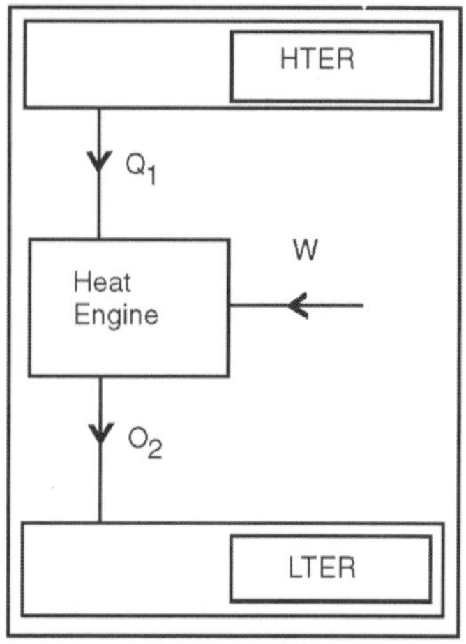

$$Q1 > W$$

which means that it is impossible to convert the whole heat input to work and

$$Q2 > 0$$

which means that a minimum of heat supply to the cold reservoir is necessary.

Reverse Heat Engine

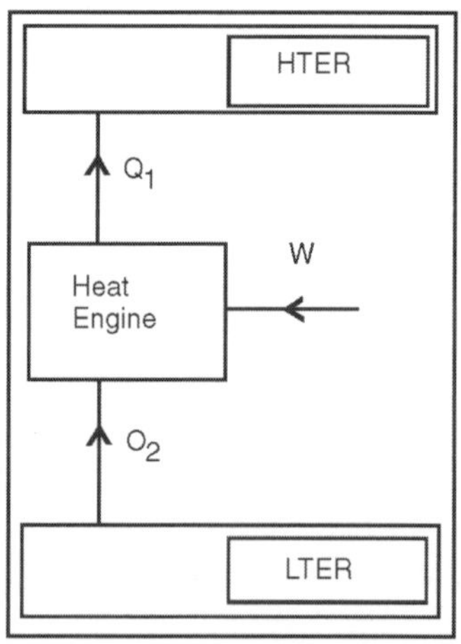

- LTER= Low Temperature Energy Reservoir
- HTER= High Temperature Energy Reservoir

A reverse heat engine has a positive work input such as heat pump and refrigerator.

Applying the first law of thermodynamics to the cycle gives:

$$-Q1 + Q2 + W = 0$$

In case of a reverse heat engine the second law of thermodynamics is as follows: It is impossible to transfer heat from a cooler body to a hotter body without any work input *i.e.*

$$W > 0$$

REVERSIBILITY

In a reversible process the state of working fluid and system's surroundings can be restored to the original ones. This requires that the working fluid goes through a continuous series of equilibrium states.

A reversible process should satisfy the following criteria:

- No internal or mechanical friction is allowed.
- The temperature and pressure difference between the working fluid and its surroundings should be infinitely small.

There are no truly reversible processes in practice. The real processes are called irreversible. However, there are some processes that can be assumed internally reversible with good approximation, such as processes in cylinders with reciprocating piston. The working fluid is always in an equilibrium state in internally reversible process. But the surroundings undergo a state change that can never be restored.

Some processes may not be assumed internally reversible, such as processes in turbo machinery. The irreversibility of these processes are due to the high degree of turbulence of the working fluid. A reversible process between two states may be shown by a continuous curve on any diagram of properties. Different points on the curve represent the intermediate states.

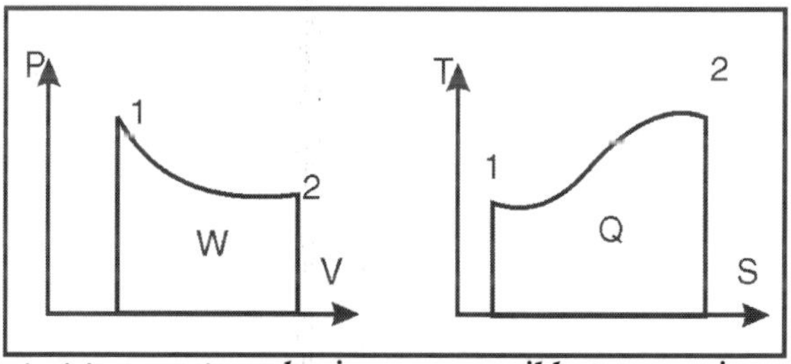

The work input to a system during a reversible process is:

W= Marked area on the P-V diagram.

and the heat supplied to a system during a reversible process is:

Q= Marked area on the T-s diagram.

Intermediate states for an irreversible process is indeterminate, therefore these processes are often shown by a dotted line joining the initial and final states.

CARNOT CYCLE

By using the second law of thermodynamics it is possible to show that no heat engine can be more efficient than a reversible heat engine working between two fixed temperature limits. This heat engine is known as Carnot cycle and consists of the following processes:

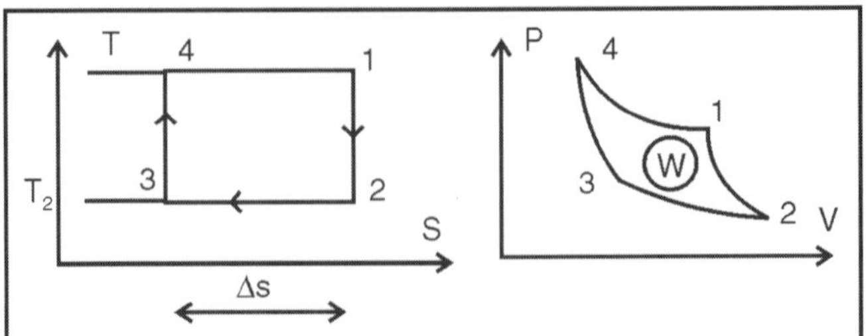

T-s and P-V diagrams for carnot cycle.

- 1 to 2: Isentropic expansion
- 2 to 3: Isothermal heat rejection
- 3 to 4: Isentropic compression
- 4 to 1: Isothermal heat supply

The supplied heat to the cycle per unit mass flow is:

$$Q1 = T1D\ s$$

The rejected heat from the cycle per unit mass flow is:

$$Q2 = T2D\ s$$

By applying the first law of thermodynamics to the cycle, we obtain:

$$Q1 - Q2 - W = 0$$

And the thermal efficiency of the cycle will be:

$$= hW/Q1 = 1 - T2/T1$$

Due to mechanical friction and other irreversiblities no cycle can achieve this efficiency. The gross work output of cycle, *i.e.* the work done by the system is:

$$Wg = W41 + W12$$

and work ratio is defined as the ratio of the net work, W, to the gross work output, Wg, *i.e.*

$$W/Wg$$

The Carnot cycle has a low work ratio. Although this cycle is the most efficient system for power generation theoretically, it can not be used in practice. There are several reasons such as low work ratio, economical aspects and practical difficulties

7

Partial Differential Equation in Fluids

PARTIAL DIFFERENTIAL EQUATION

THE REYNOLDS NUMBER

The Navier-Stokes equation is a rather complicated-looking nonlinear partial differential equation.

In order to make some progress in understanding the possible solutions, it is useful to do some dimensional analysis and to make some estimates.

For this purpose, we want to compare the inertial terms on the left hand side to the viscous terms on the right hand side. Let's assume that the flow is steady, so that $\partial v / \partial t = 0$.

Let *U* be a characteristic speed of the flow (usually taken to be the free-stream speed far from the object), and let *L* be some characteristic dimension of the flow (the typical size of an object in the fluid, say). Then for the inertial term,

$$\rho\left|(v \cdot \nabla)v\right| \sim \frac{\rho U^2}{L}.$$

For the viscous term,

$$\eta\left|\nabla^2 v\right| \sim \frac{\eta U}{L^2}.$$

Both of these quantities have the same dimension, so we can define a dimensionless number, the *Reynolds number* Re, by taking the ratio:

$$\begin{aligned} Re &= \frac{\text{intertial terms}}{\text{viscous terms}} \\ &= \frac{\rho U^2 / L}{\eta U / L^2} \\ &= \frac{UL}{v}, \end{aligned}$$

with n = η/ρ the kinematic viscosity.

To get an idea of the magnitude of the Reynolds number, consider a car moving at a speed of 55 mi/hr (= 24 m/s). An appropriate value of L would be the typical width of the car, which I would estimate to be about 1.5 m. The Reynolds number for the flow of air about the car is then,

$$\mathrm{Re} = \frac{\left(24\,\mathrm{m}\cdot\mathrm{s}^{-1}\right)\left(1.5\,\mathrm{m}\right)}{0.15\times 10^{-4}\,\mathrm{m}^2\,\mathrm{s}^{-1}} = 2.4\times 10^{6}.$$

This is fairly typical—large Reynolds number flow tends to be the rule in most situations.

VELOCITY OF A FLUID

When we deform a solid, so that it is *strained*, we know that the solid will exert a restoring force which opposes the strain; for small strains the restoring force is proportional to the strain, and we have the familiar *Hooke's Law*. Real fluids also oppose strains; however, in a fluid it is not the *amount* of strain which is important but the *rate* at which the strain is produced.

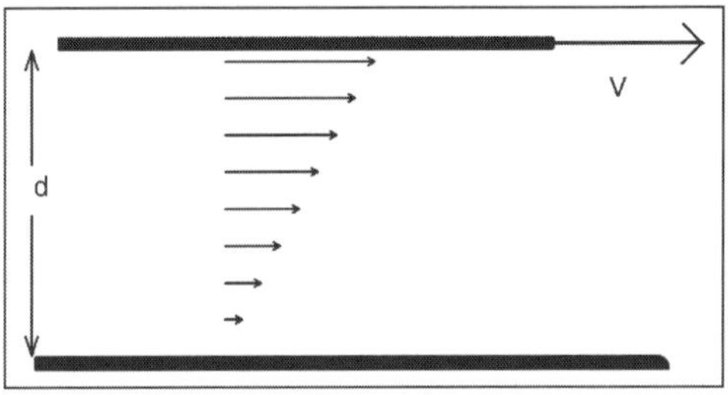

Fig. Viscous Drag between Two Plates; The Bottom Plate is Held Fixed.

For instance, if we have to draw a utensil through our favourite viscous fluid (maple syrup, say), we find that it is easy for slow motions of the utensil, but more difficult if the utensil is moved rapidly. A simple fluid flow is shown in Figure.

If the force per unit area of the top plate is measured, we would find,

$$\frac{F}{A} = \eta\frac{v}{d},$$

with d the separation between the plates. This essentially defines the viscosity η (sometimes called the *shear viscosity*). Note that we haven't derived this law—it is a consequence of experimental observation, just like Ohm's law, Fourier's law and Fick's law. A fluid which responds to a shear stress F/A in this manner is called a *Newtonian fluid* ; it has the property that the viscosity is independent of velocity. Most of the fluids that we will be interested in will be Newtonian (air, water). Non-Newtonian fluids (Silly-Putty, paint, polymeric fluids) are more complicated and beyond the scope of the present description.. Also listed in the kinematic viscosities $v \equiv \eta / \rho$. Recall that 1 pascal equals 1 $\mathrm{kg{\cdot}m^{-1}s^{-2}}$ and note that the viscosity of water is, conveniently, 1 mPa·s = 1 $\mathrm{g{\cdot}m^{-1}s^{-1}}$; it is common to quote *relative viscosities*, which are viscosities relative to water, or in mPa·s, and also to use the cgs unit of viscosity, $g \cdot cm^{-1} s^{-1}$, called the poise (after Poiseuille) and equal to 0.1 Pa·s.

Before incorporating viscosity into the equations of fluid mechanics, let's take a moment to discuss some of the properties of viscosity. Viscosity does not depend upon pressure in a significant way, but we know from personal experience that it does depend upon temperature—pancake syrup becomes less viscous after we take it out of the refrigerator and let it warm up.

Generally, the viscosity of liquids *decreases* with increasing temperature. Gases, on the other hand, have viscosities which generally *increase* with increasing temperature. Why the difference?

Viscosity is fundamentally a consequence of the intermolecular interactions in the fluid. Now viscosity is the transfer of momentum from one part of a fluid to an adjacent part, and in a dilute gas this transfer will be more effective if the collisions are very energetic. Since the mean kinetic energy of a molecule is $(3/2)k_B T$, higher temperatures imply a larger viscosity for a gas.

A liquid, on the other hand, is a strongly interacting collection of molecules, and the idea of a binary collision is meaningless. Microscopically the molecules in a liquid do have some short range order—a given molecule is surrounded by a "cage" of other molecules.

Momentum transfer in the liquid will involve the movement of these cages around one another. Increasing the temperature causes the cages to jiggle a bit more, allowing them to slip more easily past one another, thereby *reducing* the momentum transfer between adjacent bits of the fluid, and thereby reducing the viscosity of the liquid. More detail than this requires *kinetic theory*.

NAVIER-STOKES EQUATION

The velocity of a fluid will vary in a complicated way in space; however, we can still apply the above definition of viscosity to a bit of fluid of thickness Δy with an infinitesimal area ΔAy= ΔzΔx. The shear stress across the thickness Δy is,

$$\frac{\Delta F_x}{\Delta A_y} = \eta \frac{\partial v_x}{\partial_y}.$$

We can further generalize and define the *xy* component of the *viscous stress tensor* as,

$$\tau_{xy} = \eta \left(\frac{\Delta v_x}{\Delta y} + \frac{\partial v_y}{\partial x} \right).$$

The notation is made clearer in the figure below. This definition ensures that for a *uniform* rotation of the fluid there are no viscous stresses.

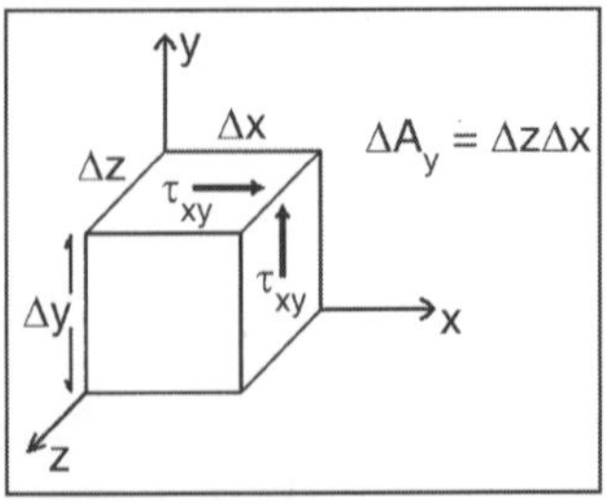

Fig. Stresses Acting on a Small Cube

We can apply the same arguments as above to other flow configurations, with the result that the viscous stress tensor τ_{ij} is given by,

$$\tau_{ij} = \eta\left(\frac{\Delta v_i}{\Delta x_i} + \frac{\partial v_j}{\partial x_i}\right),$$

where $i,j=(x,y,z)$. Now in a fluid we can have internal stresses due to the fluid pressure p in addition to the stress due to viscous forces; the total stress tensor σ_{ij} in the fluid is then,

$$\sigma_{ij} = -p\delta_{ij} + \tau_{ij},$$

where δ_{ij} is the *Kronecker delta*, defined such that $\delta_{ij} = 1$ if $i=j$, and $\delta_{ij} = 0$ otherwise.

Once we have the stress tensor, the force per unit volume of the fluid is obtained by considering the stresses acting upon an infinitesimal cube of the fluid, and calculating the net force on the cube by taking the differences in the stress acting on different faces (just like our calculation of the force arising from pressure gradients in the fluid).

This results in derivatives of the stress tensor, so that the ith component of the force per unit volume is,

$$f_i = \sum_{j=1}^{3} \frac{\partial \sigma_{ij}}{\partial x_j}$$

$$= -\frac{\partial p}{\partial x_i} + \sum_{j=1}^{3} \frac{\partial \tau_{ij}}{\partial x_j}.$$

If we work out this last sum, we find (assuming that the viscosity doesn't depend on position),

$$\sum_{j=1}^{3} \frac{\partial \tau_{ij}}{\partial x_j} = \eta\nabla^2 v_i + \eta \frac{\partial}{\partial x_i} \nabla \cdot v.$$

For an incompressible fluid $\nabla \cdot v = 0$; reinstating vector notation, we see that the sum is then just $\eta\nabla^2 v$, so that the force per unit volume is,

$$f = -\nabla p + \eta\nabla^2 v.$$

This is the force which goes into Newton's Second Law in order to obtain the equation of motion for *v*:

$$\rho = \left[\frac{\partial v}{\partial t} + \left(v \cdot \nabla \right) v \right] = \eta \nabla^2 v - \nabla p.$$

This is the *Navier-Stokes equation* for an incompressible fluid.

Along with the equation of continuity, it is the basic equation which governs the flow of Newtonian fluids. To complete the description of the flow we need to specify the boundary conditions on the velocity field at a solid surface.

In our dealings with nonviscous flow we imposed the condition that the *normal* component of the fluid velocity at a solid surface must be zero — this just ensures that no fluid flows through the surface.

Once we deal with real (viscous) fluids, we must also require that the *tangential* component of the velocity of the fluid be zero on the surface.

This condition is often called the *no-slip* boundary condition, and is the result of experimental observations.

FLUID FLOW AT LOW REYNOLDS NUMBER

At low Reynolds number the inertial terms in the Navier-Stokes equation are relatively unimportant; the flow pattern is determined by a balance of the viscous forces and the pressure gradients in the fluid. As a consequence, expressions for the velocity, drag forces, and the like do not involve the fluid density directly. One of the nice features of flow at low Reynolds numbers is that the flow is laminar—the viscosity quickly damps out any turbulent effects. We'll consider two simple low Reynolds number flows, both of which have practical applications.

VISCOUS FLUID FLOW

Poiseuille Flow

We'll start with the flow of a viscous fluid in a channel. The channel has a width in the y-direction of a, a length in the z-direction of l_z, and a length in the x-direction, the direction of flow, of .

There is a pressure drop along the length of the channel, so that the constant pressure gradient is (such a pressure gradient could be supplied by gravity, for instance). Assuming the flow to be steady, $\partial v / \partial t = 0$. Also, we'll assume that the flow is of the form ; then . The no-slip boundary condition at the top and bottom edges of the channel reads $v_x \left(y = \pm a / 2 \right) = 0$. The Navier-Stokes equation then becomes,

$$\eta \frac{\partial^2 v_x}{\partial y^2} + \frac{\Delta p}{l_x} = 0.$$

Integrating twice, we obtain,

$$v_x(y) = -\frac{1}{2\eta}\frac{\Delta p}{l_x}y^2 + C_{iy} + C_2.$$

where C_1 and C_2 are integration constants. To determine these, we impose the boundary conditions to obtain,

$$v_x(y) = \frac{1}{2\eta}\frac{\Delta p}{l_x}\left[(a/2)^2 - y^2\right].$$

We see that the velocity profile is a parabola, with the fluid in the center of the channel having the greatest speed. Once we know the velocity profile we can determine the flow rate Q, defined as the volume of fluid which passes a cross section of the channel per unit time. This is obtained by integrating the velocity profile over the cross sectional area of the channel:

$$\begin{aligned} Q &= \int_0^{l_x} dz \int_{-a/2}^{a/2} dy v_x(y) \\ &= \frac{l_z a^3}{12\eta}\frac{\Delta p}{l_x}. \end{aligned}$$

The analogous result for flow through a pipe of radius a and length l in the presence of a uniform pressure gradient $\Delta p/l$ is,

$$Q = \frac{\pi a^4 \Delta p}{8\eta l}.$$

The important feature of both of these results is the sensitive dependence upon either the channel width a or the pipe radius a. For instance, for a pipe with a fixed pressure gradient, a 20% reduction in the pipe radius leads to a 60% reduction of the flow rate!

This clearly has important physiological implications — small amounts of plaque accumulation in arteries can lead to very large reductions in the rate of blood flow.

IMPORTANT FLOW OF STOKES' FLOW

A second important flow at small Reynolds numbers is Stokes' flow—the flow of a viscous fluid around a sphere of radius a, which is moving with a speed U. The derivation of the actual velocity field and the resulting drag force D are too complicated to derive here, so I'll just quote the result:

$$D = 6\pi\eta a U.$$

This is usually referred to as Stokes' law . Note that the result is independent of the fluid density, in accordance with our earlier observation that the fluid density shouldn't influence the drag at low Reynolds numbers.

From this expression we can derive the terminal velocity of a small sphere in air, by equating the drag force with the weight, with the result that .

Note that this result requires that the Reynolds number be small, so we would need to verify that was indeed small at the end of a calculation.

Stokes' law is important in determining things such as the settling of dust (from a volcanic explosion, say), or the sedimentation of small particles (pollutants) in a river.

At this point it is useful to introduce yet another dimensionless number, the coefficient of drag C_D, which provides a measure of the drag on a solid body.

The coefficient of drag is defined as,

$$C_D = \frac{D}{\frac{1}{2}\rho U^2 A},$$

where A is the projected area of the object in the plane perpendicular to the flow.

For the Stokes' flow, $A = \pi a^2$, and the coefficient of drag is,

$$C_D = \frac{24}{Re},$$

with $Re = U(2a)/v$. In fact, dimensional analysis tells us that for the flow of any incompressible fluid about a sphere,

$$C_D = f(Re),$$

where f(Re) is some undetermined, dimensionless function of the Reynolds number; however, we know that when $Re << 1\ f(Re) \sim 24/Re$.

Table. Some Important Dimensionless Numbers in Fluid Mechanics.

Number	Symbol	Definition	Importance
Mach	Ma	*U/c*	If M<<1 for incompressible flow.
Reynolds	Re	UL/v	Measures the relative importance of viscosity.
Drag coefficient	C_D	$D/(\rho U^2 nA/2)$	Dimensionless measure of the drag.
Lift coefficient	C_L	$L/(\rho U^2 nA/2)$	Dimensionless measure of the lift.

VISCOSITY EFFECTS AT HIGH REYNOLDS NUMBERS

When the Reynolds number is large you might think that the viscosity could be ignored altogether, in which case we return to nonviscous fluid mechanics. However, we once again encounter d'Alembert's paradox—that a nonviscous fluid exerts no drag on a solid body—so we are at a loss when it comes to explaining aerodynamic drag. The important insight in resolving this paradox is due to L. Prandtl, who in 1904 suggested that the viscosity could be ignored everywhere *except* in a thin layer close to the surface of a body. Understanding the behaviour of this *boundary layer* has been crucial to the development of modern fluid mechanics and aerodynamics.

EQUATIONS OF MOTION

Let's now look at integrals of the equations of motion. If we multiply by dt and integrate, we find

$$\begin{aligned}\int f dt &= \int m(dv/dt)dt \\ &= m\int dv \\ &= mv_2 - mv_2 \\ &= \Delta mv.\end{aligned}$$

$\int f dt$ is called *impulse*, and mv is the *momentum*. Neither of these quantities has any special name, but the units are ML/T: gm-cm/s, dyne-s or lb-s. Multiplying by dx instead,

$$\begin{aligned}\int f dx &= \int (dv/dt)dx \\ &= \int (dv/dx)(dx/dt) \\ &= \int v dv \\ &= mv_2^{\,2} - mv_1^{\,2} \\ &= \Delta(mv^2/2).\end{aligned}$$

These quantities are much used: òfdx is *work*, and $mv^2/2$ is *kinetic energy* T. In cgs, the unit of mechanical energy is the erg (from the Greek for work), the dyne-cm, and in fps, it is the ft-lb. The MKS unit is joule = N-m, so 1 J = $10^5 \times 10^2 = 10^7$ erg.

The time rate of change of kinetic energy is dT/dt = mv(dv/dt) = fv, called *power*. It can be expressed as erg/s or ft-lb/s, and 10^7 erg/s = 1 W. An erg/s is 0.1 ìW. In fps, we may use the colourful horsepower, 1 hp = 550 ft-lb/s. Since 1 ft-lb = 1.356×10^7 erg, 1 hp = 745.7 W.

If f is a function of position, f = f(x,y,z), and òfdx around any closed curve is zero (as it is in the gravitational field of the earth), then it a function V(x,y,z) can be found such that $f_x = -\partial V/\partial x$, or, in vectors, f = –grad V. This procedure is very familiar from electrostatics. The function V is the *potential energy*.

Then,

$$\begin{aligned}\int f dx &= -\int (dV/dx)dx \\ &= V_1 - V_2 = -\Delta V.\end{aligned}$$

Combining this with our earlier expression, we have $\Delta V + \Delta T = \Delta(T + V) = 0$, or T + V = E = constant. This is the principle of *conservation of mechanical energy*, which proves very useful.

Energy leads the way into advanced dynamics, where it replaces the concept of force.

For example, the potential energy function $V = kx^2/2$ yields the force f = –kx, which we know gives us the harmonic oscillator. We use Newton's Law to give $-kx = md^2x/dt^2$, or $x'' + (k/m)x = 0$, where the prime indicates differentiation with respect to t.

The solution of this equation is $x = A \sin wt + B \cos wt$, with A and B constants that are determined by the initial conditions, and $\omega = \sqrt{(k/m)}$. In advanced dynamics, we form the Lagrangian $L = T - V$ and use Lagrange's equations to get the equations of motion. This is only added complication in this simple case, but the method is capable of great generalization.

Another important conservation law comes from Newton's Third Law, which states that if one body exerts a force on another, then the second body exerts an equal and opposite force on the other. This says that if,

$$f = dp_2/dt,$$

then $-f = dp_1/dt$, from which it follows that,

$$d(p_1 + p_2)/dt = 0,$$

so that $p_1 + p_2$ = constant.

This is the principle of the *conservation of momentum*. Each component of the momentum is separately conserved. The conservation of energy and momentum are powerful tools for the study of collisions. An *elastic* collision is one in which the energy is conserved. Even in an inelastic collision, the momentum is still conserved.

If the force is constant, then we can integrate the equation of motion twice and find the position as a function of time. With $(f/m) = a$, we find $x = at^2/2 + v_0t + x_0$ with two constants, the initial velocity and position. This solution is studied in every elementary physics course, and is applicable to the everyday problem of projectile motion under gravity. There are many useful consequences of this solution, such as,

$$v^2 - v_0^2 = 2a(x - x_0),$$

where the time has been eliminated between it and its first derivative. In the form $v = \sqrt{2as}$ it is very useful, even in things like hydrodynamics.

DYNAMICS IN POLAR COORDINATES

So far we have not emphasized the vector nature of Newton's Second Law, since in rectangular coordinates we can write one equation for each dimension and work entirely with components. In fact, many practical problems are one-dimensional, and the subject does not come up. When we consider motion in a plane described by polar coordinates, however, we must look more deeply. If a point is given by $P(r,\theta)$, then the unit vector r' is r/r, and the unit vector θ' is at right angles to it towards the direction of increasing è, as shown in the diagram at the right. The primes will have to do here for identifying the unit vectors. Now, since,

$$\begin{aligned} r &= rr', \\ v &= dr/dt \\ &= (dr/dt)r' + r(dr'/dt) \\ &= (dr/dt)r' + r\omega\theta', \text{ since } dr'/dt = \omega\theta', \end{aligned}$$

where ω is $d\theta/dt$.

Note also that $d\theta' = -\omega r'$, because we need this for the acceleration: $a = [d^2r/dt^2 - r\omega^2]\, r' + [2(dr/dt)\omega + r(d^2\theta/dt^2)]\theta'$. This expression has many applications.

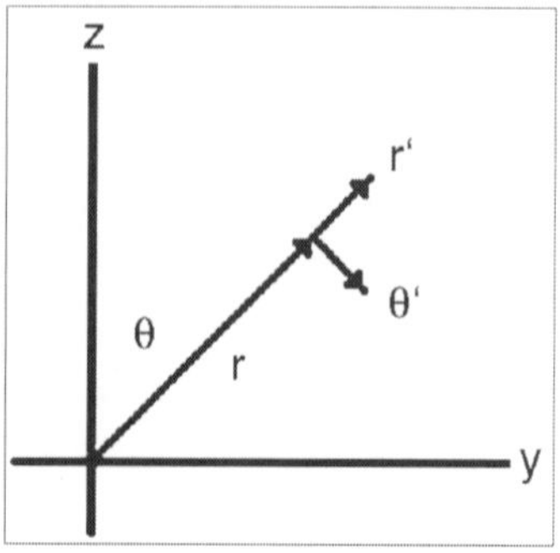

Fig: Polar Unit Vectors

Suppose first that the force acting on a particle moving in a plane is *central*, that is, directed towards the origin. Then the θ component of the acceleration must vanish, or $2(dr/dt)\omega + r(d\omega/dt) = 0$. If this is multiplied by r, we find that $d(r^2\omega)/dt = 0$, or $r^2\omega = h$, a constant.

Then, $mr^2\omega = mv_p r = H$, a constant. Here, v_p is the component of velocity perpendicular to the radius.

In vector notation, $H = mr \times v$ is called the *angular momentum*, the moment of the linear momentum about the origin. With a central force, then, the angular momentum is a constant. One consequence of this is Kepler's Area Law.

Now suppose the particle is moving in a plane with constant r, as shown in the figure. This is the case for all the points in a rigid body rotating about fixed axis normal to the plane. Then $a = -r\omega^2 r' + r\,d\omega/dt\,\theta'$. The first term is an acceleration directed towards the origin of amount v^2/r, the *centripetal acceleration* that must be supported by a force directed towards the orgin. The second term is zero unless there is an *angular acceleration*, which requires a force in the direction of dv/dt.

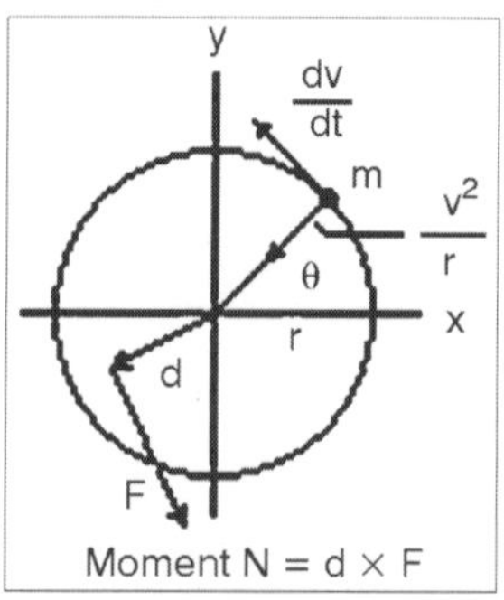

Fig: Circular Motion

This force is $f = mr\ dw/dt$, and its moment about the origin is $N = mr^2\ d^2\theta/dt^2$, which is an angular form of Newton's Second Law. If we are dealing with a rigid body, then $I = \Sigma mr^2$ is called the *moment of inertia of mass*.

We can then write $N = I\ d^2\theta/dt^2$, with N analogous to force. In the diagram, the applied moment N balances the accelerating moment. The forces do not have to have the same direction or magnitude, just the same moment.

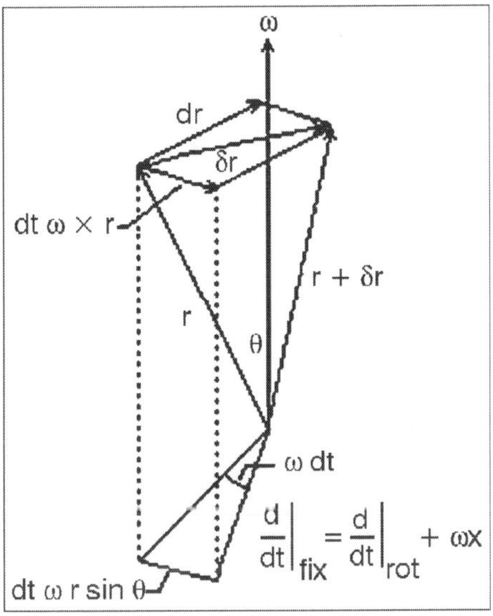

Fig: Time Derivative in a Rotating System

In vectors, if $f = dp/dt$, then $N = r \times f = r \times (dp/dt) = (d/dt)(r \times p) = dH/dt$, since $v \times p = 0$ because v and $p = mv$ are in the same direction. This shows that $N = dH/dt$ in general.

EQUATIONS OF MOTION IN TERMS OF RADIAL AND TRANSVERSE COMPONENTS

In physics, velocity is speed in a given direction. Speed describes only how fast an object is moving, whereas velocity gives both the speed and direction of the object's motion. To have a constant velocity, an object must have a constant speed and motion in a constant direction. Constant direction typically constrains the object to motion in a straight path. A car moving at a constant 20 kilometers per hour in a circular path does not have a constant velocity. The rate of change in velocity is acceleration. Velocity is a vector physical quantity; both magnitude and direction are required to define it. The scalar absolute value (magnitude) of velocity is speed, a quantity that is measured in metres per second (m/s or ms^{-1}) when using the SI (metric) system.

For example, "5 metres per second" is a scalar and not a vector, whereas "5 metres per second east" is a vector. The *average* velocity v of an object moving through a displacement (Δx) during a time interval (Δt) is described by the formula:

$$\bar{\mathbf{v}} = \frac{\Delta \mathbf{x}}{\Delta t}.$$

The rate of change of velocity is acceleration – how an object's speed or direction of travel changes over time, and how it is changing at a particular point in time.

The velocity vector v of an object that has positions x(t) at time t and x(t + Δt) at time $t + \Delta t$, can be computed as the derivative of position:

$$\mathbf{v} = \lim_{\Delta t \to 0} \frac{\mathbf{x}(t+\Delta t) - \mathbf{x}(t)}{\Delta t} = \frac{\mathrm{d}\mathbf{x}}{\mathrm{d}t}.$$

Average velocity magnitudes always smaller than or equal to average speed of a given particle. Instantaneous velocity is always tangential to trajectory. Slope of tangent of position or displacement time graph is instantaneous velocity and its slope of chord is average velocity.

The equation for an object's velocity can be obtained mathematically by evaluating the integral of the equation for its acceleration beginning from some initial period time t_0 to some point in time later t_n.

The final velocity v of an object which starts with velocity u and then accelerates at constant acceleration a for a period of time Δt is:

$$v = u + a\Delta t.$$

The average velocity of an object undergoing constant acceleration is $\frac{(\mathbf{u}+\mathbf{v})}{2}$, where u is the initial velocity and v is the final velocity. To find the position, x, of such an accelerating object during a time interval, Δt, then:

$$\Delta x = \frac{(u+v)}{2}\Delta t.$$

When only the object's initial velocity is known, the expression,

$$\Delta x = u\Delta t + \frac{1}{2}.a\Delta t^2,$$

can be used.

This can be expanded to give the position at any time t in the following way:

$$\mathbf{x}(t) = \mathbf{x}(0) + \Delta \mathbf{x} = \mathbf{x}(0) + \mathbf{u}\Delta t + \frac{1}{2}\mathbf{a}\Delta t^2,$$

These basic equations for final velocity and position can be combined to form an equation that is independent of time, also known as Torricelli's equation:

$$v^2 = u^2 + 2a\Delta x.$$

The above equations are valid for both Newtonian mechanics and special relativity. Where Newtonian mechanics and special relativity differ is in how different observers would describe the same situation. In particular, in Newtonian mechanics, all observers agree on the value of t and the transformation rules for position create a situation in which all non-accelerating observers would describe the acceleration of an object with the same values. Neither is true for special relativity. In other words only relative velocity can be calculated.

In Newtonian mechanics, the kinetic energy (energy of motion), E_K, of a moving object is linear with both its mass and the square of its velocity:

$$E_K = \frac{1}{2}mv^2.$$

The kinetic energy is a scalar quantity.

Escape velocity is the minimum velocity a body must have in order to escape from the gravitational field of the earth. To escape from the Earth's gravitational field an object must have greater kinetic energy than its gravitational potential energy. The value of the escape velocity from the Earth's surface is approximately 11100 m/s.

ANGULAR MOMENTUM OF PARTICLE

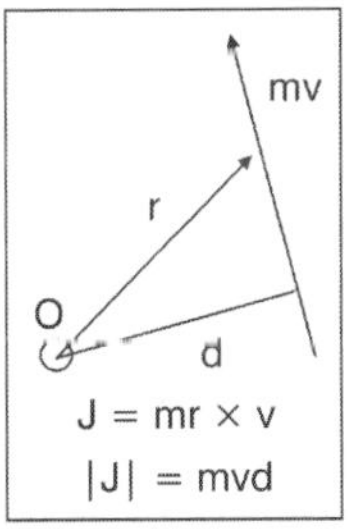

Fig. Angular Momentum

The *angular momentum* with respect to a point O is the vector J = r × p, where r is the radius vector from O to any point on the line of action of p = mv. In the figure, J points out of the page. If we use r x p instead of p in the equations of the preceding paragraph, we can prove that the total angular momentum is conserved. The conservation of linear and angular momentum gives us six scalar constants of the motion, useful in analyzing and understanding the motion. The existence of these constants is no trivial matter, but a beautiful and fundamental result of the theory.

The second law can be formally integrated to give p = òfdt = I, where I is a vector called the *impulse*, the integral of the force with respect to time. By taking moments about point O, we obtain a similar expression for the angular impulse. Of course, the limits on the integrals must be handled properly.If the second law is scalar-multiplied by v, we find mv·(dv/dt) = f·v, or (d/dt)(mv^2/2) = f·v = –dW/dt. This introduces two new scalar quantities, the *kinetic energy* T

= $mv^2/2$, and the *work* –W = ò f·dr. The minus sign, is simply means that work is being done *on* the particle, while W is conventionally the work done *by* the particle. If W is a function of position, then in vector notation f = – grad W. In one dimension, f = – dW/dx, which may be more familiar.

What we have shown is that (d/dt)(T + W) = 0, or T + W = constant, a new conservation law. If W is a *conservative* quantity (that is, its value is independent of path and depends only on position), it is called the *potential energy* V. Then, the *total energy* E = T + V is conserved in the motion. Energy is a widely used and misused quantity that could be discussed in great detail, but only its bare definition is required here, and the reader is assumed to be familiar with it.

The introduction of energy permits the use of *generalized coordinates*, and the derivation of equations of motion by the Lagrange procedure, which uses the Lagrangian function L = T– V. This facilitates the solutions of very many problems, since we are liberated from carrying around a basket of vector components. For more information, follow the link in the Introduction to the article on Lagrangian mechanics. For vector mechanics, it is convenient to have expressions for the velocity and acceleration components in polar coordinates. These are derived in the References, but will be summarized here.

$$v = (dr/dt)r' + (rd\theta/dt)\theta'.$$

r' and q' are unit vectors pointing radially outward, and tangentially in the anticlockwise sense, with respect to the position vector r, as shown in the diagram. The rectangular unit vectors i and j are also shown. Note that dr'/dθ =θ' and dθ'/dθ = –r', while the derivatives with respect to r are zero. dv/dt = $[d^2r/dt^2 - r(d\theta/dt)^2]r' + (1/r)(d/dt)(r^2 d\theta/dt)\theta'$.

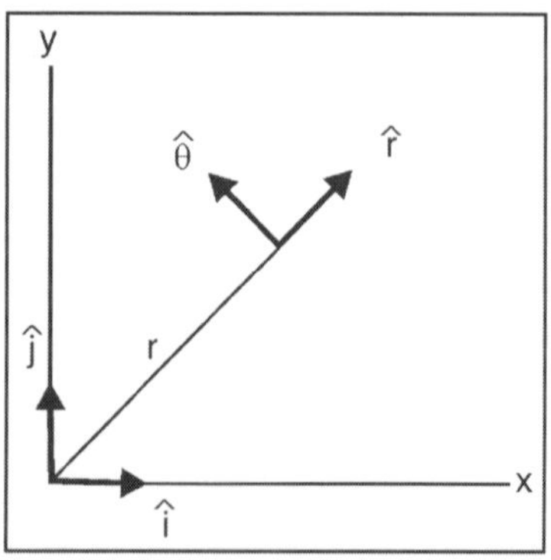

Fig. Unit Vectors

The reader should carefully note the distinction between, for example, d^2r/dt^2 and d^2r/dt^2. The first is the second derivative of a scalar quantity, a simple thing. The second is a vector, and since the direction of r can change, the complete expression is somewhat elaborate, as we have just seen. It is essential to distinguish carefully between scalars and vectors, which are arrays of scalars.

These relations are very useful. For example, suppose that the force is central, so that the tangential component of the acceleration is zero. This means that $(d/dt)(r^2dq/dt) = 0$, or r^2dq/dt = constant = h. Now, half of r times r dè is the area dA swept out by the radius vector in dt, so h is 2dA/dt. Also, mrdq/dt is the component of the momentum perpendicular to the radius vector, so $mr^2dq/dt = mh$ is the angular momentum.

We then have that the angular momentum L = mh is a constant. Also, this means that the motion takes place in a plane, and L is perpendicular to this plane. We have found out a lot here with very little effort. The constancy of areal velocity is one of the most useful properties of orbital motion, not just a curiosity. It relates the rate of change of the angle q to the radius r.

KEPLERIAN ORBITS

Kepler's three laws of planetary motion, deduced by prolonged and tedious consideration of the observed position of Mars, are:

1. The planets move in ellipses with the sun at one focus;
2. The areas swept out by the radius vector in equal time intervals are equal; and
3. The cubes of the mean distances (half the major axis of the orbit) are proportional to the squares of the periodic times. These laws are sufficient to determine the position of a planet at any later time if its position is known at one time, and the dimensions and orientation of the orbit are known.

For a deeper understanding, and the power to attack an arbitrary problem in orbital motion, such as the movement of earth satellites, we should consider the dynamics, on the basis of Newton's theory.

The force acting between two spherical, radially symmetric bodies of masses M and m a distance r apart is, by Newtonian gravitation, GMm/r^2 in magnitude, and is directed along the line joining the centres of the two bodies. G is the Newtonian gravitational constant, 6.67259×10^{-11} m^3/kg-s. Since the force is central, the angular momentum is conserved, and the bodies revolve about one another in a plane with the centre of mass (the barycentre) fixed. The motion can be analysed as the rotation of a *reduced mass* Mm/(M + m) about the centre of gravity. The potential energy V is -GMm/r, and is negative because the particles attract. For the total energy to be zero,

$$T = mv^2/2 = GMm/r, \text{ or } v = \sqrt{(2GM/r)}.$$

This is the *escape velocity* from the particle of mass M at the distance r. At the surface of the earth, the escape velocity is 11,200 m/s or 25,200 mph. This is also the velocity of a meteor that comes in from infinity. The escape velocity from the *sun* at the distance of the earth is 42,000 m/s or about 95,000 mph. It is not easy to get away from the earth; it is even harder to escape from the sun.

If we consider a circular orbit of radius a, as shown in the figure at the left, then $GMm/a^2 = [Mm/(M + m)]a(d\theta/dt)^2$, since $v = a(d\theta/dt)$ and the centripetal acceleration is v^2/a. This is the equation of motion for the relative motion, with r the distance between M and m. The center of gravity of M and m moves with constant velocity, not affected by the relative motion.

Thus,

$$G(M + m) = a^3(d\theta/dt)^2 = 4\pi^2(a^3/P^2),$$

where P is the period of the motion.

This is a derivation of Kepler's third law for a special case, but it is a general result when a is the mean distance of any orbit. The quantity $G(M + m)$ is usually called μ, so that $r^3/P^2 = \mu/4\pi^2$. If r is in astronomical units (radius of the earth's orbit, about 150×10^6 km), and the period P is in years, then $G(M + m) = 4\pi^2 = \mu$.

The quantity GM is called the Gaussian gravitational constant k^2 if time is in days instead of years, so we get approximately $k^2 = 4\pi^2/(365.25)^2$, and $k = 0.017202$.

A more precise value is 0.01720209895. Untangling G from GM means knowing the mass of the sun accurately, which is impossible, so celestial mechanics uses k instead of G. The value should be corrected for the finite mass of the earth, but since the ratio of the sun's to the earth's mass is 332,946 the correction is small.

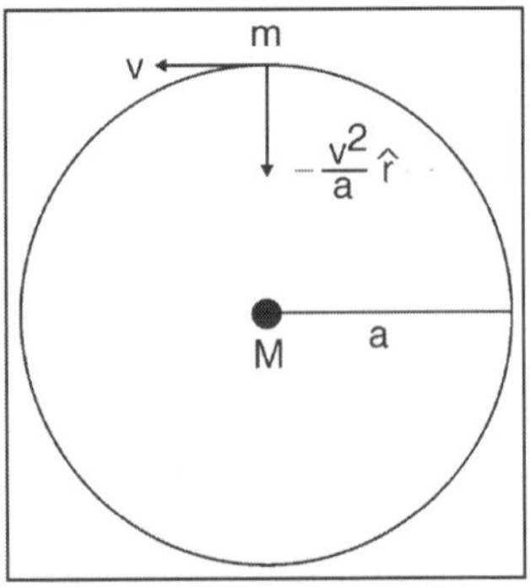

Fig. Circular Orbit

The polar equation of an ellipse, with the origin in a focus, is,

$$r = p/(1 + e \cos \theta).$$

The *eccentricity* of the ellipse is e, and p is a length called the semi-latus-rectum. In connection with orbits, the angle θ is called the *true anomaly*. In an elliptical orbit, $0 \le e < 1$. When $e = 0$, we have a circular orbit $r = p$. The same formula gives a parabola for $e = 1$, and a hyperbola for $e > 1$. In an ellipse, the maximum and minimum radii are $p/(1- e)$ and $p/(1 + e)$, respectively, and the corresponding points on the orbit are called *aphelion* and *perihelion*, respectively. The distance from perihelion to aphelion is the major axis of the ellipse, twice the mean distance a.

Therefore, $2a = p/(1 - e) + p/(1 + e) = 2p/(1- e^2)$, or $p = a(1- e^2)$. The perihelion distance is $a(1- e)$, the aphelion distance $a(1 + e)$. The distance from the center to a focus is $a- a(1- e) = ea = c$. This defines the eccentricity as $e = c/a$. Since the ellipse is the locus of points the sum of whose distances to the two foci is a constant, 2a, the length of the radius from a focus to the end of the minor axis is a. Since this is the hypotenuse of a right triangle, $a^2 = c^2 + b^2$, where b is the semiminor axis. Hence $b = a \sqrt{(1- e^2)}$. The reader can show that the average value of r over one revolution is a. This is about all we need to know about the geometry of the ellipse for present purposes.

If we know the orbit is an ellipse, and the areal velocity is constant, we should be able to prove that the force is central and varies as the inverse square of the distance. On the other hand, if we know that the force is central and inverse-square, then we should be able to prove that the orbit is an ellipse and the areal velocity is constant. Either of these things can be done fairly easily.

In the first case, differentiate the expression for r with respect to time, obtaining $dr/dt = (e/p) \sin \theta\, 2A$, where A is the areal velocity (this was called dA/dt above; now A represents the derivative). Then differentiate again to find $d^2r/dt^2 = (4A^2e/pr^2) \cos \theta$. The tangential acceleration is zero from the constancy of the areal velocity. The radial acceleration is $a_r = (4A^2e/pr^2) \cos \theta - 4A^2/r^3 = - 4A^2/pr^2$, which shows that the force is inverse square, and we have already shown that it is central. In fact, $GM = 4A^2/p$.

In the second case, the equations of motion in the plane of the orbit are $d^2r/dt^2 - r(d\theta/dt)^2 = -GM/r^2$ and $rd^2\theta/dt + 2(dr/dt)(d\theta/dt) = 0$.

Now, $(d/dt)(r^2d\theta/dt) = 2r(dr/dt)(d\theta/dt) + r^2d^2\theta/dt^2 = 0$, so the areal velocity $A = (1/2)r^2(d\theta/dt) =$ constant. Using this result, the radial equation becomes $d^2r/d\theta^2 - (2/r)(dr/d\theta)^2- r =- r^2GM/4A^2$. Making the substitution $u = 1/r$, this equation can be thrown into the form $d^2u/d\theta^2 + u = GM/4A^2$. This second-order linear equation with constant coefficients is easy to solve. We find $u = 1/r = GM/4A^2 + C \cos \theta$. This is the equation of a conic section, as we have seen above. If $p = 4A^2/GM$ and $e = pC$, we have $r = p/(1 + e \cos \theta)$, which is what we wanted to prove. Now we can find the components of the velocity. The transverse component is $rd\grave{e}/dt = 2A/r = 2A(1 + e \cos \theta)/p$, where A is the areal velocity. The radial component is $dr/dt = 2A(e/p)\sin \theta$.

The square of the velocity is the sum of the squares, or $v^2 = 4A^2[(1 + e^2 + 2e \cos \theta]/p^2] = 4A^2[(1- e^2)/p^2 + 2/pr] =-4A^2/pa + 4A^2/pr$, from which the kinetic energy per unit mass (*i.e.,* setting $m = 1$ for simplicity) is: $v^2/2 =- GM/2a + GM/r$, since $4A^2/p = GM$.

Rearranging, $v^2/2- GM/r = -GM/2a$, which is to say, $T + V = E$, with $T = v^2/2$, $V = -GM/r$ and $E = -GM/2a$.

The total energy is negative (*i.e.,* the orbiting particle is "trapped") and a function of the mean distance a only. The eccentricity is determined by the angular momentum h through $e^2 = 1 - h^2/aGM$. When $h = \sqrt{aGM}$, the orbit is circular.

The quantity a, called the "mean distance," is not the average value of the radius vector r. It is the average of the perihelion and aphelion distances, however. The time average value of 1/r turns out to be 1/a, which is its real definition.

This means that the time average potential energy is–GM/a, and the time average kinetic energy is–GM/2a + GM/a = +GM/2a. For a circular orbit, the kinetic and potential energies are constant. The time average value of r is $a(1 + e^2/2)$, and the average of r over the true anomaly is b, the minor axis of the orbit. The average of 1/r over the true anomaly is 1/p.

We obtained the relation between the mean distance a and the orbital period P for the special case of a circular orbit. A more general proof is as follows. The areal velocity dA/dt = h/2 = constant, so integrating from t = 0 to t = P we find $A = hP/2 = \pi ab$. We also know that $p = a(1 - e^2) = h^2/GM$. Therefore, $P^2 = 4\pi^2 a^2 b^2/h^2 = [4\pi^2/GM]a^3$, since $b^2/p = a$. This is the desired general result. If we know the period and mean distance of any orbit, we can calculate the gravitational constant GM, which in the general case is G(M + m).

COORDINATES AND ORBITS

The angular coordinates that are used to describe directions in space are shown in the figure at the left. The equator and ecliptic planes are shown. The dihedral angle between these two planes is ε = 23° 26' 21".412, or thereabouts.

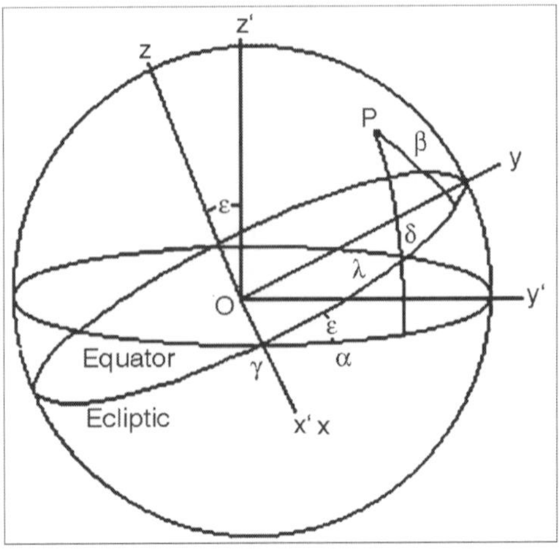

Fig: Equatorial and Ecliptic Coordinates

The equatorial coordinates are the right ascension α and the declination δ, while the analogous ecliptic coordinates are the longitude λ and the latitude β. The relations between the coordinates are best found by considering the rectangular components of the vector OP, considered as of unit length, and then performing the rotation about the x-axis that takes one system into the other. We will not require these relations at the present time.

The position of a body in the solar system can be specified in terms of a gigantic rectangular coordinate system centered on the sun. The x-axis is always taken as pointing to the vernal equinox, a direction specified by the line of intersection of the orbital plane of the earth (the ecliptic plane) with a plane parallel to the equator of the earth, the equatorial plane.

Both contain the centre of the sun. This is made interesting by the precession of the normal to the equatorial plane around the pole of the ecliptic. In the diagram, this is the movement of z' (the pole of the equator) around z (the pole of the ecliptic). Fortunately, this is not too rapid, taking about 26,500 years for one revolution. The equatorial plane has to be considered because our earth-based measurements use it as a reference, when directions are given in terms of right ascension and declination. If we know a vector whose components are referred to this plane, we can find the right ascension and declination of its direction. When the earth is located in its orbital plane on the line of intersection of the ecliptic and equatorial planes, the sun is seen in front of the vernal equinox about March 21, climbing from below the equator to above it. At this time, when we look at the sun (careful!) we are looking in the direction of the x-axis of the heliocentric coordinate system. The y-axis of the ecliptic coordinates is in the plane of the ecliptic, at right angles to the x-axis, and in the direction in which the planets move, in which a screw would advance in the direction of the north pole of the earth. The ecliptic z-axis is then perpendicular to the x- and y-axes, and makes a right-handed coordinate system with them. The equatorial rectangular coordinates are defined analogously. If we represent these coordinates by primes, x and x' are the same, while x',y' is rotated with respect to x,y by the dihedral angle between the planes. The rotation from the z' axis to the z axis is a right-handed rotation about the positive x,x' axis.

Longitudes are measured clockwise (eastwards) on the ecliptic plane from the vernal equinox, from 0° to 360°. The *longitude of perihelion*, ψ, is the angle measured clockwise from the vernal equinox to the radius passing through perihelion. (A different symbol, a kind of ù, is used conventionally, but it is not available here in HTML.) The longitude of perihelion of the earth's orbit is currently about 103°. This locates the major axis of the orbit and the direction from which the true anomaly is measured.

Other planets move on orbital planes inclined to the ecliptic by the *inclination*, i. They intersect the ecliptic on a line called the *line of nodes*. The *longitude of the ascending node*, Ω, is the angle measured clockwise from the vernal equinox to the line of nodes, on the end where the orbit is climbing above the ecliptic plane, called the *ascending node*.

The node 180° away is, not surprisingly, the *descending node*. Note in the figure that the traditional symbol for the ascending node is used, which we represent in text by Ù, which looks similar. The symbol for the descending node is inverted, with the two little loops on the top. The longitude of perihelion in this case is the longitude of the ascending node, *plus* the angle ù in the orbital plane from the node to the perihelion. This may be a little odd, but it does locate the perihelion. That is, $\Psi = \Omega + \omega$. These quantities are illustrated in the diagram.Now that the orbital plane and the direction of the perihelion have been specified by the three orbital elements i, Ω and ψ, we proceed to specify the shape of the orbit by its mean distance a and its eccentricity e. The position of the body in its orbit at a specified time is given by a sixth element, the mean longitude at the epoch.

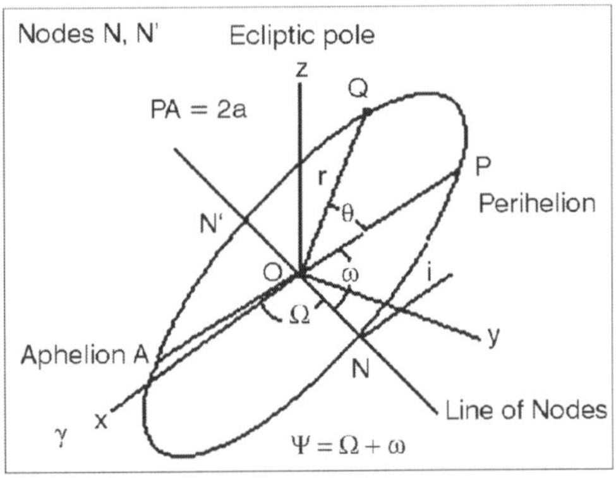

Fig: Orbital Elements

The "epoch" is just the reference time assumed, usually by its Julian date. The mean distance a determines the orbital period P through Kepler's third law. Therefore, only one of a or P need be specified. The *mean daily motion*, 360° divided by the orbital period, is often given in place of P.

The constancy of the areal velocity is used to locate the body in its orbit at a specified time. This is called Kepler's Problem, which Kepler also solved. The solution is illustrated in the figure. A reference circle of radius a is drawn around the orbital ellipse. Point Q is projected onto the auxilary circle by a perpendicular to the major axis to point Q'.

The line CQ' then defines the angle E at the centre of the circle, called the *eccentric anomaly*. We can show that it is related to the radius vector r and the true anomaly θ by $r = a(1- e \cos E)$ and $\tan(\theta/2) = \sqrt{[(1 + e)/(1- e)]} \tan E/2$. The shaded sector PFQ is the area swept out by the radius vector, and the area is proportional to the time. The area FPQ' is proportional to the area

FPQ, since it is merely magnified vertically in the ratio $a/b = 1/\sqrt{(1-e^2)}$. Indeed, the total area of the orbit is ðab, while the total area of the circle is πa^2.

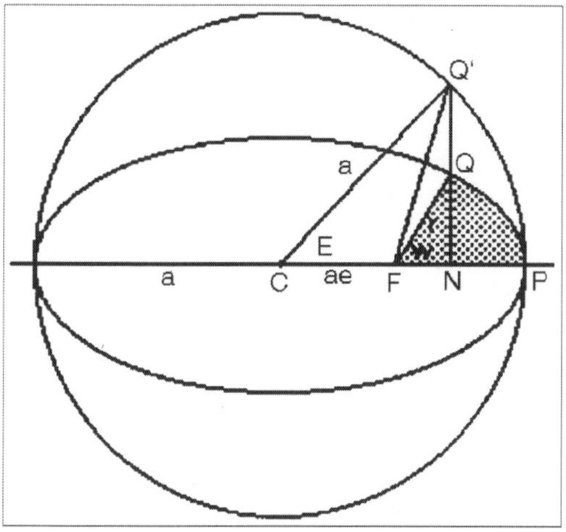

Fig. Kepler's Problem

Therefore, this larger area FPQ' increases uniformly with time as well. Since the total area is πa^2, the area swept out is given by $(\pi a^2/P)t$, where t is the time since perihelion passage. This area can also be found in terms of E, since it is the area of the whole sector PCQ' less the area of the triangle CQ'F.

The area of the sector is $\pi a^2(E/2\pi) = (a^2/2)E$. The area of the triangle is $(1/2)(a \sin E)(ae) = (a^2/2)e \sin E$. Therefore, we have $E - e \sin E = 2\pi t/P = M$, called Kepler's Equation. M is called the *mean anomaly* and increases proportionally to time after perihelion passage. This is really a beautiful result, allowing us to find the position in orbit in terms of a uniformly increasing quantity.

Kepler's Equation can be easily solved by iteration. If it is written $E = M + e \sin E$, we find the iteration equation $E_{n+1} = M + e \sin E_n$ (e is the eccentricity here, not the base of natural logarithms). Starting with $E_0 = M$, a few interations are enough to get good precision, especially when e is small.

The iterations are very easy, and can be performed on a pocket calculator. For $e = 0.1$, six iterations give 8 digits. The Newton-Raphson method can also be used. This method finds the roots of the equation $f(x) = 0$ by improving an initial guess x_0 by the formula $x_{n+1} = x_0 - f(x_n)/f'(x_n)$. A sketch will reveal the reasoning behind this formula. In this case, $f(E) = M - E + e \sin E = 0$ is the function, and $f'(E) = e \cos E - 1$. Fewer iterations are required, but each requires more work, than in the case of simple iteration. In practice, an expansion in powers of e is often used for small e, but the iteration methods are no more trouble, especially for a computer.

Although the determination of position in orbit may sound complicated, it is really quite straightforward, and after you have done it a few times, it will be easy. Let's find the position of the earth on 2003 April 2. The figures come from the *Astronomical Almanac for 2003*, p. E4. On Jan 1, JD 245 2640.5, the mean longitude of the earth was 100.2440°, and the longitude of perihelion was 103.0°. April 2 is 91 days later, so M = 100.2440– 103.0 + 91 × 0.9856 = 86.93° = 1.5172 radians. A first approximation to E is M + e sin M, and e = 0.0167, so the correction is 0.01668 radians, and E = 1.5339 radians or 87.89°. There is no need to iterate again.

The mean distance can be taken as 1.00000, so r = 0.9994 and θ = 88.84°. The earth's longitude is 103.0 + 88.84 = 191.84°. On p. C8, the solar ephemeris gives longitude 11.8506° and r = 0.9994. The solar longitude is 180° less than the earth's, or 11.84° according to our calculation. In both cases, we are using the mean ecliptic of date, and came quite close using Keplerian orbital elements instead of the more elaborate calculations used to calculate the solar ephemeris. Precise orbits require the consideration of planetary perturbations, and this is a very difficult subject indeed.

The orbit of the moon is an example, since it is subject to perturbations by the earth's spheroidal shape and the sun that make its orbit vary in a very complicated way. In the case of the earth's orbit, we really calculate the position of the earth-moon barycentre. This point is 4671 km from the centre of the earth, about two-thirds of the way to the surface.

The mean distance of the moon is 3.844×10^5 km, and its sidereal period is 27.321661 days, from which G(M + m) can be calculated and used to relate mean distance and period for other earth satellites after correcting for the mass of the moon. The moon's radius is 1738 km, and its mass 7.3483×10^{22} kg. m/M = 0.0123, which is not quite negligible. The eccentricity of the moon's orbit is about 0.0549, and its inclination 5.145396°. The plane of the orbit and the position of perigee change rather rapidly. The small eccentricity of the earth's orbit suggests that the earth-moon system did not suffer any external disturbance during its formation, and in particular any collision which created the moon. This was a one-time event, so speculations on what happened can never be proved, since no similar state is known, or is ever likely to become known.

Except for Mercury and Pluto, orbital inclinations are less than 4°, and eccentricities less than 0.0936, the value for Mars. Kepler was fortunate to have picked Mars for study because of its relatively large eccentricity, about twice those of Jupiter and Saturn. Venus's orbit is practically circular, with e = 0.0067.

Mercury not only has a large inclination, 7.005°, but an eccentric orbit, with e = 0.2056. Only Pluto has a greater inclination and eccentricity, but it is probably a special case. Observations of Mercury and Pluto were not available to Kepler, who had to work with Tycho's naked-eye observations. The angle made by r with the x-axis in our problem was 191.84°, so the coordinates will be x = 0.9994 cos 191.84° = –0.97814, and y = 0.9994 x sin 191.84° =–0.20506, in astronomical units.

In the more general case of a planet in an inclined orbital plane, the radius is first projected on the line of nodes [a = r cos (ω + θ)] and on a perpendicular (dip) line [b = r sin (ω + θ)]. Then z = b sin i, y = b cos i cos Ω + a sin Ω, x = a cos Ω– b cos i sin Ω. It is just a matter of projecting r on the coordinate axes.There is an expansion for the true anomaly directly in terms of the mean anomaly, without going through the eccentric anomaly.

To find the equatorial rectangular components, rotate about the x-axis through an angle of i. This gives y' = y cos i– z sin i and z' = y sin i + z cos i. If x', y' and z' are the components of a vector in equatorial rectangular components, then the right ascension á is given by cos α = $x'/\sqrt{(x'^2 + y'^2)}$, and the declination ä by sin ä = $z'/\sqrt{(x'^2 + y'^2 + z'^2)}$. The proper quadrants have to be determined. In this way, the direction in which a planet is seen from the earth can be found.

COMETARY ORBITS

Comets can be divided into two classes, the *short-period* comets, like Halley's Comet, with a period of 76 years, and *long-period* comets, like Hale-Bopp, which do not return for thousands of years, if at all. A period of 200 years is the conventional dividing-line. Short-period comets have elliptical orbits like the planets, except that the eccentricity is larger, the inclination can take any value, and the comets can move in a *direct* or *retrograde* direction. Retrograde motion is usually expressed by an inclination greater than 90°. Comets were the first real test for Newton's theory, which finally showed that they were normal members of the solar system, not mysterious atmospheric happenings, as had always been supposed. Positions of short-period comets are calculated in the same way as planetary positions, and the orbital elements are presented the same way. Long-period comets come from the periphery of the solar system, where they wander in the hypothetical Oort Cloud of cometary debris, normally water and carbon dioxide ice and dust. Their total energy is about zero, so the eccentricity of their orbits when they make an excursion towards the sun is about 1.

That is, their orbits are parabolic. Short-period comets are those that have suffered an energy-losing collison with a planet (usually Jupiter) and have dropped into an elliptical orbit. There must also have been comets that have gained energy, entered hyperbolic orbits, and were ejected from the solar system. Few, if any, comets have eccentricities significantly greater than 1, which would indicate that they were encountered by the solar system in its path through space, and are not part of the family.

The parabola is a considerably simpler orbit than the ellipse. Its polar formula is $r = p/(1 + \cos \theta) = (p/2) \sec^2 \theta/2)$. The areal velocity $A = \sqrt{(GMp/4)}$. Setting $x = \tan (\theta/2)$, direct integration gives Kepler's Equation as $x + x^3/3 = \sqrt{(GM/2)}(2/p)^{3/2}(t- T)$. The one real root of this equation gives the position in orbit. There are many ways to solve a cubic equation: refer to mathematics handbooks, algebra texts or mathematics programmes. The time to move from perihelion to the end of the latus rectum is $t_1 = (1/2)\sqrt{(p^3/GM)}$. $2t_1$ is a good measure of the time that the comet will spend near the sun.

The orbital elements include the inclination i, longitude of the ascending node Ω, and argument of perihelion ω. The perihelion distance p/2 is given, and the JD of perihelion passage. For Hale-Bopp, the perihelion distance was 0.91399384 and the time of perihelion passage was JD 245 0539.60742. The inclination was 89.42064850°, longitude of ascending node 282.47215310°, and argument of perihelion 130.59561740. From these elements, it was possible to find the position of the comet as seen from the earth. Cometary orbital elements are available on the internet soon after the discovery of a comet. The Jet Propulsion Laboratory supplied the elements for Hale-Bopp shortly after it was discovered.

Halley's comet was the first periodic comet to be recognized. Edmund Halley (1656–1742), friend of Newton's and later Astronomer Royal, suspected that the comets of 1531 and 1607 were the same as the comet of 1682, since they had similar orbits. He predicted the return of the comet in 1758, which duly occurred. Halley had published Newton's *Principia* at his own expense in 1678, and made an extensive study of comets using the new theory. Halley's comet is the brightest of the short-period comets; most are quite dim, and can even have orbits of small eccentricity like asteroids.

To show how to use the orbital elements given by JPL, let's look at them for Halley's comet. The time of perihelion passage, T_p is shown as 19860209.45895, which we interpret to mean 1986 February 9, at 0.45895 part of the day, or 11h 0m 53s UT. The "epoch" is the date to which the elements apply, which is given as 46480. This is a modified JD. The JD can be obtained by adding 24400000.5, or JD 2446480.5, which is 1986 February 18. The size and shape of the orbit is specified by the perihelion distance q = 0.58710374, in AU, and e = 0.96727724. The eccentricity is nearly, but not quite, unity. Near perihelion, the orbit will be indistinguishable from a parabola, so the parabolic formulas can be used for predicting its position while it is near the sun and visible. The orientation of the orbital plane is specified by the longitude of the ascending node, 58.86004°, and the inclination i = 162.24220°.

When the numbers are located, look at the symbols which may not be the usual ones because of font limitations. Since $i > 90°$, the motion of the comet in its orbit appears to be retrograde, opposite to the direction of motion of the planets, if the inclination is taken to be 180° – 162.24220° = 17.75780°.

Finally, the orientation of the orbit in its plane is given by the argument of perihelion, 111.8656°, measured from the ascending node in the direction of movement. It is not easy to comprehend the orbital position from the bare numbers; a drawing, made with some care, will show exactly what is going on.

As the perihelion distance is q = a(1 – e), it is a simple matter to determine a. In fact, a = q/(1 – e) = 0.58710374/0.03272276 = 17.941755 AU. From this, the orbital period P can be found. Since $P^2/a^3 = 1$ if P is in years (earth = 1) and a is in AU (earth = 1), P = 75.99716 years, very close to 76.0 years. The aphelion distance is 2(17.941755) – 0.58710374 = 35.296 AU.

This is outside the orbit of Neptune (a = 30) but not as far as Pluto (a = 39.5). In accordance with the law of areas, Halley's comet spends most of its time drifting in this dark region, periodically darting in to the sun to have a little more of itself boiled off into space. Although orbital elements are given to high precision, the accuracy is not necessarily as high. Especially for comets, they can be changed by planetary perturbations, sometimes by large amounts. The ejection of gases from comets also leads to reaction forces that can affect the orbit. Also, the reference coordinates may change with time, because of precession of the equinoxes and other effects, and this must be taken into account in accurate calculations. These changes are not changes in the orbit, of course.

DETERMINATION OF ORBITS

Finding the orbital elements from observations, and predicting the changes in orbital elements due to perturbations, are two of the most important problems in celestial mechanics, and have received close attention from Newton's time onwards. We cannot give any reasonable account of this work here, but we can show how orbital elements come from observed motions by a graphical analysis that is very instructive, though of little practical use where high precision is required. The reader with drawing supplies is encouraged to follow along.

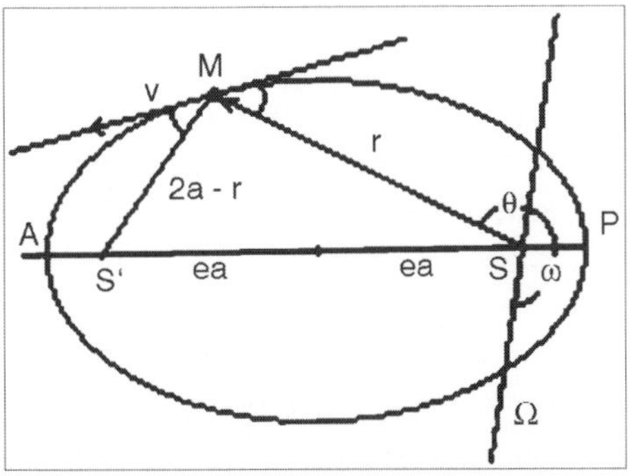

Fig: Determination of Orbital Elements

We suppose known the position r of a body M from the sun and its velocity v relative to the sun. These are six parameters that will serve to determine the six orbial elements. The orbital plane is defined by the plane of r and v, and can be represented in two views by means of orthographic projection. The line of intersection of the orbital plane with the ecliptic plane can then be found, and the dihedral angle between the two planes. Since the line of intersection will be the line of nodes, we have now found two of the orbital elements, Ω and i, which determine the plane of the orbit.

The next step is to calculate the mean distance a, using the energy relation $v^2 = GM(1/r - 1/2a)$. From a we can find the orbital period by $P^2/a^3 = 4\pi^2/GM$. The radius vector SM and the line from M to the other focus of the ellipse make equal angles with the tangent to the ellipse at M, which is in the direction of v.

In a view showing the orbital plane in true size, Lay off SM and MQ, where MQ is an arbitrary line in the direction of the empty focus. The empty focus S' is located by laying off a distance 2a – r along MQ', since the sum of the distances from the foci to M is 2a.

Now we can draw the line SS', find its center at O, and then the locations of perihelion and aphelion. The eccentricity of the orbit is e = OS/a = OS'/a. The orbit can now be drawn, and the angle from the line of nodes to the perihelion, the argument of perihelion, ù, can be measured, as well as the true anomaly θ, which is the angle PSM. We now have the five elements Ω, i, ω, e, and a.

The remaining element is the time of perihelion passage, which can be found from the true anomaly. First, we find the eccentric anomaly by $\tan(E/2) = [(1-e)/(1+e)]^{1/2} \tan(\theta/2)$, and from it the mean anomaly $M = E - e \sin E$. The mean anomaly is $M = 2\pi(t - T)/P$, where T is the time of perihelion passage, and t is the time of observation. That is, the body passed through perihelion at a time $t - T = MP/2\pi$ earlier. We have now determined all six orbital elements from the six components of the initial position and velocity.

This was easy because the distance r was one of the given quantities. In general, orbits must be determined by observations of the angular position from earth, not directly in terms of distance and velocity. The position at three different times is sufficient to determine the orbit in this case, but the analysis is more difficult than what we have done. Orbital elements vary only slowly in the solar system, except when there are near encounters, usually of light bodies with massive planets or satellites, that affect the orbits of the light bodies, but not of the heavier ones.

If we know the orbit at one instant, then we can predict the velocity and position at a later time from the orbit. We can also integrate the actual changes in position and velocity at the later time, taking into account forces exerted by bodies other than the sun. The difference will be reflected in the orbital elements, which will slowly change. This is a useful way to take perturbations into account, and is widely used. The maximum errors in using the Keplerian orbits are on the order of 5" to 30" for the inner planets, somewhat more for Jupiter and Saturn. Neptune was discovered in 1846 as a result of its perturbations of the motion of Uranus. This was a remarkable demonstration of the accuracy of Newtonian mechanics.

The earth's mean distance is decreasing by 5×10^{-8} AU per century, about 75 m per year. The eccentricity is decreasing by 3.804×10^{-5} per century. The inclination is decreasing by 46.94" per century. These are average rates, and are simply the current values of changes that may be periodic. Nevertheless, they show how gradual any changes are, and allow the calculation of the earth's position with reasonable accuracy for thousands of years either side of the present. The longitude of the node, however, moves at –3.038" per *year* because of the precession of the equinoxes, which moves the reference point.

The argument of perihelion is increasing at 11.9828" per year. It is typical for an orbit to show a relatively rapid change in longitude of the node and motion of the perihelion; the moon is an excellent example, and also earth satellites, which are perturbed by the equatorial bulge of the earth and tidal forces exerted by the sun and moon, as well as by atmospheric drag if they are in low orbits.

BERNOULLI EQUATION AND APPLICATIONS

A statement of the conservation of energy in a form useful for solving problems involving fluids. For a non-viscous, incompressible fluid in steady flow, the sum of pressure, potential and kinetic energies per unit volume is constant at any point

A special form of the Euler's equation derived along a fluid flow streamline is often called the Bernoulli Equation

$$\frac{\partial}{\partial s}\left(\frac{v}{2}+\frac{P}{\rho}+g\cdot h\right)=0 \quad (1)$$

where

v = flow speed
p = flow speed
ρ = density
g = gravity
h = height

$$\frac{v^2}{2}+\frac{p}{\rho}+g\cdot xh=(\text{Cons tan t}) \quad (2)$$

$$\frac{v^2}{2\cdot g}+\frac{p}{\gamma}+h=(\text{Cons tan t}) \quad (3)$$

where

$$\gamma = \rho \cdot g$$

$$\frac{\rho \cdot v^2}{2} + p = \text{cons tan t} \quad (4)$$

$$\frac{\rho \cdot v^2}{2} + pd \quad (5)$$

$$\frac{\rho \cdot v_1^2}{2} + p_1 = \frac{\rho \cdot v_2^2}{2} + p_2 = \text{Cons tan t} \quad (6)$$

For steady state incompressible flow the Euler equation becomes. If we integrate along the streamline it becomes. Can further be modified to by dividing by gravity.

HEAD OF FLOW

Equation (3) is often referred to the head because all elements has the unit of length.

Dynamic Pressure

(2) and (3) are two forms of the Bernoulli Equation for steady state incompressible flow. If we assume that the gravitational body force is negligible, (3) can be written as (4). Both elements in the equation have the unit of pressure and it's common to refer the flow velocity component as the dynamic pressure of the fluid flow (5).

Since energy is conserved along the streamline, (4) can be expressed as (6). Using the equation we see that increasing the velocity of the flow will reduce the pressure, decreasing the velocity will increase the pressure.

This phenomena can be observed in a venturi meter where the pressure is reduced in the constriction area and regained after. It can also be observed in a pitot tube where the stagnation pressure is measured. The stagnation pressure is where the velocity component is zero.

Example

Bernoulli Equation and Flow from a Tank through a small Orifice. Liquid flows from a tank through a orifice close to the bottom. The Bernoulli equation can be adapted to a streamline from the surface to the orifice as (e1):

$$g \cdot h_1 + \frac{p_1}{\rho} + \frac{V_1^2}{2} = g \cdot h_2 + \frac{p_2}{\rho} + \frac{V_2^2}{2} \quad \text{(ei)}$$

where,

$$h = h_1 - h_2 \text{ (e2)}$$

$$V_1 = \left(\frac{A_2}{A_1}\right) \cdot V_2 \quad \text{(e3)}$$

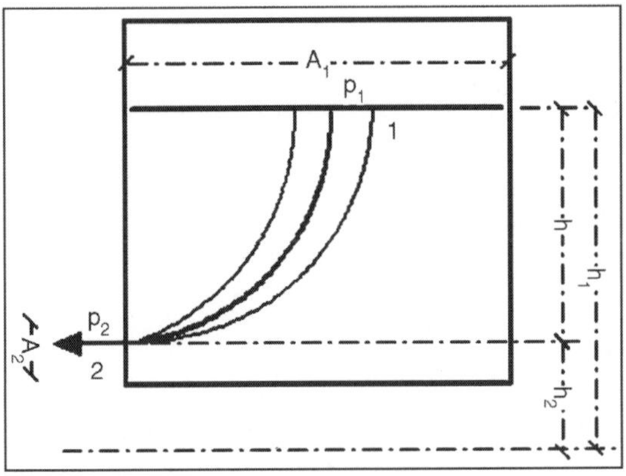

$$V_2=\sqrt{\frac{2}{\left(1-\frac{A_2^2}{A_1^2}\right)}\cdot\left(\frac{p_1-p_2}{\rho}+g\cdot h\right)} \quad \text{(e4)}$$

$$V_2=\sqrt{2\cdot g\cdot h} \quad \text{(e5)}$$

$$V_2=c\sqrt{2\cdot g\cdot h} \quad \text{(e5b)}$$

$$V_2=\sqrt{\left(\frac{2\cdot(p_1-p_2)}{\rho}\right)}$$

Since (1) and (2)'s heights from a common reference is related as (e2), and the equation of continuity can be expressed as (e3), it's possible to transform (e1) to (e4).

Vented tank

A special case of interest for equation (e4) is when the orifice area is much lesser than the surface area and when the pressure inside and outside the tank is the same - when the tank has an open surface or "vented" to the atmosphere. At this situation the (e4) can be transformed to (e5).

"The velocity out from the tank is equal to speed of a freely body falling the distance *h*." - also known as Torricelli's Theorem.

Example

Outlet velocity from a vented tank,

The outlet velocity of a tank with height *10 m* can be calculated as,

$$V_2 = (2\ (9.81\ \text{m/s}^2)\ (10\ \text{m}))^{1/2}$$
$$= 14\ \text{m/s}$$

Pressurized Tank

If the tanks is pressurized so that product of gravity and height (g h) is much lesser than the pressure difference divided by the density, (e4) can be transformed to (e6).

The velocity out from the tank depends mostly on the pressure difference.

Example

Outlet velocity from a pressurized tank.

The outlet velocity of a pressurized tank where,

$p_1 = 0.2\ MN/m^2$

$p_2 = 0.1\ MN/m^2$

$A_2/A_1 = 0.01$

$h = 10\ m$

can be calculated as:

$V_2 = ((2/(1-(0.01)^2)\ ((0.2\ 10^6\ N/m^2) - (0.1\ 10^6\ N/m^2))/(1000\ kg/m^3) + (9.81\ m/s^2)(10\ m)))^{1/2} = 19.9\ m/s$

A Pitot-static tube can measure the fluid flow velocity by converting the kinetic energy in the fluid flow into potential energy.

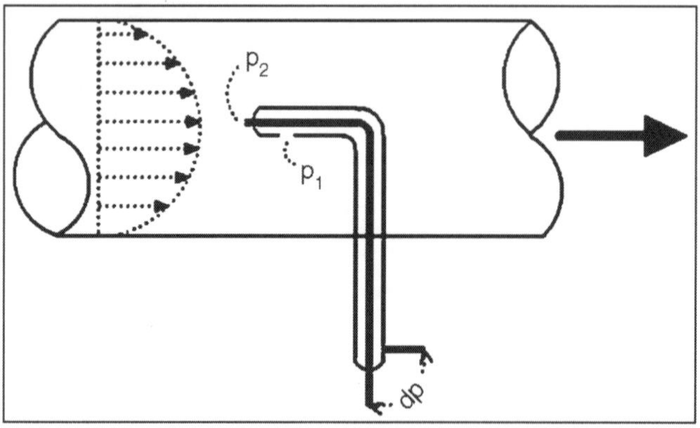

The principle is based on the Bernoulli Equation where each term can be interpreted as a form of pressure

$p + 1/2\ \rho\ v^2 + Yh$ = constant along a streamline (1)

where,

p = Static pressure (relative to the moving fluid) (Pa)

ρ = Density (kg/m³)

Y = Specific weight (kN/m³)

v = Flow velocity (m/s)

g = Acceleration of gravity (m/s²)

h = Elevation height (m)

Each term of this equation has the dimension force per unit area - N/m^2 or in imperial units *psi*, lb/ft^2.

Static Pressure

The first term - *p* - is the static pressure. It is static relative to the moving fluid and can be measured through an flat opening in parallel to the flow.

Dynamic Pressure

The second term - $1/2\ \rho\ v^2$ - is called the dynamic pressure.

Hydrostatic Pressure

The third term - Y h - is called the hydrostatic pressure. It represent the pressure due to change in elevation.

Stagnation Pressure

Since the Bernoulli Equation states that the energy along the streamline is constant, (1) can be modified to

$$p_1 + 1/2\ \rho\ v_1^2 + Y\ h_1$$
$$= p_2 + 1/2\ \rho\ v_2^2 + Y\ h_2$$
$$= \text{constant along the streamline} \quad (2)$$

where,

suffix $_1$ is a point the free flow upstream

suffix $_2$ is the stagnation point where the velocity in the flow is zero

Flow Velocity

In a measuring point we regard the hydrostatic pressure as a constant, $h_1 = h_2$ and this part can be eliminated. Since $v_2$is zero, (2) can be modified to:

$$p_1 + 1/2\ r\ v_1^2 = p_2 \quad (3)$$

or,

$$v_1 = [2\ (p_2 - p_1) / r]^{1/2} \quad (4)$$

whcrc,

$$p_2 - p_1 = dp \text{ (differential pressure)}$$

With (4) it's possible to calculate the flow velocity in point 1 - the free flow upstream - if we know the differential pressure difference $dp = p_2 - p_1$ and the density of the fluid.

PITOT TUBE

The pitot tube is a simple and convenient instrument to measure the difference between static, dynamic and total pressure (or head).

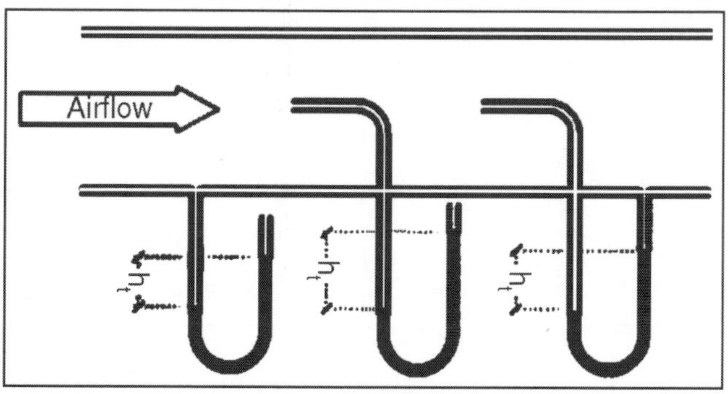

The head - *h* - (or pressure difference - *dp*) can be measured and calculated with u-tube manometers, electronic pressure transmitters or similar instrumentation. The Bernoulli equation can be applied to a great many situations not just the pipe flow we have been considering up to now. In the following sections we will see some examples of its application to flow measurement from tanks, within pipes as well as in open channels.

If a stream of uniform velocity flows into a blunt body, the stream lines take a pattern similar to this:

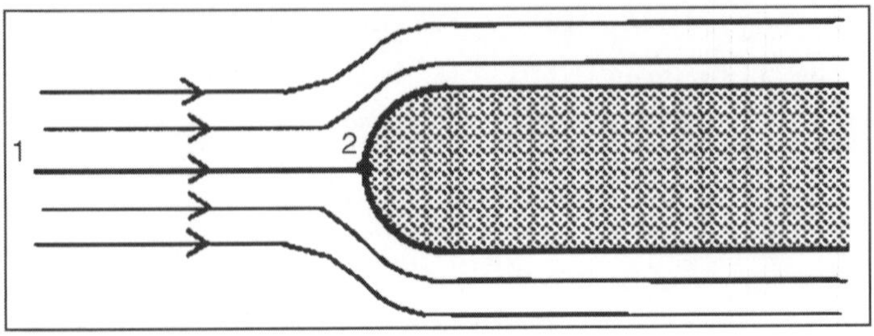

Fig. Streamlines Around a Blunt Body

Note how some move to the left and some to the right. But one, in the centre, goes to the tip of the blunt body and stops. It stops because at this point the velocity is zero - the fluid does not move at this one point. This point is known as the *stagnation point*.

From the Bernoulli equation we can calculate the pressure at this point. Apply Bernoulli along the central streamline from a point upstream where the velocity is $u_1$and the pressure p_1 to the stagnation point of the blunt body where the velocity is zero, $u_2 = 0$. Also $z_1 = z_2$.

$$\frac{p_1}{\rho g}+\frac{u_1^2}{2g}+z_1=\frac{p_2}{\rho g}+\frac{u_2^2}{2g}+z_2$$

$$\frac{p_1}{\rho}+\frac{u_1^2}{2}=\frac{p_2}{\rho}$$

$$p_2=p_1+\frac{1}{2}\rho u_1^2$$

This increase in pressure which bring the fluid to rest is called the *dynamic pressure*.

Dynamic pressure $= \frac{1}{2}\rho u_1^2$

or converting this to head (using,

$$h=\frac{p}{\rho g}).$$

Dynamic head $= \frac{1}{2g}u_1^2$.

The total pressure is know as the *stagnation pressure* (or *total pressure*).

$$\text{Stagnation pressure} = p_1 + \frac{1}{2}\rho u_1^2$$

or in terms of head.

$$\text{Stagnation head} = \frac{p_1}{\rho g} + \frac{1}{2g}u_1^2$$

The blunt body stopping the fluid does not have to be a solid. I could be a static column of fluid. Two piezometers, one as normal and one as a Pitot tube within the pipe can be used in an arrangement shown below to measure velocity of flow.

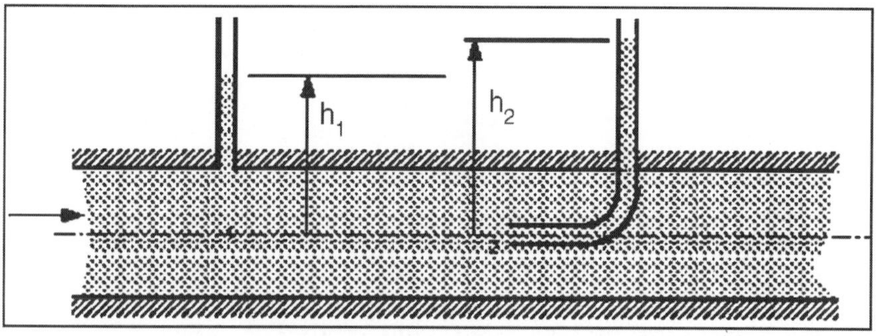

Fig. A Piezometer and a Pitot Tube

Using the above theory, we have the equation for p_2,

$$p_2 = p_1 + \frac{1}{2}\rho u_1^2$$

$$\rho g h_1 = \rho g h_1 + \frac{1}{2} r u_1^2$$

$$u = \sqrt{2g(h_2 - h_1)}$$

We now have an expression for velocity obtained from two pressure measurements and the application of the Bernoulli equation.

Pitot Static Tube

The necessity of two piezometers and thus two readings make this arrangement a little awkward. Connecting the piezometers to a manometer would simplify things but there are still two tubes.

The *Pitot static* tube combines the tubes and they can then be easily connected to a manometer. A Pitot static tube is shown below.

The holes on the side of the tube connect to one side of a manometer and register the *static head, (h_1)*, while the central hole is connected to the other side of the manometer to register, as before, the *stagnation head (h_2)*.

Consider the pressures on the level of the centre line of the Pitot tube and using the theory of the manometer:

$P_A = P_2 + pgX$

$P_B = P_1 + pg\,(X - h) + \rho_{max}gh$

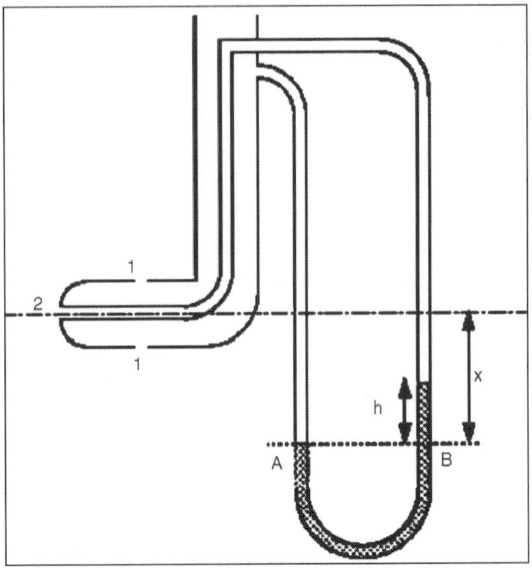

Fig. A Pitot-static tube

$$P_A = P_B$$

$$P_2 + \rho g X = p_1 + \rho g(X - h) + \rho_{max} g h$$

We know that $P_2 = P_{static} = P_1 + \frac{1}{2}\rho u_1^2$,

substituting this in to the above gives,

$$P_1 + hg(\rho_{max} - \rho) = P_1 + \frac{\rho u_1^2}{2}$$

$$u_1 = \sqrt{\frac{2gh(\rho_m - \rho)}{\rho}}$$

The Pitot/Pitot-static tubes give velocities at points in the flow. It does not give the overall discharge of the stream, which is often what is wanted. It also has the drawback that it is liable to block easily, particularly if there is significant debris in the flow.

Time for a Tank to Empty

We now have an expression for the discharge out of a tank based on the height of water above the orifice. It would be useful to know how long it would take for the tank to empty. As the tank empties, so the level of water will fall. We can get an expression for the time it takes to fall by integrating the expression for flow between the initial and final levels.

The tank has a cross sectional area of *A*. In a time *dt* the level falls by *dh* or the flow out of the tank is,

$$Q = Av$$

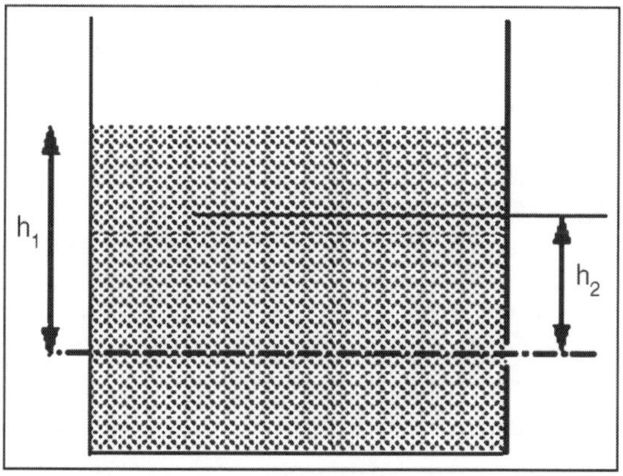

Fig. Tank Emptying from Level h_1 to h_2.

$$Q = -A\frac{\delta h}{\delta t}$$

(–ve sign as *dh* is falling), Rearranging and substituting the expression for *Q* through the orifice gives,

$$\delta h = \frac{-A}{C_d A_o \sqrt{2g}} \frac{\delta h}{\sqrt{h}}$$

This can be integrated between the initial level, h_1, and final level, h_2, to give an expression for the time it takes to fall this distance,

$$t = \frac{-A}{C_d A_o \sqrt{2g}} \int_{h_1}^{h_2} \frac{\delta h}{\sqrt{h}}$$

$$= \frac{-A}{C_d A_o \sqrt{2g}} \left[2\sqrt{h}\right]_{h_1}^{h_2}$$

$$= \frac{-2A}{C_d A_o \sqrt{2g}} \left[\sqrt{h_2} - \sqrt{h_1}\right]$$

Submerged Orifice: We have two tanks next to each other (or one tank separated by a dividing wall) and fluid is to flow between them through a submerged orifice. Although difficult to see, careful measurement of the flow indicates that the submerged jet flow behaves in a similar way to the jet in air in that it forms a vena contracta below the surface. To determine the velocity at the jet we first use the Bernoulli equation to give us the ideal velocity. Applying Bernoulli from point 1 on the surface of the deeper tank to point 2 at the centre of the orifice, gives,

$$\frac{P_1}{\rho g} + \frac{u_1^2}{2g} + z_1 = \frac{p_2}{\rho g} + \frac{u_2^2}{2g} + z_1$$

$$0 + 0 + h_1 = \frac{\rho g h_2}{\rho g} + \frac{u_2^2}{2g} + 0$$

$$u_2 = \sqrt{2g(h_1 - h_2)}$$

i.e. the ideal velocity of the jet through the submerged orifice depends on the*difference* in head across the orifice. And the discharge is given by

$$Q = C_d A_o u$$

$$= C_d A_o \sqrt{2g(h_1 - h_2)}$$

Time for Equalisation of Levels in Two Tanks:

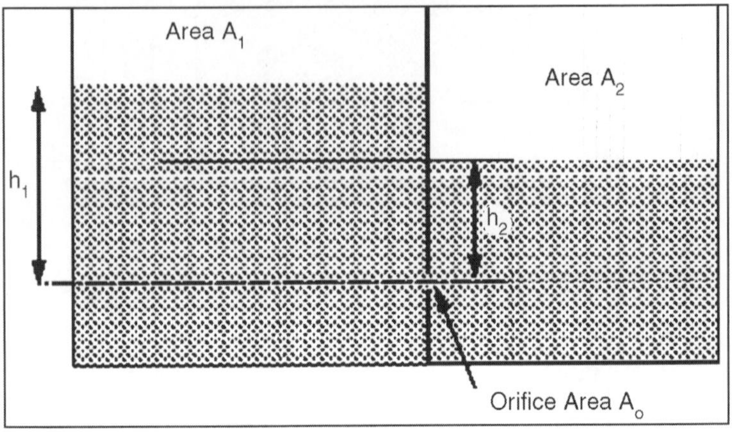

Fig. Two Tanks of Initially Different Levels Joined by an Orifice

By a similar analysis used to find the time for a level drop in a tank we can derive an expression for the change in levels when there is flow between two connected tanks.

Applying the continuity equation:

$$Q = -A_1 \frac{\delta h_1}{\delta t} = A_2 \frac{\delta h_2}{\delta t}$$

$$Q\,\delta t = -A_1\,\delta h_1 = A_2\,\delta h_2$$

Also we can write $-\,\delta h_1 + \delta h_2 = \delta h$,

So,

$$-\,A_1 dh_1 + A_2 dh_2 - A_2 dh$$

$$\delta h_1 = \frac{A_2 \delta h}{A_1 + A_2}$$

Then we get,

$$Q\ dt = -\,A_1 dh_1$$

$$C_d A_o \sqrt{2g(h_1 - h_2)}\,\delta t = \frac{A_1 A_2}{A_1 + A_2}\,\delta h$$

Re-arranging and integrating between the two levels we get,

$$\delta t = \frac{A_1 A_2}{(A_1 + A_2) C_d A_o \sqrt{2g}} \frac{\delta h}{h}$$

$$t = \frac{A_1 A_2}{(A_1 + A_2) C_d A_o \sqrt{2g}} \int_{h_{initial}}^{h_{final}} \frac{\delta h}{\sqrt{h}}$$

$$= \frac{2A_1A_2}{(A_1 + A_2)C_dA_o\sqrt{2g}}\left[\sqrt{h}\right]_{h_{initial}}^{h_{final}}$$

$$= \frac{2A_1A_2}{(A_1 + A_2)C_dA_o\sqrt{2g}}\left[\sqrt{h_{initial}} - \sqrt{h_{final}}\right]$$

(remember that *h* in this expression is the *difference* in height between the two levels ($h_2 - h_1$) to get the time for the levels to equal use $h_{initial} = h_1$ and $h_{final} = 0$).

Thus we have an expression giving the time it will take for the two levels to equal.

Flow Over Notches and Weirs

A notch is an opening in the side of a tank or reservoir which extends above the surface of the liquid. It is usually a device for measuring discharge. A weir is a notch on a larger scale - usually found in rivers. It may be sharp crested but also may have a substantial width in the direction of flow - it is used as both a flow measuring device and a device to raise water levels.

Weir Assumptions

We will assume that the velocity of the fluid approaching the weir is small so that kinetic energy can be neglected. We will also assume that the velocity through any elemental strip depends only on the depth below the free surface. These are acceptable assumptions for tanks with notches or reservoirs with weirs, but for flows where the velocity approaching the weir is substantial the kinetic energy must be taken into account (*e.g.* a fast moving river).

A General Weir Equation

To determine an expression for the theoretical flow through a notch we will consider a horizontal strip of width *b* and depth *h* below the free surface, as shown in the figure below.

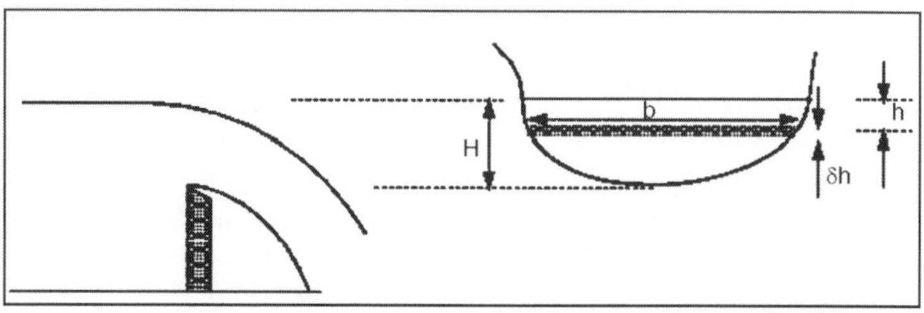

Fig. Elemental Strip of Flow Through a Notch

velocity through the strip, $u = \sqrt{2gh}$

discharge through the e strip, $\delta Q = Au = b\,\delta h\sqrt{2gh}$

integrating from the free surface, h = 0, to the weir crest, h = H gives the expression for the total theoretical discharge

$$Q_{\text{theoretical}} = \sqrt{2g}\int_{o}^{H} bh\frac{1}{2}dh$$

This will be different for every differently shaped weir or notch. To make further use of this equation we need an expression relating the width of flow across the weir to the depth below the free surface.

Rectangular Weir

For a rectangular weir the width does not change with depth so there is no relationship between b and depth h. We have the equation,

$$b = \text{constant} = B$$

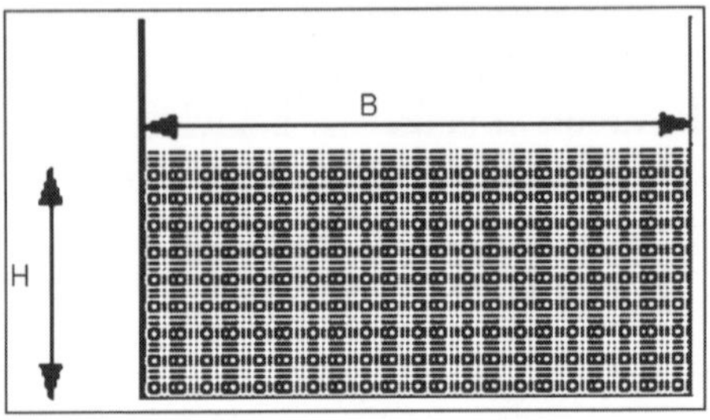

Fig. A Rectangular Weir

Substituting this into the general weir equation gives,

$$Q_{\text{theoretical}} = B\sqrt{2g}\int_{o}^{H} h\frac{1}{2}dh$$

$$= \frac{2}{3}B\sqrt{2gH^{3/2}}$$

To calculate the actual discharge we introduce a coefficient of discharge, C_d, which accounts for losses at the edges of the weir and contractions in the area of flow, giving,

$$Q_{\text{actual}} = C_d\frac{2}{3}B\sqrt{2gH^{3/2}}$$

'V' Notch Weir

For the "V" notch weir the relationship between width and depth is dependent on the angle of the "V".

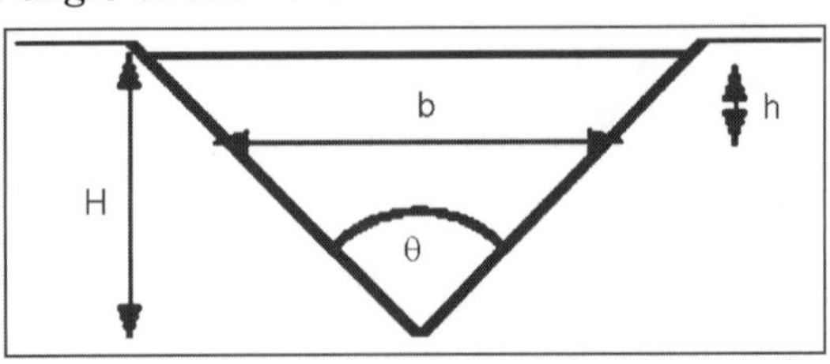

Fig. "V" Notch, or Triangular, Weir Geometry.

If the angle of the "V" is θ then the width, b, a depth h from the free surface is,

$$b = 2(H-h)\tan\left(\frac{\theta}{2}\right)$$

So the discharge is:

$$Q_{theoretical} = 2\sqrt{2g}\tan\left(\frac{\theta}{2}\right)\int_{o}^{H}(H-h)h^{1/2}\,dh$$

$$2\sqrt{2g}\tan\left(\frac{\theta}{2}\right)\left[\frac{2}{5}Hh^{3/2} - \frac{2}{5}h^{5/2}\right]_{o}^{H}$$

$$= \frac{8}{15}\sqrt{2g}\tan\left(\frac{\theta}{2}\right)H^{5/2}$$

And again, the actual discharge is obtained by introducing a coefficient of discharge,

$$Q_{actual} = C_d\,\frac{8}{15}\sqrt{2g}\tan\left(\frac{\theta}{2}\right)H^{5/2}$$

Venturi Meter

The Venturi meter is a device for measuring discharge in a pipe. It consists of a rapidly converging section which increases the velocity of flow and hence reduces the pressure. It then returns to the original dimensions of the pipe by a gently diverging 'diffuser' section. By measuring the pressure differences the discharge can be calculated. This is a particularly accurate method of flow measurement as energy loss are very small.

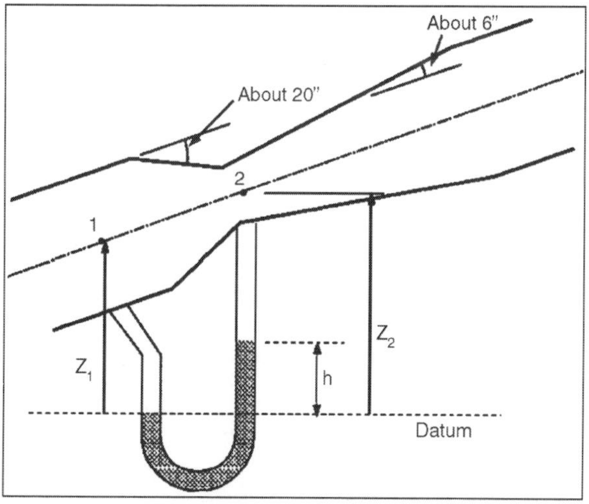

Fig. A Venturi Meter

Applying Bernoulli along the streamline from point 1 to point 2 in the narrow *throat* of the Venturi meter we have,

$$\frac{P_1}{\rho g}+\frac{u_1^2}{2g}+z_1=\frac{P_2}{\rho g}+\frac{u_2^2}{2g}+z_2$$

By the using the continuity equation we can eliminate the velocity u_2,

$$Q=u_1A_1=u_2A_2$$

$$u_2=\frac{u_1A_1}{A_2}$$

Substituting this into and rearranging the Bernoulli equation we get,

$$\frac{P_1-P_2}{\rho g}+z_1-z_2=\frac{u_1^2}{2g}\left[\left(\frac{A_1}{A_2}\right)^2-1\right]$$

$$u_1=\sqrt{\frac{2g\left[\frac{P_1-P_2}{\rho g}+z_1-z_2\right]}{\left(\frac{A_1}{A_2}\right)^2-1}}$$

To get the theoretical discharge this is multiplied by the area. To get the actual discharge taking in to account the losses due to friction, we include a coefficient of discharge,

$$Q_{ideal}=u_1A_1$$

$$Q_{ideal}=C_dQ_{ideal}=C_du_1A_1$$

$$Q_{ideal}=C_dA_1A_2\sqrt{\frac{2g\left[\frac{P_1-P_2}{\rho g}+z_1-z_2\right]}{A_1^2-A_2^2}}$$

This can also be expressed in terms of the manometer readings,

$$P_1 + rgz_1 = P_2 + r_{max}gh + rg(z_2 - h)$$

$$\frac{P_1-P_2}{\rho g}+z_1-z_2=h\left(\frac{\rho_{max}}{\rho}-1\right)$$

Thus the discharge can be expressed in terms of the manometer reading:

$$Q_{actual}=C_dA_1A_2\sqrt{\frac{2gh\left(\frac{\rho_{max}}{\rho}-1\right)}{A_1^2-A_2^2}}$$

Notice how this expression does not include any terms for the elevation or orientation (z_1 or z_2) of the Venturimeter. This means that the meter can be at any convenient angle to function.

The purpose of the diffuser in a Venturi meter is to assure gradual and steady deceleration after the throat. This is designed to ensure that the pressure rises again to something near to the original value before the Venturi meter. The angle of the diffuser is usually between 6 and 8 degrees. Wider than this and the flow might separate from the walls resulting in increased friction and energy and pressure loss.

If the angle is less than this the meter becomes very long and pressure losses again become significant. The efficiency of the diffuser of increasing pressure back to the original is rarely greater than 80%.

Flow Through A Small Orifice

We are to consider the flow from a tank through a hole in the side close to the base. The general arrangement and a close up of the hole and streamlines are shown in the figure below

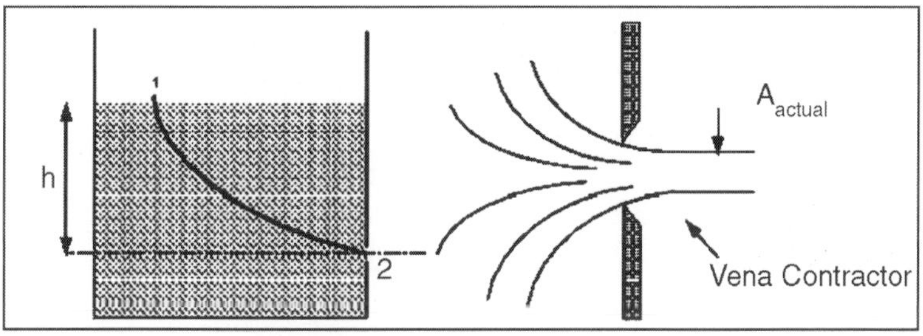

Fig. Tanket and Streamlines of Flow out of the Sharp Edged Orifice

The shape of the holes edges are as they are (sharp) to minimise frictional losses by minimising the contact between the hole and the liquid - the only contact is the very edge.

Looking at the streamlines you can see how they contract after the orifice to a minimum value when they all become parallel, at this point, the velocity and pressure are uniform across the jet.

This convergence is called the *vena contracta*. (From the Latin 'contracted vein'). It is necessary to know the amount of contraction to allow us to calculate the flow.

We can predict the velocity at the orifice using the Bernoulli equation. Apply it along the streamline joining point 1 on the surface to point 2 at the centre of the orifice. At the surface velocity is negligible ($u_1 = 0$) and the pressure atmospheric ($p_1 = 0$).At the orifice the jet is open to the air so again the pressure is atmospheric ($p_2 = 0$). If we take the datum line through the orifice then $z_1 = h$ and $z_2 = 0$, leaving

$$h = \frac{u_2^2}{2g}$$

$$u_2 = \sqrt{2gh}$$

This is the theoretical value of velocity. Unfortunately it will be an over estimate of the real velocity because friction losses have not been taken into account. To incorporate friction we use the coefficient of velocity to correct the theoretical velocity,

$$u_{actual} = C_v u_{theoretical}$$

Each orifice has its own coefficient of velocity, they usually lie in the range(0.97 – 0.99).

To calculate the discharge through the orifice we multiply the area of the jet by the velocity. The actual area of the jet is the area of the vena contracta not the area of the orifice. We obtain this area by using a coefficient of contraction for the orifice

$$A_{actual} = C_c A_{orifice}$$

So the discharge through the orifice is given by

$$\begin{aligned} Qactual &= A_{actua} u_{actual} \\ &= C_c C_v A_{orifice} u_{theoretical} \\ &= C_d A_{orifice} \sqrt{2gh} \end{aligned}$$

Where C_d is the coefficient of discharge, and $C_d = C_c\, C_v$

8

The Internal Energies in Specific Heats

INTRODUCTION

The internal energies and entropies of substances (mostly ideal gases) at some length. Unfortunately, these quantities cannot be directly measured. Instead, they must be inferred from other information. The thermodynamic property of substances which is the easiest to measure is, of course, the heat capacity, or specific heat. In fact, once the variation of the specific heat with temperature is known, both the internal energy and entropy can be easily reconstructed via

$$E(T, V) = v\int_0^T c_V(T,V)dT + E(0,V),$$

$$S(T,V) = v\int_0^T \frac{c_V(T,V)}{T}dT.$$

Here, use has been made of $dS = đQ/T$, and the third law of thermodynamics. Clearly, the optimum way of verifying the results of statistical thermodynamics is to compare the theoretically predicted heat capacities with the experimentally measured values.

Classical physics, in the guise of the equipartition theorem, says that each independent degree of freedom associated with a quadratic term in the energy possesses an average energy $(1/2)kT$ in thermal equilibrium at temperature T. Consider a substance made up of N molecules. Every molecular degree of freedom contributes $(1/2)N\ k\ T$, or $(1/2)\ vRT$, to the mean energy of the substance (with the tacit proviso that each degree of freedom is associated with a quadratic term in the energy). Thus, the contribution to the molar heat capacity at constant volume (we wish to avoid the complications associated with any external work done on the substance) is,

$$\frac{1}{v}\left(\frac{\partial E}{\partial T}\right)_V = \frac{1}{v}\frac{\partial[(1/2)vRT]}{\partial T} = \frac{1}{2}R,$$

per molecular degree of freedom. The total classical heat capacity is therefore,

$$c_V = \frac{g}{2} R,$$

where g is the number of molecular degrees of freedom. Since large complicated molecules clearly have very many more degrees of freedom than small simple molecules, the above formula predicts that the molar heat capacities of substances made up of the former type of molecules should greatly exceed those of substances made up of the latter. In fact, the experimental heat capacities of substances containing complicated molecules are generally greater than those of substances containing simple molecules, but by nowhere near the large factor predicted by previous equation.

This equation also implies that heat capacities are temperature independent. In fact, this is not the case for most substances. Experimental heat capacities generally increase with increasing temperature. These two experimental facts pose severe problems for classical physics. Incidentally, these problems were fully appreciated as far back as 1850. Stories that physicists at the end of the nineteenth century thought that classical physics explained absolutely everything are largely apocryphal.

The equipartition theorem (and the whole classical approximation) is only valid when the typical thermal energy kT greatly exceeds the spacing between quantum energy levels. Suppose that the temperature is sufficiently low that this condition is not satisfied for one particular molecular degree of freedom. In fact, suppose that kT is much less than the spacing between the energy levels.

In this situation the degree of freedom only contributes the ground-state energy, E_0, say, to the mean energy of the molecule. The ground-state energy can be a quite complicated function of the internal properties of the molecule, but is certainly not a function of the temperature, since this is a collective property of all molecules. It follows that the contribution to the molar heat capacity is,

$$\frac{1}{v}\left(\frac{\partial [N\, E_0]}{\partial T}\right)_v = 0.$$

Thus, if kT is much less than the spacing between the energy levels then the degree of freedom contributes nothing at all to the molar heat capacity. We say that this particular degree of freedom is frozen out. Clearly, at very low temperatures just about all degrees of freedom are frozen out. As the temperature is gradually increased, degrees of freedom successively “kick in,” and eventually contribute their full (1/2) R to the molar heat capacity, as kT approaches, and then greatly exceeds, the spacing between their quantum energy levels. We can use these simple ideas to explain the behaviours of most experimental heat capacities.

To make further progress, we need to estimate the typical spacing between the quantum energy levels associated with various degrees of freedom. We can do this by observing the frequency of the electromagnetic radiation emitted and absorbed during transitions between these energy levels. If the typical spacing between energy levels is ΔE then transitions between the various levels are associated with photons of frequency v, where $hv = \Delta E$.

We can define an effective temperature of the radiation via $hv = kT_{rad}$. If $T >> T_{rad}$ then $kT >>$ DE, and the degree of freedom makes its full contribution to the heat capacity. On the other hand, if $T << T_{rad}$ then $kT << \Delta E$, and the degree of freedom is frozen out. The "temperatures" of various different types of radiation. It is clear that degrees of freedom which give rise to emission or absorption of radio or microwave radiation contribute their full (1/2)R to the molar heat capacity at room temperature.

Degrees of freedom which give rise to emission or absorption in the visible, ultraviolet, X–ray, or g–ray regions of the electromagnetic spectrum are frozen out at room temperature. Degrees of freedom which emit or absorb infrared radiation are on the border line.

Table. Effective "Temperatures" of Various Types of Electromagnetic Radiation

Radiation type	Frequency (Hz)	T_{rad} (°K)
Radio	$< 10^9$	< 0.05
Microwave	$10^9 - 10^{11}$	< 0.05
Infrared	$10^{11} - 10^{14}$	5. 5000
Visible	5×10^{14}	2×10^4
Ultraviolet	$10^{15} - 10^{17}$	5×10^4 5×10^5
X-ray	$10^{17} - 10^{20}$	5×10^6 5×10^9
g-ray	$> 10^{20}$	$> 5 \times 10^9 3$

Specific Heats of Gases

Let us now investigate the specific heats of gases. Consider, first of all, translational degrees of freedom. Every molecule in a gas is free to move in three dimensions. If one particular molecule has mass m and momentum $p = mv$ then its kinetic energy of translation is,

$$K = \frac{1}{2m}\left(p_x^2 + p_y^2 + p_z^2\right).$$

The kinetic energy of other molecules does not involve the momentum P of this particular molecule. Moreover, the potential energy of interaction between molecules depends only on their position coordinates, and, thus, certainly does not involve P. Any internal rotational, vibrational, electronic, or nuclear degrees of freedom of the molecule also do not involve P.

Hence, the essential conditions of the equipartition theorem are satisfied (at least, in the classical approximation). Since equation contains three independent quadratic terms, there are clearly three degrees of freedom associated with translation (one for each dimension of space), so the translational contribution to the molar heat capacity of gases is,

$$(c_V)_{\text{translation}} = \frac{3}{2}R.$$

Suppose that our gas is contained in a cubic enclosure of dimensions L. According to Schrödinger's equation, the quantized translational energy levels of an individual molecule are given by,

$$E = \frac{\hbar^2\pi^2}{2mL^2}\left(n_1^2 + n_2^2 + n_3^2\right),$$

where n_1, n_2, and n_3 are positive integer quantum numbers. Clearly, the spacing between the energy levels can be made arbitrarily small by increasing the size of the enclosure. This implies that translational degrees of freedom can be treated classically, so that equation is always valid (except very close to absolute zero). We conclude that all gases possess a minimum molar heat capacity of (3/2) R due to the translational degrees of freedom of their constituent molecules.

The electronic degrees of freedom of gas molecules (*i.e.*, the possible configurations of electrons orbiting the atomic nuclei) typically give rise to absorption and emission in the ultraviolet or visible regions of the spectrum. It follows from electronic degrees of freedom are frozen out at room temperature.

Similarly, nuclear degrees of freedom (*i.e.*, the possible configurations of protons and neutrons in the atomic nuclei) are frozen out because they are associated with absorption and emission in the X–ray and g–ray regions of the electromagnetic spectrum. In fact, the only additional degrees of freedom we need worry about for gases are rotational and vibrational degrees of freedom. These typically give rise to absorption lines in the infrared region of the spectrum.

The rotational kinetic energy of a molecule tumbling in space can be written,

$$K = \frac{1}{2}I_x\omega_x^2 + \frac{1}{2}I_y\omega_y^2 + \frac{1}{2}I_z\omega_z^2,$$

where the x –, y–, and z–axes are the so called principle axes of inertia of the molecule (these are mutually perpendicular), w_x, w_y, and w_z are the angular velocities of rotation about these axes, and I_x, I_y, and I_z are the moments of inertia of the molecule about these axes. No other degrees of freedom depend on the angular velocities of rotation.

Since the kinetic energy of rotation is the sum of three quadratic terms, the rotational contribution to the molar heat capacity of gases is,

$$(c_V)_{\text{rotation}} = \frac{3}{2}R,$$

according to the equipartition theorem. Note that the typical magnitude of a molecular moment of inertia is md^2, where m is the molecular mass, and d is the typical interatomic spacing in the molecule. A special case arises if the molecule is linear (*e.g.* if the molecule is diatomic).

In this case, one of the principle axes lies along the line of centers of the atoms. The moment of inertia about this axis is of order ma^2, where a is a typical nuclear dimension (remember that nearly all of the mass of an atom resides in the nucleus). Since $a \sim 10^{-5}\, d$,it follows that the moment of inertia about the line of centres is minuscule compared to the moments of inertia about the other two principle axes. In quantum mechanics, angular momentum is quantized in units of $\hbar$. The energy levels of a rigid rotator are written,

$$E = \frac{\hbar^2}{2I} J(J+1),$$

where I is the moment of inertia and J is an integer. Note the inverse dependence of the spacing between energy levels on the moment of inertia.

It is clear that for the case of a linear molecule, the rotational degree of freedom associated with spinning along the line of centres of the atoms is frozen out at room temperature, given the very small moment of inertia along this axis, and, hence, the very widely spaced rotational energy levels.

Classically, the vibrational degrees of freedom of a molecule are studied by standard normal mode analysis of the molecular structure. Each normal mode behaves like an independent harmonic oscillator, and, therefore, contributes R to the molar specific heat of the gas [(1/2) R from the kinetic energy of vibration and (1/2)R from the potential energy of vibration].

A molecule containing n atoms has $n - 1$ normal modes of vibration. For instance, a diatomic molecule has just one normal mode (corresponding to periodic stretching of the bond between the two atoms). Thus, the classical contribution to the specific heat from vibrational degrees of freedom is,

$$(c_V)_{\text{vibration}} = (n-1)R.$$

The rotational and vibrational degrees of freedom actually make a contribution to the specific heats of gases at room temperature, once quantum effects are taken into consideration? We can answer this question by examining just one piece of data. The infrared absorption spectrum of Hydrogen Chloride. The absorption lines correspond to simultaneous transitions between different vibrational and rotational energy levels.

Hence, this is usually called a vibration-rotation spectrum. The missing line at about 3.47 microns corresponds to a pure vibrational transition from the ground-state to the first excited state (pure vibrational transitions are forbidden:

HCl molecules always have to simultaneously change their rotational energy level if they are to couple effectively to electromagnetic radiation).

The longer wavelength absorption lines correspond to vibrational transitions in which there is a simultaneous decrease in the rotational energy level. Likewise, the shorter wavelength absorption lines correspond to vibrational transitions in which there is a simultaneous increase in the rotational energy level.

It is clear that the rotational energy levels are more closely spaced than the vibrational energy levels. The pure vibrational transition gives rise to absorption at about 3.47 microns, which corresponds to infrared radiation of frequency 8.5×10^{11} hertz with an associated radiation "temperature" of 4400 degrees kelvin.

We conclude that the vibrational degrees of freedom of HCl, or any other small molecule, are frozen out at room temperature. The rotational transitions split the vibrational lines by about 0.2 microns.

This implies that pure rotational transitions would be associated with infrared radiation of frequency 5×10^{12} hertz and corresponding radiation "temperature" 260 degrees kelvin. We conclude that the rotational degrees of freedom of HCl, or any other small molecule, are not frozen out at room temperature, and probably contribute the classical $(1/2)R$ to the molar specific heat. There is one proviso, however. Linear molecules (like HCl) effectively only have two rotational degrees of freedom (instead of the usual three), because of the very small moment of inertia of such molecules along the line of centres of the atoms.

We are now in a position to make some predictions regarding the specific heats of various gases. Monatomic molecules only possess three translational degrees of freedom, so monatomic gases should have a molar heat capacity $(3/2)R = 12.47$ joules/degree/mole.

The ratio of specific heats $g = Cp/C_V = (C_V + R)/C_V$ should be $5/3 = 1.667$. It can be seen from both of these predictions are borne out pretty well for Helium and Argon. Diatomic molecules possess three translational degrees of freedom and two rotational degrees of freedom (all other degrees of freedom are frozen out at room temperature). Thus, diatomic gases should have a molar heat capacity $(5/2)\mathrm{R} = 20.8$ joules/degree/mole.

The ratio of specific heats should be $7/5 = 1.4$ It can be seen from these are pretty accurate predictions for Nitrogen and Oxygen. The freezing out of vibrational degrees of freedom becomes gradually less effective as molecules become heavier and more complex.

This is partly because such molecules are generally less stable, so the force constant k is reduced, and partly because the molecular mass is increased. Both these effect reduce the frequency of vibration of the molecular normal modes and, hence, the spacing between vibrational energy levels. This accounts for the obviously non-classical [*i.e.*, not a multiple of (1/2)R] specific heats of Carbon Dioxide and Ethane. In both molecules, vibrational degrees of freedom contribute to the molar specific heat (but not the full R because the temperature is not high enough).

The variation of the molar heat capacity at constant volume (in units of R) of gaseous hydrogen with temperature. The expected contribution from the translational degrees of freedom is (3/2)R (there are three translational degrees of freedom per molecule).

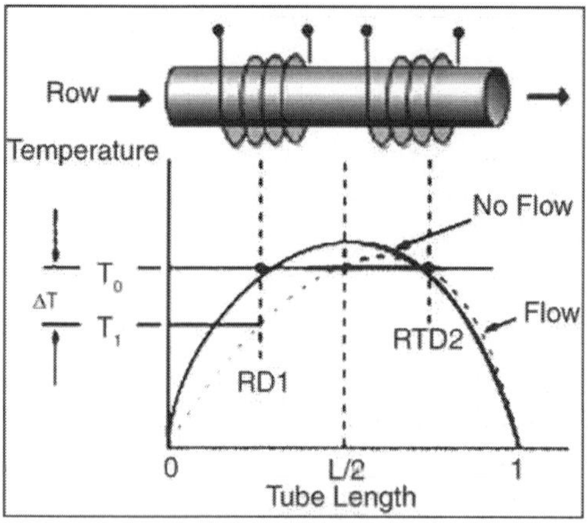

Fig. The Molar Heat Capacity at Constant Volume.

The expected contribution at high temperatures from the rotational degrees of freedom is R (there are effectively two rotational degrees of freedom per molecule). Finally, the expected contribution at high temperatures from the vibrational degrees of freedom is R (there is one vibrational degree of freedom per molecule). It can be seen that as the temperature rises the rotational, and then the vibrational, degrees of freedom eventually make their full classical contributions to the heat capacity.

THE MAXWELL DISTRIBUTION

Consider a molecule of mass m in a gas which is sufficiently dilute for the intermolecular forces to be negligible (*i.e.*, an ideal gas). The energy of the molecule is written,

$$\in = \frac{p^2}{2m} + \in^{\text{int}},$$

where p is its momentum vector, and $\hat{I}^{\text{int}}$ is its internal (*i.e.*, non-translational) energy. The latter energy is due to molecular rotation, vibration, etc.

Translational degrees of freedom can be treated classically to an excellent approximation, whereas internal degrees of freedom usually require a quantum mechanical approach.

Classically, the probability of finding the molecule in a given internal state with a position vector in the range r to $r + dr$, and a momentum vector in the range P to $P + dP$, is proportional to the number of cells (of "volume" h_0) contained in the corresponding region of phase-space, weighted by the Boltzmann factor.

In fact, since classical phase-space is divided up into uniform cells, the number of cells is just proportional to the "volume" of the region under consideration. This "volume" is written $d^3r d^3p$. Thus, the probability of finding the molecule in a given internal state s is,

$$P_s(r,p)d^3r\, d^3p \propto \exp(-\beta p^2/2m)\exp(\beta \epsilon_s^{\text{int}})d^3r\, d^3p,$$

where Ps is a probability density defined in the usual manner. The probability $P(r, p)\, d^3r\, d^3p$ of finding the molecule in any internal state with position and momentum vectors in the specified range is obtained by summing the above expression over all possible internal states.

The sum over exp $(-\mathrm{b} \epsilon_s^{\text{int}})$ just contributes a constant of proportionality (since the internal states do not depend on r or p), so,

$$P(r,p)d^3r\, d^3p \propto \exp(-\beta p^2/2m)\, d^3r\, d^3p.$$

Of course, we can multiply this probability by the total number of molecules N in order to obtain the mean number of molecules with position and momentum vectors in the specified range.

Suppose that we now want to determine: $f(r, v)\, d^3r\, d^3v$ *i.e.*, the mean number of molecules with positions between r and $r + dr$, and velocities in the range v and $v + dv$. Since $v = p/m$, it is easily seen that,

$$f(r,v)d^3r\, d^3v = C\exp(-\beta m v^2/2)\, d^3r\, d^3v,$$

where C is a constant of proportionality. This constant can be determined by the condition

$$\int_{(r)}\int_{(v)} f(r,v)d^3r\, d^3v = N:$$

i.e., the sum over molecules with all possible positions and velocities gives the total number of molecules, N. The integral over the molecular position coordinates just gives the volume V of the gas, since the Boltzmann factor is independent of position. The integration over the velocity coordinates can be reduced to the product of three identical integrals (one for v_x, one for v_y, and one for v_z), so we have,

$$CV\left[\int_{-\infty}^{\infty}\exp(-\beta m v_z^2/2)dv_z\right]^3 = N.$$

Now,

$$\int_{-\infty}^{\infty}\exp(-\beta m v_z^2/2)dv_z = \sqrt{\frac{2}{\beta m}}\int_{-\infty}^{\infty}\exp(-y^2)dy = \sqrt{\frac{2\pi}{\beta m}},$$

so $C = (N/V)\ (bm/2p)^{3/2}$. Thus, the properly normalized distribution function for molecular velocities is written,

$$f(v)d^3r\, d^3v = n\left(\frac{m}{2\pi kT}\right)^{3/2} \exp(-mv^2/2kT)d^3r\, d^3v.$$

Here, $n = N/V$ is the number density of the molecules. We have omitted the variable r in the argument of f, since f clearly does not depend on position. In other words, the distribution of molecular velocities is uniform in space. This is hardly surprising, since there is nothing to distinguish one region of space from another in our calculation.

The above distribution is called the Maxwell velocity distribution, because it was discovered by James Clark Maxwell in the middle of the nineteenth century. The average number of molecules per unit volume with velocities in the range v to $v + dv$ is obviously $f(v)\, d^3v$.

Let us consider the distribution of a given component of velocity: the z-component, say. Suppose that $g(v_z)\, dv_z$ is the average number of molecules per unit volume with the z-component of velocity in the range v_z to $v_z + dv_z$, irrespective of the values of their other velocity components. It is fairly obvious that this distribution is obtained from the Maxwell distribution by summing (integrating actually) over all possible values of v_x and v_y, with v_z in the specified range. Thus,

$$g(vz)dv_z = \int_{(v_z)}\int_{(v_y)} f(v)d^3v.$$

This gives,

$$g(v_z)dv_z = n\left(\frac{m}{2\pi kT}\right)$$

$$\int_{(v_z)}\int_{(v_y)} \exp\left[-(m/2kT)\left(v_z^2 + v_v^2 + v_z^2\right)\right]dv_x dv_y dv_z$$

$$= n\left(\frac{m}{2\pi\, kT}\right)^{3/2} \exp\left(-mv_z^2/2kT\right)\left[\int_{-\infty}^{\infty} \exp\left(-mv_x^2/2k\, T\right)\right]^2$$

$$= n\left(\frac{m}{2\pi\, kT}\right)^{3/2} \exp\left(-mv_z^2/2\, kT\right)\left(\sqrt{\frac{2\pi\, kT}{m}}\right)^2,$$

or,

$$g(v_z)\, dv_z = n\left(\frac{m}{2\pi\, kT}\right)^{1/2} \exp\left(-mv_z^2/2\, kT\right)dv_z.$$

Of course, this expression is properly normalized, so that,

$$\int_{-\infty}^{\infty} g(v_z)dv_z = n.$$

It is clear that each component (since there is nothing special about the z-component) of the velocity is distributed with a Gaussian probability distribution, centred on a mean value,

$$\overline{v_z} = 0,$$

with variance,

$$\overline{v_z^2} = \frac{kT}{m}.$$

Equation implies that each molecule is just as likely to be moving in the plus z-direction as in the minus z-direction. Equation can be rearranged to give,

$$\overline{\frac{1}{2}mv_z^2} = \frac{1}{2}kT,$$

in accordance with the equipartition theorem.

Note that equation can be rewritten,

$$\frac{f(v)d^3v}{n} = \left[\frac{g(v_x)dv_x}{n}\right]\left[\frac{g(v_y)dv_y}{n}\right]\left[\frac{g(v_z)dv_x}{n}\right],$$

where $g(v_x)$ and $g(v_y)$ are defined in an analogous way to $g(v_z)$. Thus, the probability that the velocity lies in the range v to $v + dv$ is just equal to the product of the probabilities that the velocity components lie in their respective ranges. In other words, the individual velocity components act like statistically independent variables. Suppose that we now want to calculate $F(v)\,dv$: *i.e.*, the average number of molecules per unit volume with a speed $v = |v|$ in the range v to $v + dv$. It is obvious that we can obtain this quantity by adding up all molecules with speeds in this range, irrespective of the direction of their velocities. Thus,

$$F(v)dv = \int f(v)d^3v,$$

where the integral extends over all velocities satisfying,

$$v < |v| < v + dv.$$

This inequality is satisfied by a spherical shell of radius v and thickness dv in velocity space. Since $f(v)$ only depends on $|v|$, so $f(v) \equiv f(v)$, the above integral is just $f(v)$ multiplied by the volume of the spherical shell in velocity space. So,

$$F(v)dv = 4\pi f(v)v^2dv,$$

which gives,

$$F(v)dv = 4\pi f(v)\left(\frac{m}{2\pi\, kT}\right)v^2 \exp(-mv^2/2k\,T)\,dv.$$

This is the famous Maxwell distribution of molecular speeds. Of course, it is properly normalized, so that,

$$\int_0^\infty F(v)dv = n.$$

Note that the Maxwell distribution exhibits a maximum at some non-zero value of v. The reason for this is quite simple. As v increases, the Boltzmann factor decreases, but the volume of phase-space available to the molecule (which is proportional to v^2) increases: the net result is a distribution with a non-zero maximum.

The mean molecular speed is given by,

$$\overline{v} = \frac{1}{n}\int_0^\infty F(v)v\,dv.$$

Thus, we obtain,

$$\overline{v} = 4\pi\left(\frac{m}{2\pi\,kT}\right)^{3/2}\int_0^\infty v^3 \exp(-mv^2/2kT)\,dv,$$

or,

$$\overline{v} = 4\pi\left(\frac{m}{2\pi\,kT}\right)^{3/2}\left(\frac{2k\,T}{m}\right)^2\int_0^\infty y^3 \exp(-y^2)\,dy$$

Now,

$$\int_0^\infty y^3 \exp(-y^2)\,dy = \frac{1}{2},$$

so,

$$\overline{v} = \sqrt{\frac{8}{\pi}\frac{kT}{m}}.$$

A similar calculation gives,

$$v_{\text{rms}} = \sqrt{\overline{v^2}} = \sqrt{\frac{3\,kT}{m}}.$$

However, this result can also be obtained from the equipartition theorem. Since,

$$\overline{\frac{1}{2}mv^2} = \overline{\frac{1}{2}m\left(v_z^2 + v_y^2 + v_z^2\right)} = 3\left(\frac{1}{2}kT\right),$$

then equation follows immediately. It is easily demonstrated that the most probable molecular speed (*i.e.*, the maximum of the Maxwell distribution function) is,

$$\overline{v} = \sqrt{\frac{2kT}{m}}.$$

The speed of sound in an ideal gas is given by,

$$c_s = \sqrt{\frac{\gamma\,kT}{m}},$$

where g is the ratio of specific heats.

This can also be written since $p = nkT$ and r = nm. It is clear that the various average speeds which we have just calculated are all of order the sound speed (*i.e.*, a few hundred meters per second at room temperature). In ordinary air ($\gamma = 1.4$) the sound speed is about 84% of the most probable molecular speed, and about 74% of the mean molecular speed. Since sound waves ultimately propagate via molecular motion, it makes sense that they travel at slightly less than the most probable and mean molecular speeds.

The Maxwell velocity distribution as a function of molecular speed in units of the most probable speed. Also shown are the mean speed and the root mean square speed.

It is difficult to directly verify the Maxwell velocity distribution. However, this distribution can be verified indirectly by measuring the velocity distribution of atoms exiting from a small hole in an oven.

The velocity distribution of the escaping atoms is closely related to, but slightly different from, the velocity distribution inside the oven, since high velocity atoms escape more readily than low velocity atoms. In fact, the predicted velocity distribution of the escaping atoms varies like $v^3 \exp(-mv^2/2kT)$, in contrast to the $v^3 \exp(-mv^2/2kT)$ variation of the velocity distribution inside the oven.

The measured and theoretically predicted velocity distributions of potassium atoms escaping from an oven at 157 °C. There is clearly very good agreement between the two.

SOLIDS AND SPECIFIC HEATS

Consider a simple solid containing N atoms. Now, atoms in solids cannot translate (unlike those in gases), but are free to vibrate about their equilibrium positions. Such vibrations are called lattice vibrations, and can be thought of as sound waves propagating through the crystal lattice. Each atom is specified by three independent position coordinates, and three conjugate momentum coordinates.

Let us only consider small amplitude vibrations. In this case, we can expand the potential energy of interaction between the atoms to give an expression which is quadratic in the atomic displacements from their equilibrium positions. It is alHways possible to perform a normal mode analysis of the oscillations. In effect, we can find $3N$ independent modes of oscillation of the solid.

Each mode has its own particular oscillation frequency, and its own particular pattern of atomic displacements. Any general oscillation can be written as a linear combination of these normal modes.

Let q_i be the (appropriately normalized) amplitude of the ith normal mode, and p_i the momentum conjugate to this coordinate. In normal mode coordinates, the total energy of the lattice vibrations takes the particularly simple form,

$$E = \frac{1}{2}\sum_{i=1}^{3N}(p_i^2 + \omega_i^2 q_i^2),$$

where w_i is the (angular) oscillation frequency of the ith normal mode. It is clear that in normal mode coordinates, the linearized lattice vibrations are equivalent to $3N$ independent harmonic oscillators (of course, each oscillator corresponds to a different normal mode).

The typical value of ω_i is the (angular) frequency of a sound wave propagating through the lattice. Sound wave frequencies are far lower than the typical vibration frequencies of gaseous molecules.

In the latter case, the mass involved in the vibration is simply that of the molecule, whereas in the former case the mass involved is that of very many atoms (since lattice vibrations are non-localized).

The strength of interatomic bonds in gaseous molecules is similar to those in solids, so we can use the estimate $\omega \sim \sqrt{k/m}$ (k is the force constant which measures the strength of interatomic bonds, and m is the mass involved in the oscillation) as proof that the typical frequencies of lattice vibrations are very much less than the vibration frequencies of simple molecules.

It follows from $\Delta E - \hbar\omega$ that the quantum energy levels of lattice vibrations are far more closely spaced than the vibrational energy levels of gaseous molecules. Thus, it is likely (and is, indeed, the case) that lattice vibrations are not frozen out at room temperature, but, instead, make their full classical contribution to the molar specific heat of the solid.

If the lattice vibrations behave classically then, according to the equipartition theorem, each normal mode of oscillation has an associated mean energy kT in equilibrium at temperature T [(1/2)kT resides in the kinetic energy of the oscillation, and (1/2)kT resides in the potential energy]. Thus, the mean internal energy per mole of the solid is,

$$\overline{E} = 3NkT = 3vRT.$$

It follows that the molar heat capacity at constant volume is,

$$c_V = \frac{1}{v}\left(\frac{\partial \overline{E}}{\partial T}\right)_V = 3R$$

for solids. This gives a value of 24.9 joules/mole/degree. In fact, at room temperature most solids (in particular, metals) have heat capacities which lie remarkably close to this value. This fact was discovered experimentally by Dulong and Petite at the beginning of the nineteenth century, and was used to make some of the first crude estimates of the molecular weights of solids (if we know the molar heat capacity of a substance then we can easily work out how much of it corresponds to one mole, and by weighing this amount, and then dividing the result by Avogadro's number, we can obtain an estimate of the molecular weight).

The experimental molar heat capacities *Cp* at constant pressure for various solids. The heat capacity at constant volume is somewhat less than the constant pressure value, but not by much, because solids are fairly incompressible. It can be seen that Dulong and Petite's law (*i.e.*, that all solids have a molar heat capacities close to 24.9 joules/mole/degree) holds pretty well for metals. However, the law fails badly for diamond. This is not surprising.

As is well-known, diamond is an extremely hard substance, so its intermolecular bonds must be very strong, suggesting that the force constant k is large. Diamond is also a fairly low density substance, so the mass m involved in lattice vibrations is comparatively small. Both these facts suggest that the typical lattice vibration frequency of diamond ($\omega \sim \sqrt{k/m}$) is high. In fact, the spacing between the different vibration energy levels (which scales like $\hbar\omega$) is sufficiently large in diamond for the vibrational degrees of freedom to be largely frozen out at room temperature.

This accounts for the anomalously low heat capacity of diamond.

Table. Values of (Joules/Mole/Degree)
For Some Solids at T = 298° K. K. From Reif.

Solid	*Cp*	Solid	*Cp*
Copper	24.5	Aluminium	24.4
Silver	25.5	Tin (white)	26.4
Lead	26.4	Sulphur (rhombic)	22.4
Zinc	25.4	Carbon (diamond)	6.1

Dulong and Petite's law is essentially a high temperature limit. The molar heat capacity cannot remain a constant as the temperature approaches absolute zero, since, by equation this would imply S ® ¥, which violates the third law of thermodynamics. We can make a crude model of the behaviour of C_V at low temperatures by assuming that all the normal modes oscillate at the same frequency, w, say.

According to equation the solid acts like a set of 3*N* independent oscillators which, making use of Einstein's approximation, all vibrate at the same frequency. We can use the quantum mechanical result for a single oscillator to write the mean energy of the solid in the form,

$$\overline{E} = 3N\hbar\omega\left(\frac{1}{2} + \frac{1}{\exp(\beta\hbar\omega) - 1}\right).$$

The molar heat capacity is defined,

$$c_V = \frac{1}{v}\left(\frac{\partial \overline{E}}{\partial T}\right)_V = \frac{1}{v}\left(\frac{\partial \overline{E}}{\partial T}\right)_V \frac{\partial \beta}{\partial T} = -\frac{1}{vkT^2}\left(\frac{\partial \overline{E}}{\partial T}\right)_V,$$

giving,

$$c_V = -\frac{3\,N_A\hbar\omega}{kT^2}\left[-\frac{\exp(\beta\hbar\omega)}{[\exp(\beta\hbar\omega) - 1]^2}\right],$$

which reduces to,

$$c_V = 3R\left(\frac{\theta_E}{T}\right)^2 \frac{\exp(\theta_E/T)}{[\exp(\theta_E/T)-1]^2}.$$

Here,

$$\theta_E = \frac{\hbar\omega}{k}$$

is called the Einstein temperature. If the temperature is sufficiently high that $T >> \theta_E$ then $kT >> \hbar\omega$, and the above expression reduces to $C_V = 3R$,, after expansion of the exponential functions. Thus, the law of Dulong and Petite is recovered for temperatures significantly in excess of the Einstein temperature. On the other hand, if the temperature is sufficiently low that $T << \theta_E$ then the exponential factors in equation become very much larger than unity, giving,

$$C_V \sim 3R\left(\frac{\theta_E}{T}\right)\exp(-\theta_E/T).$$

So, in this simple model the specific heat approaches zero exponentially as $T \to 0$.

In reality, the specific heats of solids do not approach zero quite as quickly as suggested by Einstein's model when $T \to 0$. The experimentally observed low temperature behaviour is more like $C_V \mu\, T^3$. The reason for this discrepancy is the crude approximation that all normal modes have the same frequency. In fact, long wavelength modes have lower frequencies than short wavelength modes, so the former are much harder to freeze out than the latter (because the spacing between quantum energy levels, $\hbar\omega$, is smaller in the former case).

The molar heat capacity does not decrease with temperature as rapidly as suggested by Einstein's model because these long wavelength modes are able to make a significant contribution to the heat capacity even at very low temperatures.

A more realistic model of lattice vibrations was developed by the Dutch physicist Peter Debye in 1912. In the Debye model, the frequencies of the normal modes of vibration are estimated by treating the solid as an isotropic continuous medium. This approach is reasonable because the only modes which really matter at low temperatures are the long wavelength modes: *i.e.*, those whose wavelengths greatly exceed the interatomic spacing. It is plausible that these modes are not particularly sensitive to the discrete nature of the solid: *i.e.*, the fact that it is made up of atoms rather than being continuous.

Consider a sound wave propagating through an isotropic continuous medium. The disturbance varies with position vector *r* and time *t* like exp[–*i* (k.r – w*t*)], where the wave-vector *k* and the frequency of oscillation w satisfy the dispersion relation for sound waves in an isotropic medium:

$$\omega = kC_s.$$

Here, C_S is the speed of sound in the medium. Suppose, for the sake of argument, that the medium is periodic in the x–, y–, and z–directions with periodicity lengths L_x, L_y, and L_z, respectively. In order to maintain periodicity we need

$$k_x(x + L_x) = k_x + 2\pi n_x,$$

where n_x is an integer. There are analogous constraints on k_y and k_z. It follows that in a periodic medium the components of the wave-vector are quantized, and can only take the values

$$k_x = \frac{2\pi}{L_x} n_x,$$

$$k_y = \frac{2\pi}{L_y} n_y,$$

$$k_z = \frac{2\pi}{L_z} n_z,$$

where n_x, n_y, and n_z are all integers. It is assumed that L_x, L_y, and L_z are macroscopic lengths, so the allowed values of the components of the wave-vector are very closely spaced. For given values of k_y and k_z, the number of allowed values of k_x which lie in the range k_x to $k_x + dk_x$ is given by,

$$\Delta n_x = \frac{L_x}{2\pi} 2k_x.$$

It follows that the number of allowed values of k (*i.e.*, the number of allowed modes) when k_x lies in the range k_z to $k_z + dk_z$, $k_z + k_y$ lies in the range k_y to $k_y + dk_y$, and k_z lies in the range k_z to $k_z + dk_z$, is,

$$\rho d^3k = \left(\frac{L_x}{2\pi} dk_x\right)\left(\frac{L_y}{2\pi} dk_y\right)\left(\frac{L_z}{2\pi} dk_z\right) = \frac{V}{(2\pi)^3} dk_x dk_y dk_z,$$

where $V = L_x L_y L_z$ is the periodicity volume, and d^3k ° $dk_x dk_y dk_z$. The quantity r is called the density of modes. Note that this density is independent of k, and proportional to the periodicity volume.

Thus, the density of modes per unit volume is a constant independent of the magnitude or shape of the periodicity volume. The density of modes per unit volume when the magnitude of k lies in the range k to $k + dk$ is given by multiplying the density of modes per unit volume by the "volume" in k-space of the spherical shell lying between radii k and $k + dk$. Thus,

$$\rho_k dk = \frac{4\pi k^2 \, dk}{(2\pi)} = \frac{V^2}{2\pi^2} dk.$$

Consider an isotropic continuous medium of volume V. According to the above relation, the number of normal modes whose frequencies lie between ω and $\omega + d\omega$ (which is equivalent to the number of modes whose k values lie in the range ω/C_s to $\omega/C_s + d\omega/C_s$) is,

$$\sigma_C(\omega)\,d\omega = 3\frac{k^2 V}{2\pi^2}dk = 3\frac{V^2}{2\pi^2 C_S{}^3}\omega^2 d\omega.$$

The factor of 3 comes from the three possible polarizations of sound waves in solids. For every allowed wavenumber (or frequency) there are two independent torsional modes, where the displacement is perpendicular to the direction of propagation, and one longitudinal mode, where the displacement is parallel to the direction of propagation. Torsion waves are vaguely analogous to electromagnetic waves (these also have two independent polarizations). The longitudinal mode is very similar to the compressional sound wave in gases. Of course, torsion waves can not propagate in gases because gases have no resistance to deformation without change of volume.

The Debye approach consists in approximating the actual density of normal modes s(w) by the density in a continuous medium s_C(w), not only at low frequencies (long wavelengths) where these should be nearly the same, but also at higher frequencies where they may differ substantially.

Suppose that we are dealing with a solid consisting of N atoms. We know that there are only $3N$ independent normal modes. It follows that we must cut off the density of states above some critical frequency, w_D say, otherwise we will have too many modes. Thus, in the Debye approximation the density of normal modes takes the form,

$$\sigma_D(\omega) = \sigma_C(\omega) \text{ for } \omega < \omega_D$$
$$\sigma_D(\omega) = 0 \text{ for } \omega < \omega_D$$

Here, ω_D is the Debye frequency. This critical frequency is chosen such that the total number of normal modes is $3N$, so,

$$\int_0^\infty \sigma_C(\omega)\,d\omega = \int_0^{\omega_D} \sigma_C(\omega)\,d\omega = 3N.$$

Substituting equation into the previous formula yields,

$$\frac{3V}{2\pi^2 C_s^3}\int_0^{\omega_D} d\omega = \frac{3V}{2\pi^2 C_s^3}\omega_D^3 = 3N.$$

This implies that,

$$\omega_D = C_s\left(6\pi^2\frac{N}{V}\right)^{1/3}.$$

Thus, the Debye frequency depends only on the sound velocity in the solid and the number of atoms per unit volume. The wavelength corresponding to the Debye frequency is $2\pi\, C_s/\omega_D$, which is clearly on the order of the interatomic spacing $a \sim (V/N)^{1/3}$. It follows that the cut-off of normal modes whose frequencies exceed the Debye frequency is equivalent to a cut-off of normal modes whose wavelengths are less than the interatomic spacing. Of course, it makes physical sense that such modes should be absent.

Compares the actual density of normal modes in diamond with the density predicted by Debye theory. Not surprisingly, there is not a particularly strong resemblance between these two curves, since Debye theory is highly idealized. Nevertheless, both curves exhibit sharp cut-offs at high frequencies, and coincide at low frequencies. Furthermore, the areas under both curves are the same. This is sufficient to allow Debye theory to correctly account for the temperature variation of the specific heat of solids at low tem.peratures.

We can use the quantum mechanical expression for the mean energy of a single oscillator, equation to calculate the mean energy of lattice vibrations in the Debye approximation. We obtain

$$\bar{E} = \int_0^\infty \sigma_D(\omega)\hbar\omega\left(\frac{1}{2}+\frac{1}{\exp(\beta\hbar\omega)-1}\right)d\omega.$$

According to equation the molar heat capacity takes the form,

$$C_V = \frac{1}{\upsilon kT^2}\int_0^\infty \sigma_D(\omega)\hbar\omega\left[\frac{\exp(\beta\hbar\omega)\hbar\omega}{[\exp(\beta\hbar\omega)-1]^2}\right]d\omega.$$

Substituting in equation we find that,

$$C_V = \frac{k}{\upsilon}\int_0^{\omega_D}\frac{\exp(\beta\hbar\omega)(\beta\hbar\omega)^2}{[\exp(\beta\hbar\omega)-1]^2}\frac{3V}{2\pi^2C_s^3}\omega^2 d\omega,$$

giving,

$$C_V = \frac{3Vk}{2\pi^2\upsilon(C_s\beta\hbar)^3}\int_0^{\beta\hbar\omega_D}\frac{\exp x}{(\exp x-1)^2}x^4 dx,$$

in terms of the dimensionless variable $x = \beta\hbar\omega$. According to equation the volume can be written,

$$V = 6\pi^2 N\left(\frac{Cs}{\omega_D}\right)^3,$$

so the heat capacity reduces to,

$$C_V = 3Rf_D(\beta\hbar\omega_D) = 3Rf_D(\theta_D/T),$$

where the Debye function is defined,

$$f_D(y) \equiv \frac{3}{y^3}\int_0^y \frac{\exp x}{(\exp x-1)^2}x^4 dx.$$

We have also defined the Debye temperature θ_D as,

$$k\theta_D = \hbar\omega_D.$$

Consider the asymptotic limit in which $T >> \theta_D$. For small y, we can approximate $\exp x$ as $1 + x$ in the integrand of equation so that,

$$f_D(y) \to \frac{3}{y^3}\int_0^y x^2 dx = 1.$$

Thus, if the temperature greatly exceeds the Debye temperature we recover the law of Dulong and Petite that $C_V = 3R$. Consider, now, the asymptotic limit in which $T << \theta_D$. For large y,

$$\int_0^y \frac{\exp x}{(\exp x - 1)^2} x^4 dx \simeq \int_0^\infty \frac{\exp x}{(\exp x - 1)^2} x^4 dx = \frac{4\pi^4}{15}.$$

Thus, in the low temperature limit,

$$f_D(y) \to \frac{4\pi^4}{5}\frac{1}{y^3}.$$

This yields,

$$c_V \simeq \frac{12\pi^4}{5} R \left(\frac{T}{\theta_D}\right)^3$$

in the limit $T << \theta_D$: *i.e.*, c_V varies with temperature like T^3.

Table. Comparison of Debye Temperatures (In Degrees Kelvin)

Solid	θ_D from low temp.	θ_D from sound speed
NaCl	308	320
KCl	230	246
Ag	225	216
Zn	308	305

The fact that c_V goes like T^3 at low temperatures is quite well verified experimentally, although it is sometimes necessary to go to temperatures as low as 0.02 θ_D to obtain this asymptotic behaviour.

Theoretically, θ_D should be calculable from equation in terms of the sound speed in the solid and the molar volume. A comparison of Debye temperatures evaluated by this means with temperatures obtained empirically by fitting the law to the low temperature variation of the heat capacity. It can be seen that there is fairly good agreement between the theoretical and empirical Debye temperatures.

This suggests that the Debye theory affords a good, thought not perfect, representation of the behaviour of c_V in solids over the entire temperature range.

Finally, the actual temperature variation of the molar heat capacities of various solids as well as that predicted by Debye's theory. The prediction of Einstein's theory is also show for the sake of comparison. Note that 24.9 joules/mole/degree is about 6 calories/gram-atom/degree (the latter are chemist's units).

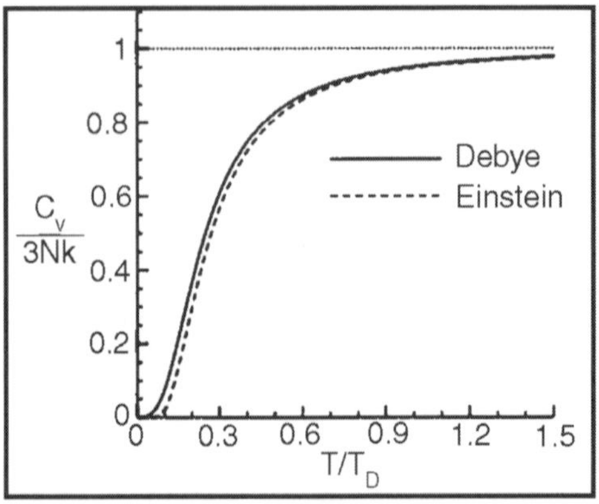

Fig. The Molar Heat Capacity of Various Solids.

SPECIFIC HEATS

In general,

$$Q = C\Delta T$$

where C is a constant that depends on the substance.

We can determine the constant for any substance if we know how much heat is transferred. Since heat is path dependent, however, we must specify the process, *i.e.*, the path, to find C .

We will call the specific heat at constant pressure C_p, and that at constant volume C_v, or C_p and C_v per unit mass.

Specific Heat at Constant Volume

Remember that if we specify any two properties of the system, then the state of the system is fully specified.

In other words we can write

$u = u(T,\nu)$ $u = u(p,\nu)$ $u = u(p,T)$, or

Consider the form u=u(T,v), and use the chain rule to write how u changes with respect to T and v :

$$du = \left(\frac{\partial u}{\partial T}\right)_\nu dT + \left(\frac{\partial u}{\partial \nu}\right)_T d\nu$$

For a constant volume process, the second term is zero since there is no change in volume, dv = 0.

Now if we write the First Law for a quasi-static process, with ,

$$du = \delta q - pd\nu$$

We see that again the second term is zero if the process is also constant volume.

Equating equations with dv canceled in each,

$$\delta q = \left(\frac{\partial u}{\partial T}\right)_{\nu} dT$$

and rearranging

$$\left(\frac{\partial u}{\partial T}\right)_{\nu} = \left(\frac{\partial q}{\partial T}\right)_{\nu}$$

In this case, any energy increase is due only to energy transfer as heat. We can therefore use our definition of specific heat from Equation to define the specific heat for a constant volume process,

$$C_{\nu} \equiv \left(\frac{\partial u}{\partial T}\right)_{\nu}$$

The Specific Heat at Constant Pressure

If we write h = h(T,p) and consider a constant pressure process, we can perform a similar derivation to the one above and show that

$$C_p \equiv \left(\frac{\partial h}{\partial T}\right)_p$$

In the derivation of c_v, we considered only a constant volume process, hence the name, "specific heat at constant volume."

It is more useful, however, to think of c_v in terms of its definition as a certain partial derivative, which is a thermodynamic property, rather than as a quantity related to heat transfer in a special process.

The names "specific heat at constant volume" and "specific heat at constant pressure" are therefore unfortunate misnomers; c_v and C_p are thermodynamic properties of a substance, and by definition depend only the state.

To recap:

$$C_p \equiv \left(\frac{\partial h}{\partial T}\right)_p \quad \text{and} \quad C_{\nu} = \left(\frac{\partial u}{\partial T}\right)_{\nu}$$

or,

$$C_p \equiv \left(\frac{\partial H}{\partial T}\right)_p \quad \text{and} \quad C_{\nu} = \left(\frac{\partial U}{\partial T}\right)_{\nu}$$

SPECIFIC HEATS AND CALORIMETRY

The specific heat of a substance is the heat required in calories to raise the temperature of 1 gram by 1 degree Celsius. As Fahrenheit continues his measurements of heat flow, it quickly became evident that for different materials, the amount of heat needed to raise the temperature of one gram by one degree could be quite different. It was natural to measure specific heats relative to that of water, the simplest and most readily available substance for calorimetric experiments.

There were some surprises. For example, it had been widely thought before the measurements were made, that one cc of Mercury, being a lot heavier than one cc of water, would take more heat to raise its temperature by one degree. This proved not to be the case—Fahrenheit himself made the measurement. In an insulating container, called a "calorimeter" he added 100ccs of water at 100°F to 100ccs of mercury at 150°F, and stirred so they quickly reached thermal equilibrium.

Question: what do you think the final temperature was?

Answer: The final temperature was, surprisingly, about 120°F. 100 cc of water evidently "contained more heat" than 100 cc of mercury, despite the large difference in weight.

The technique, called calorimetry, was widely used to find specific heats of many different substances, and at first no clear pattern emerged. It was puzzling that the specific heat of mercury was so low compared with water. As the experiments progressed, it gradually became evident that heavier substances, paradoxically, had lower specific heats.

Meanwhile, this quantitative approach to scientific observation had spread to chemistry. Towards the end of the 1700's, Lavoisier weighed chemicals involved in reactions before and after the reaction.

This involved weighing the gases involved, so had to be carried out in closed containers, so that, for example, the weight of oxygen used and the carbon dioxide, etc., produced would accounted for in studying combustion. The big discovery was that mass was neither created nor destroyed. This had not been realized before because no one had weighed the gases involved. It made the atomic theory suddenly more plausible, with the idea that maybe chemical reactions were just rearrangements of atoms into different combinations.

Lavoisier also clarified the concept of an element, an idea that was taken up in about 1800 by John Dalton, who argues that a given compound consisted of identical molecules, made up of elementary atoms in the same proportion, such as H2O (although that was thought initially to be HO). This explained why, when substances reacted chemically, such as the burning of hydrogen to form water, it took exactly eight grams of oxygen for each gram of hydrogen. (produce H2O2 under the right conditions, with exactly sixteen grams of oxygen to one of hydrogen, but the simple ratios of amounts of oxygen needed for the two reactions were simply explained by different molecular structures, and made the atomic hypothesis even more plausible.)

Much effort was expended carefully weighing the constituents in many chemical reactions, and constructing diagrams of the molecules. The important result of all this work was that it became possible to list the relative weights of the atoms involved.

For example, the data on H2O and H2O2 led to the conclusion that an oxygen atom weighed sixteen times the weight of a hydrogen atom.

It must be emphasized, though, that these results gave no clue as to the actual weights of atoms! All that was known was that atoms were too small to see in the best microscopes. Nevertheless, knowing the relative weights of some atoms in 1820 led to an important discovery.

Two professors in France, Dulong and Petit, found that for a whole series of elements the product of atomic weight and specific heat was the same.

Element	Specific Heat (cals/gm/degree C)	Relative weights of the atoms	Product of relative atomic weight and specific heat
Lead	0.0293	12.95	0.3794
Tin	0.0514	7.35	0.3779
Zinc	0.0927	4.03	0.3736
Sulphur	0.188	2.011	0.378

The significance of this, as they pointed out, was that the "specific heat", or heat capacity, of each atom was the same—a piece of lead and a piece of zinc having the same number of atoms would have the same heat capacity. So heavier atoms absorbed no more heat than lighter atoms for a given rise in temperature.

This partially explained why mercury had such a surprisingly low heat capacity. Of course, having no idea how big the atoms might be, they could go no further. And, indeed, many of their colleagues didn't believe in atoms anyway, so it was hard to convince them of the significance of this discovery.

LATENT HEAT

One of Black's experiments was to set a pan of water on a steady fire and observe the temperature as a function of time. He found it steadily increased, reflecting the supply of heat from the fire, until the water began to boil, whereupon the temperature stayed the same for a long time. The steam coming off was at the same (boiling) temperature as the water. So what was happening to the heat being supplied?

Black correctly concluded that heat needed to be supplied to change water from its liquid state to its gaseous state, that is, to steam.

In fact, a lot of heat had to be supplied: 540 calories per gram, as opposed to the mere 100 calories per gram needed to bring it from the freezing temperature to boiling.

He also discovered that it took 80 calories per gram to melt ice into water, with no rise in temperature. This heat is released when the water freezes back to ice, so it is somehow "hidden" in the water. He called it latent heat, meaning hidden heat.

THE GAS LAW

Coefficients of Expansion

Almost all materials expand on heating—the most famous exception being water, which contracts as it is warmed from 0 degrees Celsius to 4 degrees. This is actually a good thing, because as freezing weather sets in, the coldest water, which is about to freeze, is less dense than slightly warmer water, so rises to the top of a lake and the ice begins to form there. For almost all other liquids, solidification on cooling begins at the bottom of the container. So, since water behaves in this weird way, ice skating is possible! Also, as a matter of fact, life in lakes is possible—the ice layer that forms insulates the rest of the lake water from very cold air, so fish can make it through the winter.

LINEAR EXPANSION

The coefficient of linear expansion α of a given material, for example a bar of copper, at a given temperature is defined as the fractional increase in length that takes place on heating through one degree:

$$L \to L + \Delta L = (1+\alpha)\, L \; when T \to T + 1°C.$$

Of course, α might vary with temperature (it does for water, as we just mentioned) but in fact for most materials it stays close to constant over wide temperature ranges.

For copper, $\alpha = 17 \times 10^{-6}$

VOLUME EXPANSION

For liquids and gases, the natural measure of expansion is the coefficient of volume expansion, β.

$$V \to V + \Delta V = (1+\beta)V \text{ when} T \to T + 1°C.$$

Of course, on heating a bar of copper, clearly the volume as well as the length increases—the bar expands by an equal fraction in all directions (this could be experimentally verified, or you could just imagine a cube of copper, in which case all directions look the same).

The volume of a cube of copper of side L is $V = L3$. Suppose we heat it through one degree. Putting together the definitions of α, β above,

$$V \to (1+\beta)V, L \to (1+\alpha)L,\; L^3 \to (1+\alpha^3)\, L^3 or V \to (1+a)^3 V.$$

So $(1 + \beta) = (1 + \alpha)^3$. But remember α is very, very small—so even though $(1 + \alpha)^3 = 1 + 3\alpha + \alpha^2 + \alpha^3$, the last two terms are completely negligible so to a fantastically good approximation:

$$\beta = 3\alpha$$

The coefficient of volume expansion is just three times the coefficient of linear expansion.

GAS PRESSURE INCREASE WITH TEMPERATURE

In 1702, Amontons discovered a linear increase of P with T for air, and found P to increase about 33 per cent from the freezing point of water to the boiling point of water. That is to say, he discovered that if a container of air were to be sealed at 0°C, at ordinary atmospheric pressure of 15 pounds per square inch, and then heated to 100°C but kept at the same volume, the air would now exert a pressure of about 20 pounds per square inch on the sides of the container. Remarkably, Amontons discovered, if the gas were initially at a pressure of thirty pounds per square inch at 0°C, on heating to 100°C the pressure would go to about 40 pounds per square inch—so the percentage increase in pressure was the same for any initial pressure: on heating through 100°C, the pressure would always increase by about 33 per cent .

Furthermore, the result turned out to be the same for different gases! Finding a Natural Temperature Scale.In class, we plotted air pressure as a function of temperature for a fixed volume of air, by making several measurements as the air was slowly heated (to give it a chance to all be at the same temperature at each stage).

We found a straight line. On the graph, we extended the line backwards, to see how the pressure would presumably drop on cooling the air.

We found the remarkable prediction that the pressure should drop to zero at a temperature of about –273°C. In fact, if we'd done the cooling experiment, we would have found that air doesn't actually follow the line all the way down, but condenses to a liquid at around –200°C.

However, helium gas stays a gas almost to –270°C, and follows the line closely. The physics of gases, and the interpretation of this, much more fully in a couple of lectures. For now, the important point is that this suggests a much more natural temperature scale than the Celsius one: we should take -273°C as the zero of temperature! For one thing, if we do that, the pressure/ temperature relationship for a gas becomes beautifully simple:

$$P \propto T.$$

This temperature scale, in which the degrees have the same size as in Celsius, is called the Kelvin or absolute scale. Temperatures are written 300K. To get from Celsius to Kelvin, just add 273.15.

AN IDEAL GAS

Physicists at this point introduce the concept of an "Ideal Gas". This is like the idea of a frictionless surface: it doesn't exist in nature, but it is a very handy approximation to some real systems, and makes problems much easier to handle mathematically. The ideal gas is one for which $P \propto T$ for all temperatures, so helium is close to ideal over a very wide range, and air is close to ideal at ordinary atmospheric temperatures and above.

THE GAS LAW

We say earlier in the course that for a gas at constant temperature PV = constant (Boyle's Law). Now at constant volume,

$$P \propto T$$

We can put these together in one equation to find a relationship between pressure, volume and temperature:

$$PV = CT$$

where C is a constant. Notice, by the way, that we can immediately conclude that at fixed pressure, $V \propto T$, this is called Charles' Law. But what is C? Obviously, it depends on how much gas we have—double the amount of gas, keeping the pressure and temperature the same, and the volume will be doubled, so C will be doubled.

But notice that C will not depend on what gas we are talking about: if we have two separate one-liter containers, one filled with hydrogen, the other with oxygen, both at atmospheric pressure, and both at the same temperature, then C will be the same for both of them. One might conclude from this that C should be defined for one liter of gas at a specified temperature and pressure, such as 0°C and 1 atmosphere, and that could be a consistent scheme. It might seem more natural, though, to specify a particular mass of gas, since then we wouldn't have to specify a particular temperature and pressure in the definition of C.

But that idea brings up a further problem: one gram of oxygen takes up a lot less room than one gram of hydrogen. Since we've just seen that choosing the same volume for the two gases gives the same constant C for the two gases, evidently taking the same mass of the two gases will give different C's.

AVOGADRO'S HYPOTHESIS

The resolution to this difficulty is based on a remarkable discovery the chemists made two hundred years or so ago: they found that one liter of nitrogen could react with exactly one liter of oxygen to produce exactly two liters of NO, nitrous oxide, all volume measurements being at the same temperature and pressure. Further, one liter of oxygen combined with two liters of hydrogen to produce two liters of steam.

These simple ratios of interacting gases could be understood if one imagined the atoms combining to form molecules, and made the further assumption, known as Avogadro's Hypothesis :

Equal volumes of gases at the same temperature and pressure contain the same number of molecules

One could then understand the simple volume results by assuming the gases were made of diatomic molecules, H2, N2, O_2 and the chemical reactions were just molecular recombinations given by the equations $N_2 + O_2 = 2NO$, $2H_2 + O_2 = 2H_2O$, etc. Of course, in 1811 Avogadro didn't have the slightest idea what this number of molecules was for, say, one liter, and nobody else did either, for another fifty years. So no-one knew what an atom or molecule weighed, but assuming that chemical reactions were atoms combining into molecules, or rearranging from one molecular pairing or grouping to another, they could figure out the relative weights of atoms, such as an oxygen atom had mass 16 times that of a hydrogen atom—even though they had no idea how big these masses were!

This observation led to defining the natural mass of a gas for setting the value of the constant C in the gas law to be a "mole" of gas: hydrogen was known to be H_2 molecules, so a mole of hydrogen was 2 grams, oxygen was O_2, so a mole of oxygen was 32 grams, and so on.

With this definition, a mole of oxygen contains the same number of molecules as a mole of hydrogen: so at the same temperature and pressure, they will occupy the same volume. At 0°C, and atmospheric pressure, the volume is 22.4 liters. So, for one mole of a gas (for example, two grams of hydrogen), we set the constant C equal to R, known as the universal gas constant, and equal to 8.3 J/(mol.K) and PV = RT.

For n moles of a gas, such as 2n grams of hydrogen, the law is:

$$PV = nRT$$

and this is the standard form of the Gas Law.

By the late 1700's, the experiments of Fahrenheit, Black and others had established a systematic, quantitative way of measuring temperatures, heat flows and heat capacities—but this didn't really throw any new light on just what was flowing.

This was a time when the study of electricity was all the rage, led in America by Benjamin Franklin, who had suggested in 1747 that electricity was one (invisible) fluid (it had previously been suggested that there were two fluids, corresponding to the two kinds of electrical charging observed).

SPECIFIC HEAT AT CONSTANT PRESSURE

The specific heat of gases is expressed as molar specific heats. For a monoatomic ideal gas the internal energy is in the form of kinetic energy.

The expression for the internal energy is:

$$U = nN_A KE_{avg} = nN_A \frac{3}{2} KT = \frac{3}{2} nRT$$

Two specific heats are defined for gases, one for constant volume (C_V) and one for constant pressure (C_P).

For a constant volume process with a monoatomic ideal gas the first law of thermodynamics gives:

$$Q = c_V n\Delta T$$

$$Q = \Delta U + P\Delta v = \Delta U$$

$$c_{\forall} = \frac{1\Delta U}{n\Delta T} = \frac{3}{2}R$$

Further application of the ideal gas law and first law gives the relationship,

$$c_p = c_v + R$$

The ratio of the specific heats $\gamma = C_P/C_V$ is a factor in adiabatic engine processes and in determining the speed of sound in a gas. This ratio $\gamma = 1.66$ for an ideal monoatomic gas and $\gamma = 1.4$ for air, which is predominantly a diatomic gas.

CONSTANT PRESSURE

The molar specific heat at constant pressure is defined by,

$$Q = nc_p\Delta T$$

Using the first law of thermodynamics, this can be in the form,

$$\Delta U + P\Delta V = nc_p\Delta T$$

From the ideal gas law (PV=nRT) under constant pressure:

$$P\Delta V = nR\Delta T$$

So that

$$\frac{\Delta U}{\Delta T} + nR = nc_p$$

Since the constant volume specific heat is,

$$c_V - \frac{1}{n}\frac{\Delta U}{\Delta T}$$

it follows that,

$$C_p = C_v + R$$

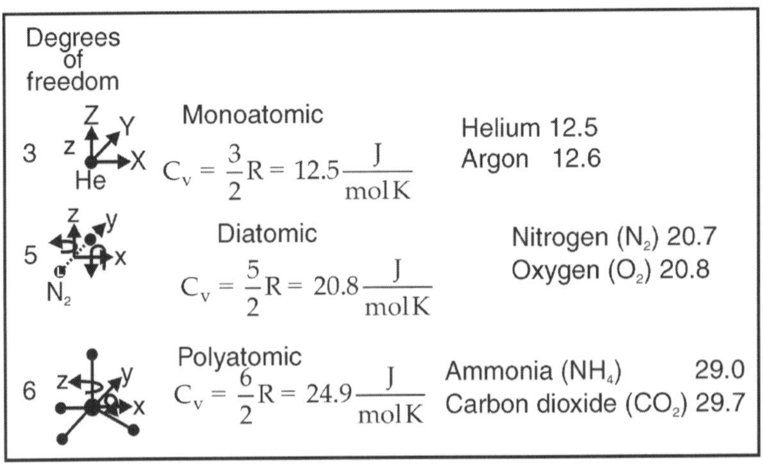

For an ideal monoatomic gas,

$$C_p = \frac{5}{2}R = 20.8\frac{J}{mol\,K}$$

The constant pressure specific heat is related to the constant volume value by $C_P = C_V + R$. The ratio of the specific heats $\gamma = C_P/C_V$ is a factor in adiabatic engine processes and in determining the speed of sound in a gas.

CONSTANT VOLUME SPECIFIC HEAT

The molar specific heat at constant volume is defined by $Q = nC_v\Delta T$ (Constant volume)

Using the first law of thermodynamics this can be put in the form,

$$\Delta U + P\Delta V = nC_v\Delta T$$

but since $\Delta V = 0$, the expression for C_V becomes,

$$C_v = \frac{1}{n}\frac{\Delta U}{\Delta T}$$

For a monoatomic ideal gas,

$$U = \frac{3}{2}nRT \text{ so } C_v = \frac{3}{2}R = 12.5\frac{J}{mol\,K}$$

The equipartition of energy predicts,

$$U = \frac{f}{2}nRT \text{ and } C_v = \frac{f}{2}R$$

Where f is the number of degrees of freedom in the molecular motion.

9

Ideal Gas

INTRODUCTION TO IDEAL GASES

An ideal gas is a substance possessing very simple thermodynamic properties to which actual gases and vapours appear to approximate indefinitely at low pressures and high temperatures. It has the characteristic equation pv = Re, and obeys Boyle's law at all temperatures. The coefficient of expansion at constant pressure is equal to the coefficient of increase of pressure at constant volume. The difference of the specific heats by equation is constant and equal to R. The isothermal elasticity – v(dp/dv) is equal to the pressure p. The adiabatic elasticity is equal to y p, where-y is the ratio S/s of the specific heats. The heat absorbed in isothermal expansion from $v_0$to v at a temperature 0 is equal to the work done by equation (since d_0 = o, and 0(dp/d_0)dv =pdv), and both are given by the expression RO log e (v/vo).

The energy E and the total heat F are functions of the temperature only and their variations take the form dE = sdO, d F = Sd_0. The specific heats are independent of the pressure or density by equations. If we also assume that they are constant with respect to temperature (which does not necessarily follow from the characteristic equation, but is generally assumed, and appears from Regnault's experiments to be approximately the case for simple gases), the expressions for the change of energy or total heat from 00 to 0 may be written E – E0 = s(0 – 0 0), F – Fo = S(0 - 00). In thiscase the ratio of the specific heats is constant as well as the difference, and the adiabatic equation takes the simple form, pv v = constant, which is at once obtained by integrating the equation for the adiabatic elasticity, – v(dp/dv) = yp.

DEVIATIONS OF ACTUAL GASES FROM THE IDEAL STATE

Since no gas is ideally perfect, it is most important for practical purposes to discuss the deviations of actual gases from the ideal state, and to consider how their properties may be thermodynamically explained and defined. The most natural method of procedure is to observe the deviations from Boyle's law by measuring the changes of pv at various constant temperatures. It is found by experiment that the change of pv with pressure at moderate pressures is nearly proportional to the change of p, in other words that the coefficient d(pv)/dp is to a first approximation a function of the temperature only.

This coefficient is sometimes called the " angular coefficient," and may be regarded as a measure of the deviations from Boyle's law, 'which may be most simply expressed at moderate pressures by formulating the variation of the angular coefficient with temperature. But this procedure in itself is not sufficient, because, although it would be highly probable that a gas obeying Boyle's law at all temperatures was practically an ideal gas, it is evident that Boyle's law would be satisfied by any substance having the characteristic equation pv = f (0), where f (0) is any arbitrary function of 0, and that the scale of temperatures given by such a substance would not necessarily coincide with the absolute scale.

A sufficient test, in addition to Boyle's law, is the condition dE/dv = o at constant temperature.

This gives by equation the condition Odp/d0 = p, which is satisfied by any substance possessing the characteristic equation p/0 = f(v), where f(v) is any arbitrary function of v. This test was applied by Joule in the well-known experiment in which he allowed a gas to expand from one vessel to another in a calorimeter without doing external work.

Under this condition the increase of intrinsic energy would be equal to the heat absorbed, and would be indicated by fall of temperature of the calorimeter. Joule failed to observe any change of temperature in his apparatus, and was therefore justified in assuming that the increase of intrinsic energy of a gas in isothermal expansion was very small, and that the absorption of heat observed in a similar experiment in which the gas was allowed to do external work by expanding against the atmospheric pressure was equivalent to the external work done. But owing to the large thermal capacity of his calorimeter, the test, though sufficient for his immediate purpose, was not delicate enough to detect and measure the small deviations which actually exist.

JOULE–THOMSON EFFECT

In physics, the Joule–Thomson effect or Joule–Kelvin effect describes the temperature change of a gas or liquid when it is forced through a valve or porous plug while kept insulated so that no heat is exchanged with the environment.

This procedure is called a throttling process or Joule-Thomson process. At room temperature, all gases except hydrogen, helium and neon cool upon expansion by the Joule-Thomson process.

The effect is named for James Prescott Joule and William Thomson, 1st Baron Kelvin who discovered it in 1852 following earlier work by Joule on Joule expansion, in which a gas undergoes free expansion in a vacuum.

Description

The adiabatic (no heat exchanged) expansion of a gas may be carried out in a number of ways. The change in temperature experienced by the gas during expansion depends on the initial and final pressure, but also on the manner in which the expansion is carried out.

- If the expansion process is reversible, meaning that the gas is in thermodynamic equilibrium at all times, it is called an isentropic expansion. In this scenario, the gas does positive work during the expansion, and its temperature decreases.
- In a free expansion, on the other hand, the gas does no work and absorbs no heat, so the internal energy is conserved. Expanded in this manner, the temperature of an ideal gas would remain constant, but the temperature of a real gas may either increase or decrease, depending on the initial temperature and pressure.
- The method of expansion in which a gas or liquid at pressure P1 flows into a region of lower pressure P2 via a valve or porous plug under steady state conditions and without change in kinetic energy, is called the Joule–Thomson process. During this process, enthalpy remains unchanged.

Temperature change of either sign can occur during the Joule-Thomson process. Each real gas has a Joule–Thomson (Kelvin) inversion temperature above which expansion at constant enthalpy causes the temperature to rise, and below which such expansion causes cooling. This inversion temperature depends on pressure; for most gases at atmospheric pressure, the inversion temperature is above room temperature, so most gases can be cooled from room temperature by isenthalpic expansion.

Physical Mechanism

As a gas expands, the average distance between molecules grows. Because of intermolecular attractive forces expansion causes an increase in the potential energy of the gas. If no external work is extracted in the process and no heat is transferred, the total energy of the gas remains the same because of the conservation of energy. The increase in potential energy thus implies a decrease in kinetic energy and therefore in temperature.

A second mechanism has the opposite effect. During gas molecule collisions, kinetic energy is temporarily converted into potential energy. As the average intermolecular distance increases, there is a drop in the number of collisions per time unit, which causes a decrease in average potential energy. Again, total energy is conserved, so this leads to an increase in kinetic energy (temperature).

Below the Joule–Thomson inversion temperature, the former effect (work done internally against intermolecular attractive forces) dominates, and free expansion causes a decrease in temperature. Above the inversion temperature, gas molecules move faster and so collide more often, and the latter effect (reduced collisions causing a decrease in the average potential energy) dominates: Joule-Thomson expansion causes a temperature increase.

The Joule–Thomson (Kelvin) Coefficient

The rate of change of temperature T with respect to pressure P in a Joule–Thomson process (that is, at constant enthalpy H) is the Joule–Thomson (Kelvin) coefficient ìJT. This coefficient can be expressed in terms of the gas's volume V, its heat capacity at constant pressure Cp, and its coefficient of thermal expansion á as:

$$\mu_{JT} \equiv \left(\frac{\partial T}{\partial P}\right)_H = \frac{V}{C_p}(\alpha T - 1).$$

The value of mJT is typically expressed in °C/bar (SI units: K/Pa) and depends on the type of gas and on the temperature and pressure of the gas before expansion.

All real gases have an inversion point at which the value of mJT changes sign. The temperature of this point, the Joule–Thomson inversion temperature, depends on the pressure of the gas before expansion.

In a gas expansion the pressure decreases, so the sign of ¶*P* is always negative. With that in mind, the following table explains when the Joule–Thomson effect cools or warms a real gas:

If the gas temperature is	Then mJT is	Since ∂P is	Thus∂P must be	So the gas
Below the inversion temperature	Positive	Always negative	Negative	Cools
Above the inversion temperature	Negative	Always negative	Positive	Warms

Helium and hydrogen are two gases whose Joule–Thomson inversion temperatures at a pressure of one atmosphere are very low (*e.g.*, about 51 K (–222 °C) for helium). Thus, helium and hydrogen warm up when expanded at constant enthalpy at typical room temperatures. On the other hand nitrogen and oxygen, the two most abundant gases in air, have inversion temperatures of 621 K (348 °C) and 764 K (491 °C) respectively: these gases can be cooled from room temperature by the Joule–Thomson effect.

For an ideal gas, ìJT is always equal to zero: ideal gases neither warm nor cool upon being expanded at constant enthalpy.In practice, the Joule–Thomson effect is achieved by allowing the gas to expand through a throttling device (usually a valve) which must be very well insulated to prevent any heat transfer to or from the gas.

No external work is extracted from the gas during the expansion. The effect is applied in the Linde technique as a standard process in the petrochemical industry, where the cooling effect is used to liquefy gases, and also in many cryogenic applications (*e.g.* for the production of liquid oxygen, nitrogen, and argon).

Only when the Joule–Thomson coefficient for the given gas at the given temperature is greater than zero can the gas be liquefied at that temperature by the Linde cycle. In other words, a gas must be below its inversion temperature to be liquefied by the Linde cycle. For this reason, simple Linde cycle liquefiers cannot normally be used to liquefy helium, hydrogen, or neon.

Proof that Enthalpy Remains Constant in a Joule-Thomson Process

In a Joule-Thomson process the enthalpy remains constant. To prove this, the first step is to compute the net work done by the gas that moves through the plug. Suppose that the gas has a volume of V_1 in the region at pressure P_1 and a volume of V_2 when it appears in the region at pressure P_2. Then the work done on the gas by the rest of the gas in region 1 is P_1V_1. In region 2 the amount of work done by the gas is $P_2 V_2$. So, the total work done by the gas is,

$$P_2V_2 - P_1V_1$$

The change in internal energy plus the work done by the gas is, by the first law of thermodynamics, the total amount of heat absorbed by the gas (here it is assumed that there is no change in kinetic energy). In the Joule-Thompson process the gas is kept insulated, so no heat is absorbed. This means that,

$$E_2 - E_1 + P_2V_2 - P_1V_1 = 0$$

where E_1 and E_2 denote the internal energy of the gas in regions 1 and 2, respectively.

The above equation then implies that:

$$H_1 = H_2$$

where H_1 and H_2 denote the enthalpy of the gas in regions 1 and 2, respectively.

Derivation of the Joule-Thomson (Kelvin) coefficient

A derivation of the formula,

$$m_{JT} \equiv \left(\frac{\partial T}{\partial P}\right)_H = \frac{V}{C_P}(\alpha T - 1)$$

for the Joule–Thomson (Kelvin) coefficient. The partial derivative of T with respect to P at constant H can be computed by expressing the differential of the enthalpy dH in terms of dT and dP, and equating the resulting expression to zero and solving for the ratio of dT and dP.

It follows from the fundamental thermodynamic relation that the differential of the enthalpy is given by:

$dH = TdS + VdP$ (here, S is the entropy of the gas).

Expressing dS in terms of dT and dP gives:

$$dH = T\left(\frac{\partial S}{\partial T}\right)_P dT + \left[V + T\left(\frac{\partial S}{\partial P}\right)_T\right] dP$$

Using,

$$C_P = T\left(\frac{\partial S}{\partial T}\right)_P, \text{ we can write:}$$

$$dH = C_P dT + \left[V + T\left(\frac{\partial S}{\partial P}\right)_T\right]dP.$$

The remaining partial derivative of S can be expressed in terms of the coefficient of thermal expansion via a Maxwell relation as follows. From the fundamental thermodynamic relation, it follows that the differential of the Gibbs energy is given by:

$$dG = -SdT + VdP.$$

The symmetry of partial derivatives of G with respect to T and P implies that:

$$\left(\frac{\partial S}{\partial P}\right)_T = -\left(\frac{\partial V}{\partial T}\right)_P = -V\alpha$$

where a is the coefficient of thermal expansion. Using this relation, the differential of H can be expressed as,

$$dH = C_p dT + V(1 - T\alpha)dP.$$

Equating dH to zero and solving for dT/dP then gives:

$$\left(\frac{\partial T}{\partial P}\right)_H = \frac{V}{C_P}(\alpha T - 1).$$

CONCEPT OF IDEAL GAS

An ideal gas is defined as gas in which all collisions between atoms or molecules are elastic and has no intermolecular attractive forces. In such a gas, all internal energy in form of kinetic energy is accompanied by a change in temperature.

An ideal gas can be characterized by three state variables: absolute pressure (P), volume (V), and absolute temperature (T).

The relationship between them may be deduced from kinetic theory and is called the,

Ideal gas law: PV=nRT=NkT

- n = number of moles
- R = universal gas constant = 8.3145 J/mol K
- N = number of molecules
- k = Boltzmann constant = 1.38066×10^{-23} J/K = 8.617385×10^{-5} eV/K
- k = R/N_A
- N_A = Avogadro's number = 6.0221×10^{23} /mol

The ideal gas law arises from the kinetic pressure of gas molecules which are colliding with the walls of a container. But there is also a statistical element in the determination of the average kinetic energy of those molecules.

ENTHALPY OF IDEAL GAS

Under conditions of *constant pressure* the heat absorbed or released is termed as *enthalpy*. We cannot measure enthalpy *directly*, but can calculate the *change* in enthalpy DH.

The change in enthalpy, DH, equals the heat, q_p, added to or lost by the system when the process occurs under constant pressure:

$$DH = q_p$$

DH represents the difference between the enthalpy of the system at the beginning of the reaction compared to what it is at the end of the reaction:

$$DH = H_{final} - H_{initial}$$

We are considering the enthalpic state of the *system*. Thus:

- If the *system* has higher enthalpy at the end of the reaction, then it *absorbed* heat from the surroundings (*endothermic* reaction)
- If the system has a lower enthalpy at the end of the reaction, then it *gave off heat* during the reaction (*exothermic* reaction)

Therefore:

- For *endothermic* reactions $H_{final} > H_{initial}$ *and DH is positive (+DH)*
- For exothermic reactions $H_{final} < H_{initial}$ *and DH is negative (–DH)*

For an ideal gas the enthalpy - *h* - is function of temperature and the change in enthalpy can be expressed as,

$$dh = c_p \, dT$$

where,

dh = change in enthalpy

c_p = *specific heat capacity for the gas in a constant pressure process*

c_p can within a moderate temperature change be regarded as constant.

The enthalpy in a fluid is defined as:

$$h = u + p / \tilde{n}$$

where,

h = enthalpy

u = internal energy

p = absolute pressure

ñ = density

Combining previous equation and the Ideal Gas Law gives:

$$h = u + R\,T$$

where,

R = the individual gas constant

The change in enthalpy can be expressed by differentiating:

$$dh = du + R\,dT$$

Dividing previous equation with dT gives:

$$(dh / dT) - (du / dT) = R$$

Modifying equations:

$$c_p - c_v = R$$

The difference c_p - c_v is constant for an ideal gas.

SPECIFIC HEAT CAPACITIES OF IDEAL GAS

If we are dealing with a gas, it is most convenient to use forms of the thermodynamics equations based on the enthalpy of the gas. From the definition of enthalpy:

h = e + p * v

where h in the specific enthalpy, p is the pressure, v is the specific volume, and e is the specific internal energy. During a process, the values of these variables change. Let's denote the change by the Greek letter delta which looks like a triangle. So "delta h" means the change of "h" from state 1 to state 2 during a process.

delta h = delta e + p * delta v

The enthalpy, internal energy, and volume are all changed, but the pressure remains the same. From our derivation of enthalpy equation, the change of specific enthalpy is equal to the heat transfer for a constant pressure process:

delta h = cp * delta T

where delta T is the change of temperature of the gas during the process, and c is the specific heat capacity. We have added a subscript "p" to the specific heat capacity to remind us that this value only applies to a constant pressure process.

The equation of state of a gas relates the temperature, pressure, and volume through a gas constant R . The gas constant used by aerodynamicists is derived from the universal gas constant, but has a unique value for every gas.

p * v = R * T

If we have a constant pressure process, then:

p * delta v = R * delta T

Now let us imagine that we have a constant volume process with our gas that produces exactly the same temperature change as the constant pressure process that we have been discussing. Then the first law of thermodynamics tells us:

delta e = delta q - delta w

where q is the specific heat transfer and w is the work done by the gas. For a constant volume process, the work is equal to zero. And we can express the heat transfer as a constant times the change in temperature. This gives:

delta e = cv * delta T

where delta T is the change of temperature of the gas during the process and c is the specific heat capacity.

If we substitute the expressions for "delta e", "p * delta v", and "delta h" into the enthalpy equation we obtain:

cp * delta T = cv * delta T + R * delta T

dividing by "delta T" gives the relation:

cp = cv + R

We can define an additional variable called the specific heat ratio, which is given the Greek symbol "gamma", which is equal to cp divided by cv:

gamma = cp / cv

"Gamma" is just a number whose value depends on the state of the gas. For air, gamma = 1.4 for standard day conditions. "Gamma" appears in many fluids equations including the equation relating pressure, temperature, and volume during a simple process, the equation for the speed of sound, and all the equations for isentropic flows, and shock waves.

P – V – T SURFACE OF AN IDEAL GAS

For a fixed number of molecules the Ideal Gas Law forms the surface of a three-dimensional graph where the axis are Pressure, Volume, and Temperature.

Lines of constant pressure, constant volume, and constant temperature form a coordinate system labeling the location of an ideal gas.

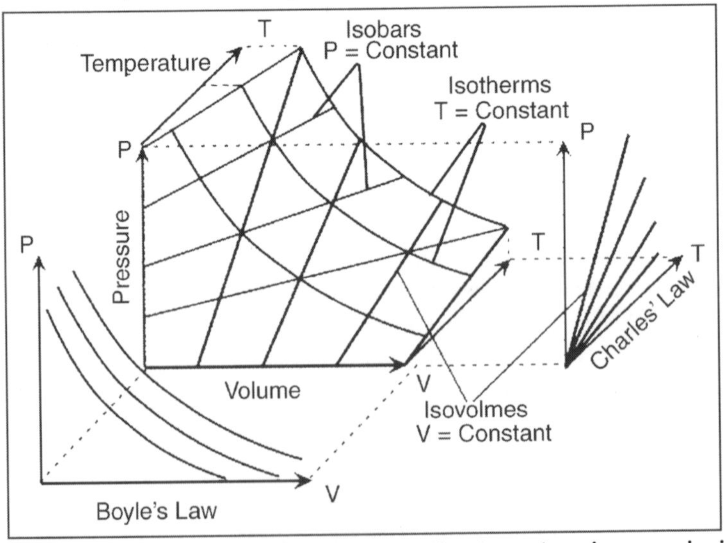

Boyle law showed that the pressure of a low-density gas is inversely proportional to its volume, when the temperature is constant.

$P \, \alpha_1/V$ for constant temperature.

Charles and Gay-Lussac law showed that the pressure of a low-density gas is proportional to its temperature, when the volume is constant.

$P \, \alpha \, T$ for constant volume.

The ideal gas law is only valid for low-density gas. There exists a PVT surface for high-density gases only.

PVT SURFACE FOR SUBSTANCE WHICH CONTRACTS ON FREEZING

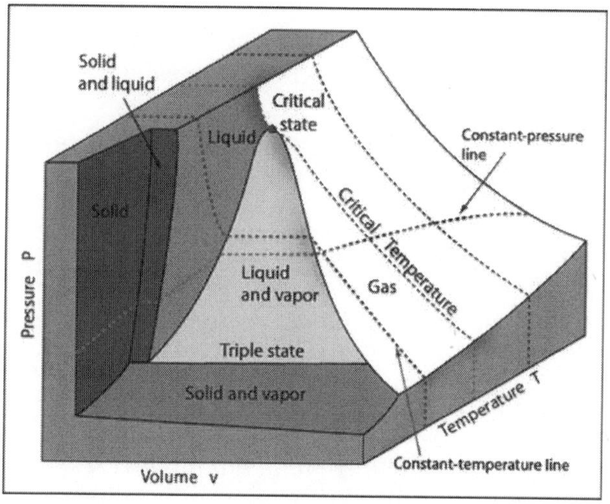

The equilibrium state of a simple, compressible substance can be specified in terms of its pressure, volume and temperature. If any two of these state variables is specified, we can determine the third. The states of the substance can be represented as a surface in three dimensional PVT space. PVT surface above represents a substance which contracts on freezing. PVT surface image is shown in figure.

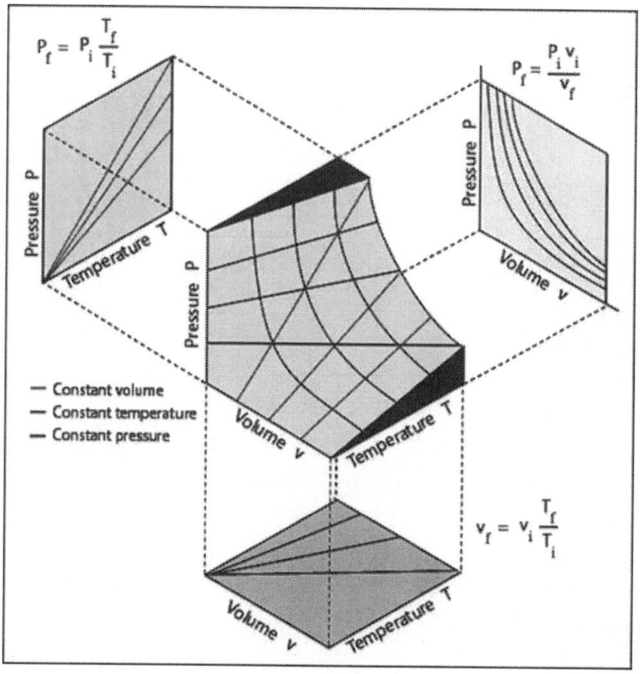

In this the solid, liquid and gas phases can be determined by regions on the surface. A point lying on a line between a single-phase and a two-phase region represents a "saturation state". The line between the liquid and liquid-vapour regions is called as liquid-saturation line. A point on the boundary between the vapour and liquid-vapour regions is called as a saturated-vapour state.

All the possible states of an ideal gas can be represented by a PVT surface as shown in figure.

TRIPLE POINT

Triple Point of a substance is defined as temperature and pressure at which three phases such as gas, liquid, and solid of that substance coexist in equilibrium.

For example, triple point of mercury occurs at a temperature of -38.8344 °C and a pressure of 0.2 mPa. In addition to the triple point between solid, liquid, and gas, there can be triple points involving more than one solid phase, for substances with multiple polymorphs. Helium-4 is a special case that presents a triple point involving two different fluid phases.

IDEAL MONATOMIC GASES

Let us now practice calculating thermodynamic relations using the partition function by considering an example with which we are already quite familiar: *i.e.*, an ideal monatomic gas. Consider a gas consisting of N identical monatomic molecules of mass m enclosed in a container of volume V.

Let us denote the position and momentum vectors of the ith molecule by r_i and p_i, respectively.

Since the gas is ideal, there are no interatomic forces, and the total energy is simply the sum of the individual kinetic energies of the molecules:

$$E = \sum_{i=1}^{N} \frac{p_i^{\,2}}{2m},$$

where $p_i^2 = p_i.\, p_i$.

Let us treat the problem classically. In this approach, we divide up phase-space into cells of equal volume h_0^f.

Here, f is the number of degrees of freedom, and h_0 is a small constant with dimensions of angular momentum which parameterizes the precision to which the positions and momenta of molecules are determined. Each cell in phase-space corresponds to a different state. The partition function is the sum of the Boltzmann factor exp $(-\beta E_r)$ over all possible states, where E_r is the energy of state t. Classically, we can approximate the summation over cells in phase-space as an integration over all phase-space. Thus,

$$Z = \int \cdots \int \exp(-\beta E) \frac{d^3 r_1 \cdots d^3 r_N \, d^3 p_1 \cdots d^3 p_N}{h_0^{3N}},$$

where $3N$ is the number of degrees of freedom of a monatomic gas containing N molecules. Making use of equationthe above expression reduces to,

$$Z = \frac{V^N}{h_0^{3N}} \int \cdots \int \exp[-\beta/2m) p_1^2] \, d^3 p_1 \cdots \exp[-(\beta/2m) p_N^2] d^3 p_N.$$

Note that the integral over the coordinates of a given molecule simply yields the volume of the container, V, since the energy E is independent of the locations of the molecules in an ideal gas. There are N such integrals, so we obtain the factor V^N in the above expression. Note, also, that each of the integrals over the molecular momenta in equation are identical: they differ only by irrelevant dummy variables of integration. It follows that the partition function Z of the gas is made up of the product of N identical factors: *i.e.*,

$$Z = \zeta^N,$$

where,

$$\zeta = \frac{V}{h_0^3} \int \exp[-(\beta/2m) p^2] d^3 p$$

is the partition function for a single molecule. Of course, this result is obvious, since we have already shown that the partition function for a system made up of a number of weakly interacting subsystems is just the product of the partition functions of the subsystems.

The integral in equation is easily evaluated: $\int \exp[-(\beta/2m) p^2] d^3 p$

$$= \int_{-\infty}^{\infty} \exp[-(\beta/2m) \, p_x^2] dp_x \int_{-\infty}^{\infty} \exp[-(\beta/2m) \, p_y^2] dp_y$$

$$\times \int_{-\infty}^{\infty} \exp[-(\beta/2m) \, p_z^2] dp_z = \left(\sqrt{\frac{2\pi m}{\beta}} \right)^3,$$

where use has been made of equation. Thus,

$$\zeta = V \left(\frac{2\pi m}{h_0^2 \beta} \right)^{3/2},$$

and,

$$\ln Z = N \ln \zeta = N \left[\ln V - \frac{3}{2} \ln \beta + \frac{3}{2} \ln \left(\frac{2\pi m}{h_0^2} \right) \right].$$

The expression for the mean pressure yields,

$$\bar{p} = \frac{1}{\beta}\frac{\partial \ln Z}{\partial V} = \frac{1}{\beta}\frac{N}{V},$$

which reduces to the ideal gas equation of state,

$$pV = NkT = \upsilon RT,$$

where use has been made of $N = \upsilon N_A$ and $R = N_A k$. According to equation the mean energy of the gas is given by,

$$\bar{E} = -\frac{\partial \ln Z}{\partial \beta} = \frac{3}{2}\frac{N}{\beta} = \upsilon\frac{3}{2}RT.$$

Note that the internal energy is a function of temperature alone, with no dependence on volume. The molar heat capacity at constant volume of the gas is given by,

$$c_\upsilon = \frac{1}{\upsilon}\left(\frac{\partial \bar{E}}{\partial T}\right)_V = \frac{3}{2}R,$$

so the mean energy can be written,

$$\bar{E} = \upsilon\, c_\upsilon T.$$

We have seen all of the above results before. Let us now use the partition function to calculate a new result. The entropy of the gas can be calculated quite simply from the expression,

$$S = k(\ln Z + \beta\bar{E}) .$$

Thus,

$$S = \upsilon R\left[\ln V - \frac{3}{2}\ln\beta + \frac{3}{2}\ln\left(\frac{2\pi m}{{h_0}^2}\right) + \frac{3}{2}\right],$$

or,

$$S = \upsilon R\left[\ln V + \frac{3}{2}\ln T + \sigma\right],$$

where,

$$\sigma = \frac{3}{2}\ln\left(\frac{2\pi mk}{{h_0}^2}\right) + \frac{3}{2}.$$

The above expression for the entropy of an ideal gas is certainly new. Unfortunately, it is also quite obviously incorrect.

GIBB'S PARADOX

What has gone wrong? First of all, let us be clear why equation is incorrect.

We can see that $S \to -\infty$ as $T \to 0$, which contradicts the third law of thermodynamics. However, this is not a problem. Equation was derived using classical physics, which breaks down at low temperatures. Thus, we would not expect this equation to give a sensible answer close to the absolute zero of temperature.

Equation is wrong because it implies that the entropy does not behave properly as an extensive quantity. Thermodynamic quantities can be divided into two groups, extensive and intensive. Extensive quantities increase by a factor a when the size of the system under consideration is increased by the same factor. Intensive quantities stay the same. Energy and volume are typical extensive quantities. Pressure and temperature are typical intensive quantities. Entropy is very definitely an extensive quantity. We have shown that the entropies of two weakly interacting systems are additive. Thus, if we double the size of a system we expect the entropy to double as well. Suppose that we have a system of volume V containing v moles of ideal gas at temperature T. Doubling the size of the system is like joining two identical systems together to form a new system of volume $2V$ containing $2v$ moles of gas at temperature T. Let,

$$S = vR\left[\ln V + \frac{3}{2}\ln T + \sigma\right]$$

denote the entropy of the original system, and let,

$$S' = 2vR\left[\ln 2V + \frac{3}{2}\ln T + \sigma\right]$$

denote the entropy of the double-sized system. Clearly, if entropy is an extensive quantity (which it is!) then we should have,

$$S' = 2S.$$

But, in fact, we find that,

$$S' - 2S = 2vR\ln 2.$$

So, the entropy of the double-sized system is more than double the entropy of the original system.

Where does this extra entropy come from? Well, let us consider a little more carefully how we might go about doubling the size of our system. Suppose that we put another identical system adjacent to it, and separate the two systems by a partition. Let us now suddenly remove the partition. If entropy is a properly extensive quantity then the entropy of the overall system should be the same before and after the partition is removed. It is certainly the case that the energy (another extensive quantity) of the overall system stays the same. However, according to above equation the overall entropy of the system increases by $2vR \ln 2$ after the partition is removed.

Suppose, now, that the second system is identical to the first system in all respects except that its molecules are in some way slightly different to the molecules in the first system, so that the two sets of molecules are distinguishable. In this case, we would certainly expect an overall increase in entropy when the partition is removed. Before the partition is removed, it separates type 1 molecules from type 2 molecules. After the partition is removed, molecules of both types become jumbled together.

This is clearly an irreversible process. We cannot imagine the molecules spontaneously sorting themselves out again. The increase in entropy associated with this jumbling is called entropy of mixing, and is easily calculated. We know that the number of accessible states of an ideal gas varies with volume like $\Omega \propto V^N$. The volume accessible to type 1 molecules clearly doubles after the partition is removed, as does the volume accessible to type 2 molecules. Using the fundamental formula $S = k \ln \Omega$, the increase in entropy due to mixing is given by,

$$S = 2k \ln \frac{\Omega_f}{\Omega_i} = 2Nk \ln \frac{V_f}{V_i} = 2v\, R \ln 2.$$

It is clear that the additional entropy $2vR \ln 2$, which appears when we double the size of an ideal gas system by joining together two identical systems, is entropy of mixing of the molecules contained in the original systems. But, if the molecules in these two systems are indistinguishable, why should there be any entropy of mixing? Well, clearly, there is no entropy of mixing in this case.

At this point, we can begin to understand what has gone wrong in our calculation. We have calculated the partition function assuming that all of the molecules in our system have the same mass and temperature, but we have never explicitly taken into account the fact that we consider the molecules to be indistinguishable.

In other words, we have been treating the molecules in our ideal gas as if each carried a little license plate, or a social security number, so that we could always tell one from another. In quantum mechanics, which is what we really should be using to study microscopic phenomena, the essential indistinguishability of atoms and molecules is hard-wired into the theory at a very low level. Our problem is that we have been taking the classical approach a little too seriously. It is plainly silly to pretend that we can distinguish molecules in a statistical problem, where we do not closely follow the motions of individual particles. A paradox arises if we try to treat molecules as if they were distinguishable. This is called Gibb's paradox, after the American physicist Josiah Gibbs who first discussed it. The resolution of Gibb's paradox is quite simple: treat all molecules of the same species as if they were indistinguishable. In our previous calculation of the ideal gas partition function, we inadvertently treated each of the N molecules in the gas as distinguishable.

Because of this, we overcounted the number of states of the system. Since the $N!$ possible permutations of the molecules amongst themselves do not lead to physically different situations, and, therefore, cannot be counted as separate states, the number of actual states of the system is a factor $N!$ less than what we initially thought. We can easily correct our partition function by simply dividing by this factor, so that,

$$Z = \frac{\zeta^N}{N!}.$$

This gives,

$$\ln Z = N \ln \zeta - N!,$$

or,

$$\ln Z = N \ln \zeta - N \ln N + N,$$

using Stirling's approximation. Note that our new version of ln Z differs from our previous version by an additive term involving the number of particles in the system. This explains why our calculations of the mean pressure and mean energy, which depend on partial derivatives of ln Z with respect to the volume and the temperature parameter β, respectively, came out all right. However, our expression for the entropy S is modified by this additive term. The new expression is,

$$S = vR\left[\ln V - \frac{3}{2}\ln \beta + \frac{3}{2}\ln\left(\frac{2\pi\, mk}{h_0^{\;2}}\right) + \frac{3}{2}\right] + k(-N \ln N + N).$$

This gives,

$$S = vR\left[\ln \frac{V}{N} + \frac{3}{2}\ln T + \sigma_0\right]$$

where,

$$\sigma_0 = \frac{3}{2}\ln\left(\frac{2\pi\, mk}{h_0^{\;2}}\right) + \frac{5}{2}.$$

It is clear that the entropy behaves properly as an extensive quantity in the above expression: *i.e.*, it is multiplied by a factor a when v, V, and N are multiplied by the same factor.

THERMODYNAMICS OF IDEAL GAS

The Kinetic Theory picture of a gas is often called the Ideal Gas Model. It ignores interactions between molecules, and the finite size of molecules. In fact, though, these only become important when the gas is very close to the temperature at which it become liquid, or under extremely high pressure. In this lecture, we will be analysing the behaviour of gases in the pressure and temperature range corresponding to heat engines, and in this range the Ideal Gas Model is an excellent approximation. Essentially, our progamme here is to learn how gases absorb heat and turn it into work, and vice versa. This heat-work interplay is called thermodynamics.

Julius Robert Mayer was the first to appreciate that there is an equivalence between heat and mechanical work. The tortuous path that led him to this conclusion is described in an earlier lecture, but once he was there, he realised that in fact the numerical equivalence—how many Joules in one calorie in present day terminology—could be figured out easily from the results of some measurements of gas specific heat by French scientists. The key was that they had measured specific heats both at constant volume and at constant pressure. Mayer realised that in the latter case, heating the gas necessarily increased its volume, and the gas therefore did work in pushing to expand its container. Having convinced himself that mechanical work and heat were equivalent, evidently the extra heat needed to raise the temperature of the gas at constant pressure was exactly the work the gas did on its container.

The simplest way to see what's going on is to imagine the gas in a cylinder, held in by a piston, carrying a fixed weight, able to move up and down the cylinder smoothly with negligible friction. The pressure on the gas is just the total weight pressing down divided by the area of the piston, and this total weight, of course, will not change as the piston moves slowly up or down: the gas is at constant pressure.

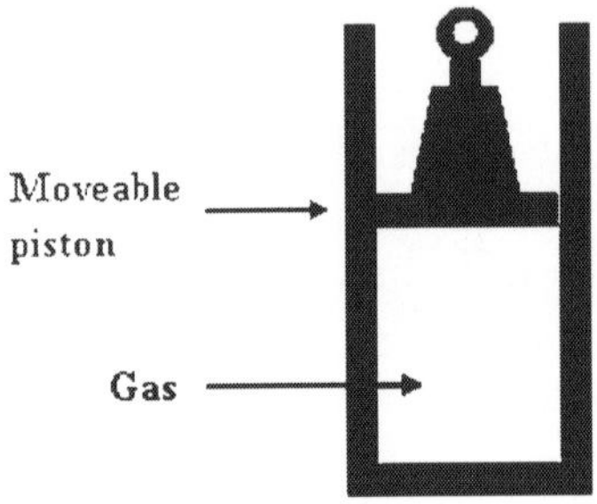

THE GAS SPECIFIC HEATS CV AND CP

Consider now the two specific heats of this same sample of gas, let's say one mole:

- Specific heat at constant volume, C_V (piston glued in place),
- Specific heat at constant pressure, C_P (piston free to rise, no friction).

In fact, we already worked out C_V in the Kinetic Theory lecture: at temperature T, recall the average kinetic energy per molecule is $\frac{3}{2}kT$, so one mole of gas—Avogadro's number of molecules—will have total kinetic energy, which we'll label internal energy,

$$E_{\text{int}}\ \tfrac{3}{2}kT.N_A\ \tfrac{3}{2}RT.$$

(In this simplest case, we are ignoring the possibility of the molecules having their own internal energy: they might be spinning or vibrating—we'll include that shortly).

That the internal energy is $\frac{3}{2}kT$ per mole immediately gives us the specific heat of a mole of gas in a fixed volume,

$$C_V = \tfrac{3}{2}R$$

that being the heat which must be supplied to raise the temperature by one degree. However, if the gas, instead of being in a fixed box, is held in a cylinder at constant pressure, experiment confirms that more heat must be supplied to raise the gas temperature by one degree. As Mayer realised, the total heat energy that must be supplied to raise the temperature of the gas one degree at constant pressure is $\tfrac{3}{2}k$ per molecule plus the energy required to lift the weight. The work the gas must do to raise the weight is the force the gas exerts on the piston multiplied by the distance the piston moves. If the area of piston is A, then the gas at pressure P exerts force PA. If on heating through one degree the piston rises a distance D*h*, the gas does work

$$PA.\Delta h = P\Delta V.$$

Now, for one mole of gas, $PV = RT$, so at constant P,

$$P\Delta V = P\Delta T.$$

Therefore, the work done by the gas in raising the weight is just RDT, the specific heat at constant pressure, the total heat energy needed to raise the temperature of one mole by one degree,

$$C_P - C_V + R.$$

In fact, this relationship is true whether or not the molecules have rotational or vibrational internal energy. (It's known as Mayer's relationship.) For example, the specific heat of oxygen at constant volume,

$$C_V(\mathrm{O}_2) = \tfrac{5}{2}R.$$

and this is understood as a contribution of $\tfrac{3}{2}R$ from kinetic energy, and R from the two rotational modes of a dumbbell molecule (just why there is no contribution form rotation about the third axis can only be understood using quantum mechanics). The specific heat of oxygen at constant pressure,

$$C_P(O_2) = \tfrac{3}{2}R.$$

It's worth having a standard symbol for the ratio of the specific heats:

$$\frac{C_P}{C_V} = \gamma.$$

ISOTHERMS AND ADIABATS

An ideal gas in a box has three thermodynamic variables: P, V, T. But if there is a fixed mass of gas, fixing two of these variables fixes the third from $PV = nRT$ (for n moles). In a heat engine, heat can enter the gas, then leave at a different stage. The gas can expand doing work, or contract as work is done on it. To track what's going on as a gas engine transfers heat to work, say, we must follow the varying state of the gas. We do that by tracing a curve in the

(P, V) plane. Supplying heat to a gas which consequently expands and does mechanical work is the key to the heat engine. But just knowing that a gas is expanding and doing work is not enough information to follow its path in the (P, V) plane. The route it follows will depend on whether or not heat is being supplied (or taken away) at the same time. There are, however, two particular ways a gas can expand reversibly—meaning that a tiny change in the external conditions would be sufficient for the gas to retrace its path in the (P, V) plane backwards. It's important to concentrate on reversible paths, because as Carnot proved and we shall discuss later, they correspond to the most efficient engines. The two sets of reversible paths are the isotherms and the adiabats.

Isothermal behaviour: The gas is kept at constant temperature by allowing heat flow back and forth with a very large object (a "heat reservoir") at temperature T. From $PV = nRT$, it is evident that for a fixed mass of gas, held at constant T but subject to (slowly) varying pressure, the variables P, V will trace a hyperbolic path in the (P, V) plane.

This path, $PV = nRT_1$, *say is called the isotherm at temperature* T_1. *Here are two examples of isotherms*:

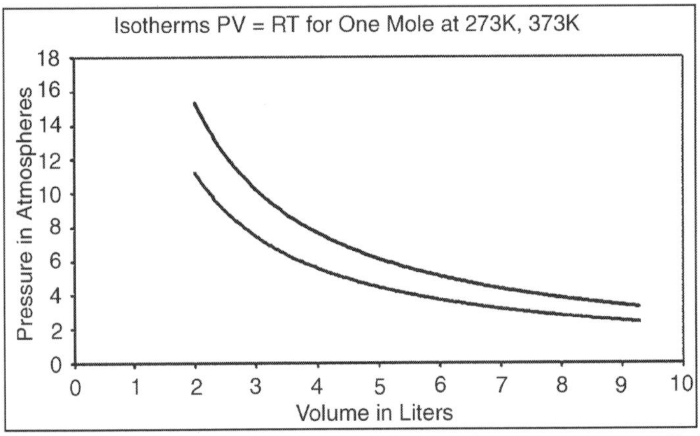

Adiabatic behaviour: "adiabatic" means "nothing gets through", in this case no heat gets in or out of the gas through the walls. So all the work done in compressing the gas has to go into the internal energy E_{int}.

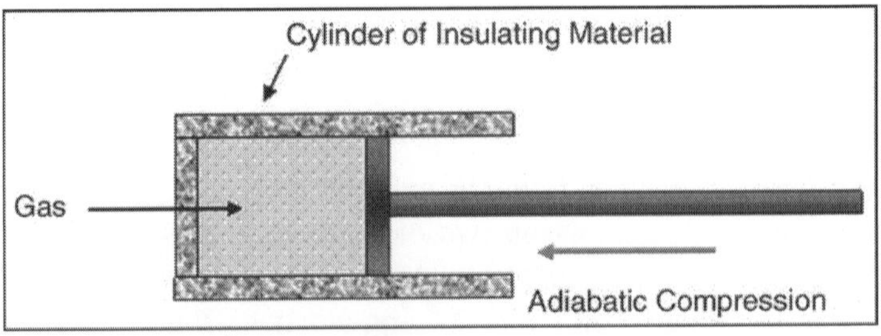

As the gas is compressed, it follows a curve in the (P, V) plane called an adiabat. To see how an adiabat differs from an isotherm, imagine beginning at some point on the blue 273 K isotherm on the above graph, and applying pressure so the gas moves to higher pressure and lower volume. Since the gas's internal energy is increasing, but the number of molecules is staying the same, its temperature is necessarily rising, it will move towards the red curve, then above it.

This means the adiabats are always steeper than the isotherms.

EQUATION FOR AN ADIABAT

What equation for an adiabat corresponds to $PV = nRT_1$ for an isotherm?

On raising the gas temperature by DT, the change in the internal energy—the sum of molecular kinetic energy, rotational energy and vibrational energy,

$$\Delta E_{\text{int}} = C_V \Delta T.$$

This is always true: whether or not the gas is changing volume is irrelevant, all that counts in E_{int} is the sum of the energies of the individual molecules (assuming as we do here that attractive or repulsive forces between molecules are negligible). In adiabatic compression, all the work done by the external pressure goes into this internal energy, so,

$$-P\Delta V = C_V \Delta T.$$

(Compressing the gas of course gives negative ΔV, positive ΔEint.)

To find the equation of an adiabat, we take the infinitesimal limit

$$-PdV = C_V dT.$$

Divide the left-hand side by PV, the right-hand side by RT (since $PV = RT$, that's OK) to find,

$$-\frac{R}{C_V}\frac{dV}{V} = \frac{dT}{T}.$$

Recall now that $C_P = C_V + R$, and $C_P/C_V =$ g. It follows that,

$$\frac{R}{C_V} = \frac{C_P - C_V}{C_V} = \gamma - 1.$$

Hence,

$$-(\gamma - 1)\int \frac{dV}{V} = \int \frac{dT}{T}$$

and integrating,

$$\ln T + (\gamma - 1) \ln V = \text{const.}$$

from which the equation of an adiabat is,

$$TV^{-1} = \text{const}.$$

From $PV - RT$, the P, V equation for an adiabat can be found by multiplying the left-hand side of this equation by the constant PV/T, giving,

PV^{g} = const. for an adiabat,

where $\gamma = {}^5/_3$ for a monatomic gas, ${}^7/_5$ for a diatomic gas.

THE CARNOT CYCLE

All standard heat engines (steam, gasoline, diesel) work by supplying heat to a gas, the gas then expands in a cylinder and pushes a piston to do its work. The catch is that the heat and/or the gas must somehow then be dumped out of the cylinder to get ready for the next cycle.

We examine the first step, the expansion, then go on to the full cycle—Carnot's analysis.

Carnot's aim was to figure out how to maximize the efficiency of a heat engine, and then work out what that efficiency was, that is, how much of the heat supplied was actually converted into the mechanical work done by the engine.

Remember that he had in mind the analogy of the water wheel, at that time still a main driving force of industry. He knew that the most efficient water wheels were those that operated smoothly, the water went into the buckets at the top from the same level, it didn't fall through any height, and didn't splash around.

In the limit of a frictionless wheel, with gentle flow on and off the wheel, the machine would be reversible—turning it in reverse to raise the water back would take the same amount of work the wheel had delivered as the water fell.

This was clearly perfect efficiency, so these were to conditions to emulate in the heat engine.

The analog to having the water flow into buckets at the same height, with no wasteful drop, is to have the heat from the heat supply flow into the gas at the same temperature. There must of course be a slight drop in temperature for the heat to flow at all, but this must be minimized.

This means that as the heat is supplied and the gas expands, the temperature of the gas stays the same as that of the heat supply (the "heat reservoir") and the gas is expanding isothermally.

ISOTHERMAL EXPANSION

So the first question is: how much work is done by an isothermally expanding gas? Taking the temperature of the heat reservoir to be TH (H for hot), the expanding gas follows the isothermal path $PV = nRT_H$ in the (P, V) plane.

The work done by the gas in a small volume expansion DV is just PDV, the area under the curve.Hence the work done in expanding isothermally from volume Va to Vb is the total area under the curve between those values,

$$\text{work done isothermally} = \int_{V_a}^{V_b} PdV = \int_{V_a}^{V_b} \frac{nRV_H}{V} dV = nRT_H \ln \frac{V_b}{V_a}.$$

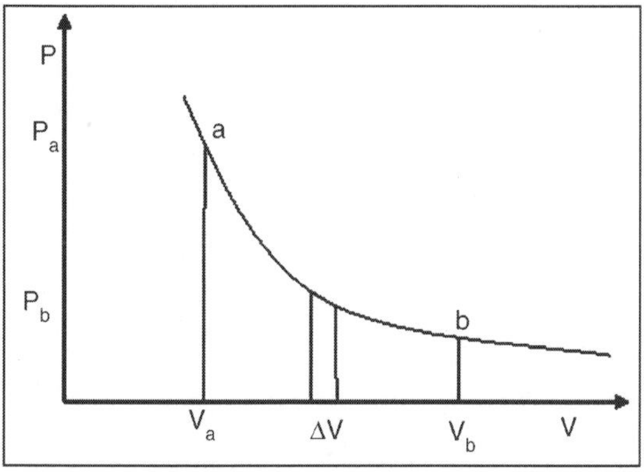

Since the gas is at constant temperature TH, there is no change in its internal energy during this expansion, so the total heat supplied must be nRT_H $\ln \frac{V_b}{V_a}$, the same as the external work the gas has done.

In fact, this isothermal expansion is only the first step: the gas is at the temperature of the heat reservoir, hotter than its other surroundings, and will be able to continue expanding even if the heat supply is cut off. To ensure that this further expansion is also reversible, the gas must not be losing heat to the surroundings. That is, after the heat supply is cut off, there must be no further heat exchange with the surroundings, the expansion must be adiabatic.

ADIABATIC EXPANSION

The work done in an adiabatic expansion is like that done in allowing a compressed spring to expand against a force—equal to the work needed to compress the spring in the first place, for a perfect spring, and an adiabatically enclosed gas is essentially perfect in this respect. In other words, adiabatic expansion is reversible.

To find the work the gas does in expanding adiabatically from Vb to Vc, say, the above analysis is repeated with the isotherm $PV = nRT_H$ replaced by the adiabat $PV^g = P_b V_b^g$, work done adiabatically W_{adiabat}

$$\int_{V_a}^{V_c} PdV = P_b V_b^{\gamma} \int_{V_b}^{V_c} \frac{dV}{V^{\gamma}} = P_b V_b^{\gamma} \frac{V_c^{1-\gamma} - V_b^{1-\gamma}}{1-\gamma}.$$

Again, this is the area under the curve, in this case under the adiabat, from b to c in the (P, V) plane.

Since points b, c are on the same adiabat, $P_b V_b^g = P_c V_c^g$, and the expression can be written more neatly:

$$W_{\text{adiabat}} = \frac{P_c V_c - P_b V_b}{1-\gamma}.$$

This is a useful expression for the work done since we are plotting in the (P, V) plane, but note that from the gas law $PV = nRT$, the numerator is just $nR(Tc - T_b)$, and from this $W_{\text{adiabat}} = nC_V(T_c - T_b)$, as of course it must be—this is the loss of internal energy that has been expended by the gas on expanding against external pressure. We've looked in detail at the work a gas does in expanding as heat is supplied (isothermally) and when there is no heat exchange (adiabatically). These are the two initial steps in a heat engine, but it is equally necessary for the engine to get back to where it began, for the next cycle. The general idea is that the piston drives a wheel, which continues to turn and pushes the gas back to the original volume.

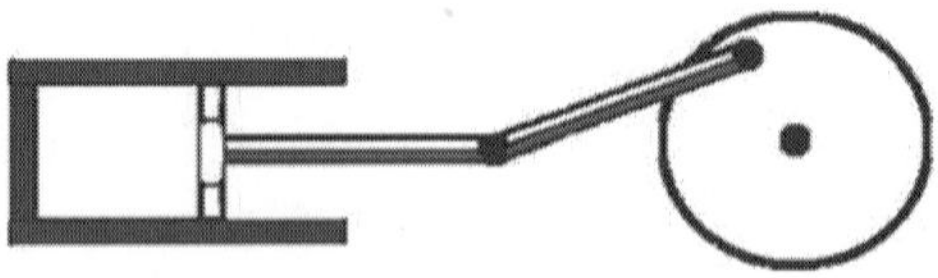

But it is also essential for the gas to be as cold as possible on this return leg, because the wheel is now having to expend work on the gas, and we want that to be as little work as possible—it's costing us. The colder the gas, the less pressure the wheel is pushing against.

To ensure that the engine is as efficient as possible, this return path to the starting point (P_a, V_a)must also be reversible. We can't just retrace the path taken in the first two legs, that would take all the work the engine did along those legs, and leave us with no net output. Now the gas cooled during the adiabatic expansion from b to c, from TH to TC, say, so we can go some distance back along the reversible colder isotherm TC. But this won't get us back to (P_a, V_a), because that's on the TH isotherm. The simplest option—the one chosen by Carnot—is to proceed back along the cold isotherm to the point where it intersects the adiabat through a, then follow that isotherm back to a. (One could follow a more complicated path: provided it was composed of segments each being adiabatic or isothermal, it would be reversible.) Carnot's cycle is around that curved quadrilateral having these four curves as its sides.

Efficiency of the Carnot Engine

In a complete cycle of Carnot's heat engine, the gas traces the path abcd. The important question is: what fraction of the heat supplied from the hot reservoir (along the red top isotherm) is turned into mechanical work? This fraction is called the efficiency of the engine.

The work output along any curve in the (P, V) plane is just $\int PdV$—the area under the curve, but it will be negative if the volume is decreasing! So the work done by the engine during the hot isothermal segment is the area abfh, then the adiabatic expansion adds the area bcef, but as the gas is compressed back, the wheel has to do work on the gas equal to the area cdge as heat is dumped into the cold reservoir, then dahg as the gas is recompressed to the starting point.

The bottom line is that the total work done by the gas is the area bounded by the four paths: the curved "parallelogram" in the picture above. We could compute this area by finding $\int PdV$ for each segment, but that is unnecessary—on completing the cycle, the gas is back to its initial temperature, so has the same internal energy. Therefore, the work done by the engine must be just the difference between the heat supplied at TH and that dumped at TC.

Now the heat supplies along the initial hot isothermal path ab, equal to the work done along that leg, is (from the paragraph above on isothermal expansion):

$$Q_H = nRT_H \ln\frac{V_b}{V_a}$$

and the heat dumped into the cold reservoir along cd is,

$$Q_C = nRT_C \ln\frac{V_c}{V_d}.$$

The difference between these two is the net work output. This can be simplified using the adiabatic equations for the other two sides of the cycle:

$$T_H V_b^{\gamma-1} = T_C V_c^{\gamma-1}$$

$$T_H V_a^{\gamma-1} = T_C V_d^{\gamma-1}.$$

Dividing the first of these equations by the second,

$$\left(\frac{V_b}{V_a}\right) = \left(\frac{V_c}{V_d}\right)$$

and using that in the preceding equation for QC,

$$Q_C = nRT_C \ln\frac{V_a}{V_b} = \frac{T_C}{T_H} Q_H.$$

The work done can now be written simply:

$$W = Q_H - Q_C = \left(1 - \frac{T_C}{T_H}\right) Q_H.$$

Therefore the efficiency of the engine, defined as the fraction of the ingoing heat energy that is converted to available work, is,

$$\text{efficiency} = \frac{W}{Q_H} = 1 - \frac{T_C}{T_H}.$$

These temperatures are of course in degrees Kelvin, so for example the efficiency of a Carnot engine having a hot reservoir of boiling water and a cold reservoir ice cold water will be 1– (273/373) = 0.27, just over a quarter of the heat energy is transformed into useful work.After all the effort to construct an efficient heat engine, making it reversible to eliminate "friction" losses, etc., it is perhaps somewhat disappointing to find this figure of 27 per cent efficiency when operating between 0 and 100 degrees Celsius. Surely we can do better than that? After all, the heat energy of hot water is the kinetic energy of the moving molecules, can't we find some device to channel all that energy into useful work? Well, we can do better than 27 per cent , by having a colder cold reservoir, or a hotter hot one. But there's a limit: we can never reach 100 per cent efficiency, because we cannot have a cold reservoir at $T_C = 0K$, and even if we did after the first cycle the heat dumped into it would warm it up!

The Second Law of Thermodynamics states that we cannot devise an engine, working in a cycle, that simply extracts heat from a hot reservoir and delivers mechanical work.

This means any engine that takes heat and delivers work also dumps out some of the initial heat to a reservoir at a lower temperature.

It's important to note that the First Law of Thermodynamics, the conservation of total energy including heat, would not be violated by an engine that powered a ship by extracting heat energy from the surrounding water. This Second Law is saying something new. And, this Second Law does not follow from the First by logical deduction—it comes (like the First) from experiment and observation.

THE MOST EFFICIENT ENGINE

An important consequence of the second Law is that no engine can be more efficient than the Carnot cycle. Essentially, this is because a "super efficient" engine, if one existed, could be used to drive a Carnot cycle in reverse, which would pump back to the hot reservoir the heat the super efficient engine dumped in the cold reservoir, and the net effect of the two coupled engines would be to take heat from the hot reservoir and do work, contradicting the Second Law.

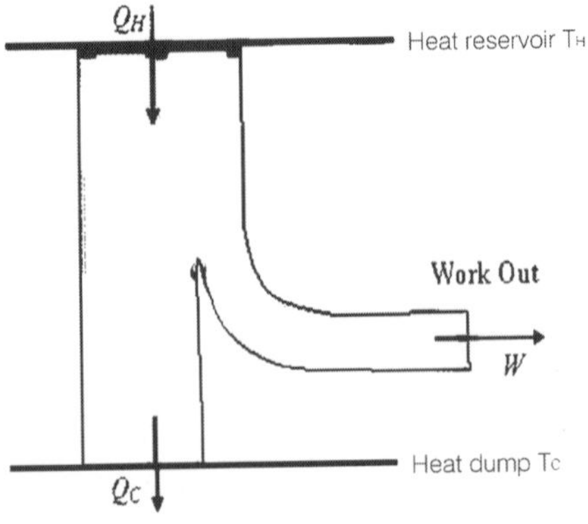

To see this, we plot the heat/energy flow for the Carnot cycle:

Here $Q_H = Q_C + W$ (all expressed in Joules, of course).

Since the engine is reversible, it can also be run backwards (this would be a refrigerator: outside work is supplied, and heat is extracted from a cold reservoir and dumped into a hot reservoir:

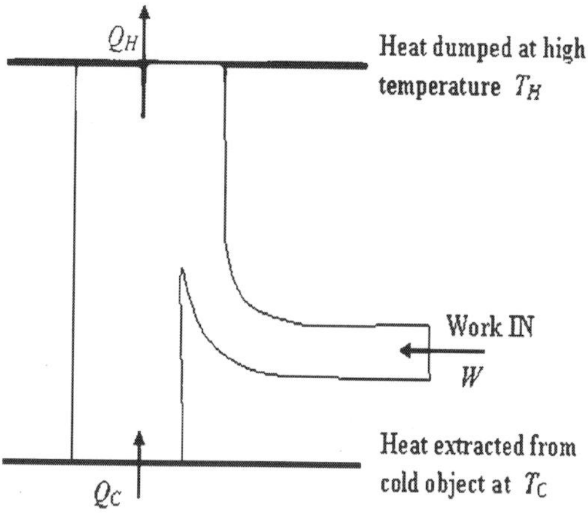

Suppose now we have a super efficient engine, represented by the first diagram above, and dumping the same heat per cycle QC into the cold reservoir, but taking in more heat energy Q_H + Δ Joules from the hot reservoir, and performing work W + Δ. Now, we hook up our super efficient engine to the "Carnot refrigerator" in the diagram above. The refrigerator sucks out of the cold reservoir all the heat the super efficient engine dumped there, and needs W Joules of work per cycle to do it.

The super efficient engine can provide this, and there are still Δ Joules of work to spare. Of course, the Carnot refrigerator has also dumped QH Joules of heat in the hot reservoir. But the bottom line is that between them, the super efficient engine and the Carnot refrigerator have extracted Δ Joules of heat from the hot reservoir and performed Δ Joules of work—contradicting the Second Law.

The Second Law therefore forces the conclusion that no amount of machine design will produce an engine more efficient than the Carnot cycle. The rather low ultimate efficiencies this dictated came as a shock to nineteenth century engineers.

MOLECULAR COLLISIONS

In analysing the gas so far, we've ignored collisions between molecules, and in fact for air at ordinary temperatures the relationship between pressure, volume and temperature came out correctly. Furthermore, Maxwell's speed distribution can be used to find what fraction of the molecules in a planet's atmosphere are moving at above escape velocity, so we can predict what gases will remain surrounding a planet, given the gravitational force near the surface, and the temperature. But there are other phenomena for which an understanding of collisions is all-important.

For example, if two different gases, say oxygen and nitrogen, in a container at the same pressure and temperature are separated by a partition, how quickly will they mix once the partition is removed? Assuming room temperature, the molecules will be moving at hundreds of meters per second, so one might imagine the mixing will be over in hundredths of a second. But that is not the case at all—observationally, it might take an hour, for a box holding a few liters. This surprisingly slow penetration of one gas by another is called diffusion. The reason it takes the gases so long to mix becomes evident on tracking one molecule as it enters the other gas. Think of an oxygen molecule moving into nitrogen. We'll take O_2 and N_2 to be little spheres of diameter d. So now visualize the little O_2 sphere shooting into this space where all these other spheres are moving around. Temporarily, for ease of visualization, let's imagine all the other spheres to be at rest.

How far can we expect the O_2 to get before it hits an N_2? The average distance before a collision is called the mean free path. Let's try to picture how much room there is to fly between these fixed N_2 spheres. We do know that if it were liquid nitrogen, there would be very little room: liquids are just about incompressible, so the molecules must be touching. Roughly speaking, a molecule of diameter d will occupy a cubical volume of about d^3 (there has to be some space left over—we can pack cubes to fill space, but not spheres).

We also know that liquid nitrogen weighs about 800 kg per cubic meter, whereas N_2 gas at room temperature (and pressure) weighs about 1.2 kg per cubic meter, a ratio of 670. This means that on average each molecule in the gas has 670 times more room—that is, it has a space 670 times the volume d^3 we gave it in the liquid. So in the gas, the average center-to-center separation of the molecules will be the cube root of 670, which is about 8.75d.

So the picture is a gas of spheres of diameter d, placed at random, but separated on average by distances of order 10d. It's clear that shooting an oxygen molecule into this it will get quite a way.

We now try to estimate just how far an O_2 will get, on average, as it shoots into this forest of spheres. Picture the motion of the center of the oxygen molecule. Before any collision, it will be moving on a straight-line path. Just how close does the O_2 center have to get to an N_2 center for a hit? Taking both O_2, N_2 to be spheres of diameter d, if an N_2 center lies within d of the O_2 center's path, there will be a hit. So we can think of the O_2 as sweeping out a volume, a cylinder of radius d centered on its path, hitting and deflecting if it encounters an N_2 centered within that cylinder. So how far will it get, on average, before a hit? In traveling a distance x, it sweeps out a volume $\pi d^2 x$.

Now picture it going through the gas for some considerable length of time, so there are many collisions.

The volume swept out will look like a stovepipe, long straight cylindrical sections connected by elbows at the collisions. The total volume of this stovepipe (ignoring tiny corrections from the elbows) will be just pd^2L, L being the total length, that is, the total distance the molecule traveled. If the density of the nitrogen is n molecules per cubic meter, the number of N_2's in this stovepipe volume will be pd^2Ln, in other words, this will be the number of collisions. Therefore, the average distance between collisions, the mean free path l, is given by:

$$\text{mean free path } l = \frac{\text{total distance traveled}}{\text{number of collisions}} = \frac{L}{\pi d^2 Ln} = \frac{L}{\pi d^2 n}.$$

So what is n? We estimated above that each molecule has space $670d^3$ to itself, so n is just how many of those volumes there are in one cubic meter, that is, $n = 1/670d^3$.

Therefore, the mean free path is given by,

$$l = \frac{1}{\pi d^2 n} = \frac{670d^3}{\pi d^3} = 200d.$$

Notice that this derivation of the mean free path in terms of the molecular diameter depends only on knowing the ratio of the gas density to the liquid density—it does not depend on the actual size of the molecules! But it does mean that if we can somehow measure the mean free path, by measuring how fast one gas diffuses into another, for example, we can deduce the size of the molecules, and historically this was one of the first ways the size of molecules was determined, and so Avogadro's number was found.

Let us now put in some numbers to find this mean free path: for O_2, N_2, d = 0.3 nm, so the mean free path l = 60nm, or 6×10^{-8}m. The speed of the molecules at room temperature v is approximately 500 meters per sec., so the molecule has of order 1010 collisions per second!

Actually, there is one further correction we should make. We took the N_2 molecules to be at rest, whereas in fact they're moving as fast as the oxygen molecule, approximately. This means that even if the O_2 is temporarily at rest, it can undergo a collision as an N_2 comes towards it. Clearly, what really counts in the collision rate is the relative velocity of the molecules. Defining the average velocity as the root mean square velocity, if the O_2 has velocity $\vec{v}_1$ and the N_2 $\vec{v}_2$, then the square of the relative velocity $\overline{(\vec{v}_1 - \vec{v}_2)^2} = \overline{\vec{v}_1^2} - \overline{2\vec{v}_1 - \vec{v}_2}^{\,2} + \overline{\vec{v}_2^2} = \overline{\vec{v}_1^2} + \overline{\vec{v}_2^2}$, since $\overline{\vec{v}_1 . \vec{v}_2}$ must average to zero, the relative directions being random. So the average square of the relative velocity is twice the average square of the velocity, and therefore the average root-mean-square velocity is up by a factor –2, and the collision rate is increased by this factor. Consequently, the mean free path is decreased by a factor of –2 when we take into account that all the molecules are moving.

Our final result, then, is that the mean free path

$$1 = \frac{1}{\sqrt{2}\pi d^2 n}.$$

Finding the mean free path is—literally—the first step in figuring out how rapidly the oxygen atoms will diffuse into the nitrogen gas, and of course vice versa.

What we really want to know is just how much we can expect the gases to have intermingled after a given period of time. We'll just follow the one molecule, and estimate how far it gets. To begin with, let's assume for simplicity that it tales steps all of the same length l, but after each collision it bounces off in a random direction. So after N steps, it will have moved to a point,

$$\vec{L} = \vec{l}_1 + \vec{l}_2 + \vec{l}_3 + \cdots + \vec{l}_N,$$

where each vector $\vec{l}_j$ has length l, but the vectors all point in random different directions.

If we now imagine many of the oxygen molecules following random paths like this, how far on average can we expect them to have drifted after N steps? (note that they could with equal likelihood be going backwards!) The appropriate measure is the root-mean-square distance,

$$\overline{\vec{L}^2} = \overline{\left(\vec{l}_1 + \vec{l}_2 + \vec{l}_3 \ldots \vec{l}_N\right)^2} = Nl^2 + \sum_{i.j} \overline{\vec{l}_i . \vec{l}_j}$$

Since the direction after each collision is completely random, $\overline{\vec{l}_i . \vec{l}_j} = 0$, and the root-mean-square distance $\sqrt{\overline{\vec{L}}^2} = \sqrt{N}l.$

If we allow steps of different lengths, the same argument works, but now l is the root-mean-square path length.

The important factor here is the $\sqrt{N}$. Recall from above that l = 60 nm, or 6 × 10^{-8} m., and there are of order 10^{10} collisions per second. This means that the average distance diffused in one second is $\sqrt{10^{10}}l = 10^5 l$, say half a centimeter. The average distance in one hour would be only 60 times this, or 30 cm., one foot, and in a day about five feet—the average distance traveled is only increasing as the square root of the time elapsed!

This is a very general result. For example, suppose we have a gas in which the mean free path is l and the average speed of the molecules is v. Then the average time between collisions $\tau = l/n$. The number of collisions in time t will be t/t, so the average distance a molecule moves in time t will be $r = l\sqrt{t/\tau}$.

A famous mystery cleared up by arguments like this was that Newton predicted the speed of sound would be given by c^2 = B/r, as we discussed earlier in the course, with B the bulk modulus. But when B was measured carefully by slowly compressing air, the result was in error by about 30 per cent ! The speed of sound predicted a higher (stiffer) bulk modulus.

The explanation turned out to be that in a slow measurement of the bulk modulus, the gas stays at the same temperature—the heating caused by slow compression leaks away. But if the compression is rapid, the gas heats up and so the pressure goes up more than if it had stayed at the same temperature. So the question is whether the compression and decompression as a sound wave passes through is so rapid that the heated-up gas doesn't have time to spread to the cooled regions. For sound at say 1000 Hz, the wavelength is 34 cm. If compression heats gas locally, the hot molecules will diffuse away in a similar manner to that discussed above. They will be slightly faster than the average molecules. In 1/1000 th of a second, they will have 10^7 collisions, so will travel about $\sqrt{10^7}\, l = 30C, l = 0.2$ mm. This tiny distance compared with the wavelength of the sound wave means that during the compression/ decompression cycles as the wave passes through, the heat has no chance to dissipate—so, effectively, it's like compressing a gas in an insulated container, it's harder to compress than it would be if the heat generated could flow away, and the bulk modulus is higher by an amount (around 30 per cent) we shall work out in a forthcoming lecture.

The simple kinetic picture of a gas leads to the Gas Law, $PV - nRT$, with the macroscopic variable T now seen as proportional to the average energy of a molecule. Putting heat into a gas simply means increasing the molecules' average kinetic energy (plus rotational energy, etc., for more complicated molecules) and the work done by an expanding gas is just a transfer of energy from this internal energy to whatever the gas is pushing.

The First Law of Thermodynamics is nothing but conservation of total energy, now that we know heat is really microscopic kinetic energy for an ideal gas (and rotational energy, etc.). By applying these findings to the Carnot Cycle, we discovered that this reversible ideal engine had efficiency $1 - T_C/T_H$. Furthermore, we found that no heat engine could be more efficient, at least if we granted the truth of the Second Law of Thermodynamics, which states that it is impossible to construct an engine that just takes heat energy out of the air, say, and converts it into work, without any available "cold reservoir" to dump waste heat into.

The incredible thing is that Carnot, in the 1820's, who believed heat was a conserved caloric fluid, and therefore didn't know the First Law (that heat is just microscopic energy, which is not conserved by itself, but can be transformed into other forms of energy) correctly found the efficiency of his cycle to be 1 –

T_C/T_H, and then argued—also correctly—that this set the absolute limit on heat engine efficiency! How did he do that? It's worth a brief examination of his ideas, because they give a clue about something we haven't mentioned so far—entropy.

Remember, Carnot saw his cycle as a water wheel, the caloric fluid being the water, the temperature difference being the height difference from the top to the bottom of the waterfall. But he did know one difference from a waterfall: he knew there was an absolute zero, a level below which nothing could "fall". (This was from the well known results on the contraction of a gas with cooling. He took the absolute zero to be -267 Celsius, only a few degrees off the correct value). This meant that one could assign to the caloric fluid an absolute value of "potential energy" in this imaginary water wheel scenario: at temperature T, an amount of fluid F would have "potential energy" FT. This is the total possible amount of heat energy in the fluid, and to extract it all you would have to "drop" the fluid all the way down the temperature scale to the absolute zero. But in our real (if idealized) heat engine, it can only drop as far as the lowest temperature available, that of the cold reservoir TC. Hence the efficiency factor $1-T_C/T_H$: it's just the ratio of how far the caloric fluid is able to fall in our engine to how far it would have to fall (down from TH to absolute zero) to give up all its heat energy.

In this picture, the heat energy delivered from the hot reservoir $Q_H = FT_H$, so the amount of actual "caloric fluid" must be $F = Q_H/T_H$. But the fluid is conserved, in Carnot's picture, so the waste heat is $Q_C = FT_C$, with the same F, and therefore $Q_H/T_H = Q_C/T_C$. This equation is the important one: even though the argument he used is incorrect—there is no conserved caloric fluid—the equation is right, and relates directly to the efficiency of his engine: the heat utilized is $Q_H - Q_C = Q_H\,(1-T_C/T_H)$. But how did he know no engine could do better? Ironically, although he didn't know the First Law of Thermodynamics, he had figured out the Second Law, and he presented an argument exactly like that given in the last lecture to demonstrate that no engine could outperform his reversible one.

Let's go back to the Carnot cycle and consider a bit more how the gas gains heat from a reservoir as it follows a path in the (P, V) plane. We begin with the first half of the cycle, from a to c:

ISOTHERM

This does not mean that we can say the gas at (P_C, V_C) has QH more heat than the gas at (P_a,V_a). Why not? Because we could equally well have gone from a to c by a route which is the second half of the Carnot cycle traveled backwards:

This is a perfectly well defined reversible route, ending at the same place, but with quite a different amount of heat supplied. So we cannot say that a gas at a given (P, V) contains a definite amount of heat. It does of course have a definite internal energy, but that energy can be increased by adding a mix of external work and supplied heat, and the two different routes from a to c have the same total energy supplied to the gas, but with more heat and less work along the top route.

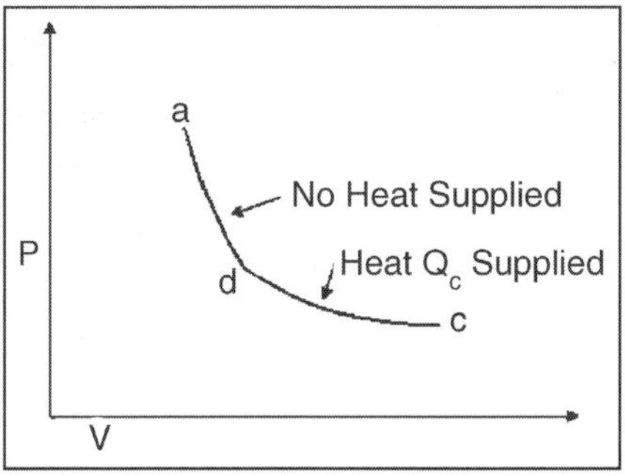

However, notice that one thing is the same over the two routes in the diagrams above: the ratio of the heat supplied to the temperature at which is was delivered: $Q_H/T_H = Q_C/T_C$.

Of course, we've chosen two particular reversible routes from a to c, but it turns out that for any reversible route from a to c, the integral of increments of heat supplied divided by the temperature of delivery, $\int_a^c dQ/T$, is the same! To see how this can be, consider cutting a corner in the previous route:

If we follow the path aefgc instead of adc, where ef is an isotherm and fg an adiabat, how does that affect $\int_a^c dQ/T$? The answer is it doesn't: look at the little Carnot cycle efgd. We've just changed from the bottom route to the top route around this cycle from e to g, so from the previous argument about the original big Carnot cycle, $\int_e^g dQ/T$ is the same.

But we can now cut corners on the corners: any zigzag route from a to c, with the zigs isotherms and the zags adiabats, in other words, any reversible route, can be constructed by adding little Carnot cycles to the original route. In fact, any path you can draw in the plane from a to c can be approximated arbitrarily well by a reversible route made up of little bits of isotherms and adiabats.

The value of $\int_a^c dQ/T$ is evidently the same along any reversible path from a to c: in contrast to $\int_a^c dQ$, the heat change, which was different for different reversible routes.

We can define a new state variable, S, such that the difference

$$S(P_2, V_2) = S(P_1, V_1) + \int_{(P_1,V_1)}^{(P_2,V_2)} \frac{dQ}{T}$$

where the integral is understood to be along a reversible path. Given the value of S at a single point (P_1,V_1), we can find its value everywhere.

Bibliography

A K Sharma and A S Sharma : *Fluid Dynamics*, Discovery Pub., Delhi, 2008.

Anil Date : *Introduction to Computational Fluid Dynamics*, Cambridge Univ Press.

Arun Solanki : *Thermodynamics*, Oxford Book Company, 2008.

Astarita : *Thermodynamics*, Springer, Publication, 2004

D Jolly : *Thermodynamics and Statistical Mechanics*, Ivy Pub, Delhi, 2006.

D.R. Olander : *General Thermodynamics*, CRC Press, 2011.

Donald A. McQuarrie and John D. Simon : *Molecular Thermodynamics*, Viva Books, 2010.

Greiner : *Thermodynamics and Statistical Mechanics*, Springer, Publication 2004.

H Versteeg and W Malalsakra : *An Introduction To Computational Fluid Dynamics: The Finite Volume Method*, Dorling Kindersley, 2012.

Hilary D Brewster : *Heat And Thermodynamics*, Oxford Book Company, 2009

John O'Connell and J.M. Haile : *Thermodynamics: Fundamentals for Applications*, Cambridge Univ Press, 2004.

M Yadav : *Thermodynamics*, Arise Pub, Delhi, 2008.

M.D. Raisinghania : *Fluid Dynamics*, S. Chand Publisher , Delhi, 2010.

Manish Saxena : *Thermodynamics and Heat Engines*, Anmol, Publication, 2010.

Michael Abbott and Hendrick Van Ness : *Schaum's Outline of Thermodynamics with Chemical Applications*, Tata Mcgrawhill, 2004.

N.B.L. Mathur : *Fundamentals of Classical Thermodynamics*, Anmol, Publication, 2010.

Nishant Patel and Guatham Sharma : *Experiments in Fluid Dynamics*, Campus Books, 2010.

P R Manjunath : *Thermodynamics*, RBSA Pub, Delhi, 2007.

P. Chattopadhyay : *Engineering Thermodynamics*, Oxford University Press, 2011.

PK Nag, S Pati and T K Jana: *Engineering Thermodynamics and Fluid Mechanics*, Tata Mcgraw-Hill, 2010.

R K Singh : *Thermodynamics*, Shree Pub, Delhi, 2005.

Robert Balmer : *Thermodynamics*, Jaico Publishing, 2004.

S K Aggarwal : *Advances In Fluid Dynamics*, ALP Books, 2009.

S K Aggarwal : *Basic Concept Of Fluid Dynamics*, ALP Books, 2009.

S K Bose and S K Roy : *Principles of Metallurgical Thermodynamics*, Universities Press, 2014.

S.A. Khan : *Textbook of Thermodynamics*, Sonali Publications, 2011.

S.K. Agarwal : *Fundamentals of Fluid Dynamics*, A.K. Publications, Delhi, 2011.

Sidharth Sharma : *Heat and Thermodynamics*, Mohit Books International, 2012.

Sonntag, Borgnakke : *Fundamentals of Thermodynamics*, Van Wylen, Wiley, 2002

Srinivasan Ayenger : *Thermodynamics*, Anmol Publication, Delhi, 2007.

V K Srivastava : *Motion Studies and Thermodynamics*, ABD Pub, 2006.

V.K. Bansal : *Thermodynamics*, ALP Books, 2010.

V.N. Dass : *Heat and Thermodynamics*, Dominant, Publication, 2005.

Venkata Chalapathi Rau : *Heat and Thermodynamics*, Adhyayan Pub, 2008.

Wassim M. Haddad, Vijay Sekhar Chellaboina and Sergey G. Nersesov : *Thermodynamics A Dynamical Systems Approach*, New Age International , 2010.

Y.V.C. Rao : *Engineering Thermodynamics*, Universities Press, 2003.

Index